# Knowledge Spaces

Jean-Claude Falmagne • Dietrich Albert
Christopher Doble • David Eppstein • Xiangen Hu
Editors

# Knowledge Spaces

## Applications in Education

Springer

*Editors*
Jean-Claude Falmagne
School of Social Sciences,
Dept. Cognitive Sciences
University of California, Irvine
Irvine, CA, USA

Dietrich Albert
Department of Psychology
University of Graz
Graz, Austria

Christopher Doble
ALEKS Corporation
Irvine, CA, USA

David Eppstein
Donald Bren School of Information
& Computer Sciences
University of California, Irvine
Irvine, CA, USA

Xiangen Hu
Department of Psychology
University of Memphis
Memphis, TN, USA

ISBN 978-3-642-35328-4 ISBN 978-3-642-35329-1 (eBook)
DOI 10.1007/978-3-642-35329-1
Springer Heidelberg New York Dordrecht London

Library of Congress Control Number: 2013942001



Printed on acid-free paper

Springer is part of Springer Science+Business Media (www.springer.com)

# Preface

As can be gathered from the database maintained by Cord Hockemeyer at the University of Graz[1], the literature on Knowledge Spaces is expanding rapidly. The recent book by Falmagne and Doignon (2011) gives a comprehensive account of the theory up to its date of publication. However there have been important developments after that. We thought that a volume gathering some of these developments would be timely.

This volume has two parts. Part I describes a number of chosen empirical works. Part II deals with recent theoretical results. Two chapters play a special role. The first chapter in Part I, called 'Overview', gives an informal, intuitive presentation of all the important concepts of learning space theory, which is an important special case of knowledge space theory. This chapter is intended for readers primarily interested in the applications of learning space theory. Chapter 8, the first chapter in Part II, gives a technical presentation of the theory, including all the main theorems (without the proofs) and a description of some important algorithms. This chapter may serve as the basic reference for the works described in Part II.

In Part I, Chapters 2 to 6 describe some applications of the ALEKS system in education. The ALEKS system is the most elaborate application of learning space theory to date. It is equipped with an *assessment module* and a *learning module*. This system is bilingual (English and Spanish) and currently covers all of K–12 mathematics[2] (excluding calculus), and beginning chemistry. Using extensive assessment and learning data, Chapter 2 investigates various statistical measures of the validity of the assessment and learning modules of the ALEKS system. Chapter 3 reports an application of the ALEKS system at the University of Illinois. The chapter demonstrates that the use of ALEKS as a placement test replacing the SAT and the ACT resulted in a substantial decrease of F grades and withdrawals. Chapter 4 describes two studies evaluating whether online individualized instruction by the ALEKS system will increase student scores on a standardized high-stakes test. (The answer is 'Yes.') The students were elementary and middle school students of the Memphis area schools. The ALEKS system was used at the University of Memphis to teach a statistics course to students majoring in psychology or social sciences. Chapter 5 compares the results for black and white students in this course with those obtained in a traditional lecture type course. While the white students, on the

[1]See: http://liinwww.ira.uka.de/bibliography/Ai/knowledge.spaces.html.
[2]Some of these subjects are also used in colleges and universities.

average, do much better than the black students in the lecture type course—which is consistent with traditional results—the discrepancy disappears in the online course. Chapter 6 describes an application of the ALEKS system to the teaching of General Chemistry to college students. The chapter presents a selection of learning and assessment data for the course, and gives interpretations of those data in the framework of knowledge space theory.

Chapter 7 analyzes the ability of students to make logical connections between fundamental chemical principles and the various representations of chemical phenomena. The ALEKS system is not used in this study. The authors build their own knowledge structures.

As mentioned earlier, the introductory chapter of Part II, Chapter 8, gives a condensed description of the most important mathematical results. Distinguishing between behavioral performance and its underlying skills and competencies, Chapters 9 and 10 deal with performance while Chapters 11 and 12 focus on competencies; all of them go well beyond Albert and Lukas (1999). Chapter 9 describes recent extensions in knowledge space theory (multiple answer alternatives, relations between sets/tests), relationships between knowledge space theory and other theoretical approaches (formal concept analysis, FCA; latent class analysis, LCA; item response theory, IRT) as well as methods for data driven generation of knowledge structures, their empirical validation (item tree analysis, ITA; inductive item tree analysis, IITA; measures and indices of t) and respective software resources. Methodological considerations and applications in Chapter 10 exemplify empirical research dealing with generating and validating knowledge structures for sets of items or tests. The different skill- and competence-oriented approaches have been developed independently. Thus, Chapter 11 for the first time relates systematically these approaches to each other by presenting a united framework which allows for identifying their commonalities and differences. These approaches are further developed in Chapter 12, which asks how to deal with distributed information, how to formulate a probabilistic approach, how to link observable navigation and problem solving behavior to cognitive and competence states, how to support self-regulated learning behavior, and how to assess competencies in educational games noninvasively. Furthermore, respective applications in technology enhanced learning and competence management are described.

Chapters 13 and 14 describe a data structure, the learning sequence, that can be used to efficiently implement learning-space based systems. In Chapter 13, learning sequences are applied to the tasks of generating the states of a learning space and using learning spaces to assess the knowledge of a learner. Chapter 14 discusses the use of learning sequences to project learning spaces onto smaller sets of concepts (important for the efficiency of assessment in large learning spaces) and to modify learning spaces by adding or removing states.

We are most grateful to all the referees whose reports led to improvements of the presentation of the works described in this volume. We thank in particular Eric Cosyn, Cornelia Dowling, Yung-Fong Hsu, Mathieu Koppen, Jeff Matayoshi, Alexander Nussbaumer, Martin Schrepp, Luca Stefanutti, Reinhard Suck, and Nicolas Thiéry. We also thank Brian Junker and Don Laming for their useful reactions to a presentation of the material in Chapter 2.

The Editors
March 11, 2013.

# Contents

## Part I LEARNING IN A KNOWLEDGE SPACE

## Part II RECENT THEORETICAL PROGRESS

# Part I

# LEARNING IN A KNOWLEDGE SPACE

# 1

# Overview

Jean-Claude Falmagne[1]

Christopher Doble[2]

## 1.1 Introduction

A student facing a computer terminal answers questions displayed on the screen by manipulating the keyboard. The questions probe the student's competence in a scholarly subject, such as arithmetic, elementary algebra (called Algebra I in many U.S. schools), or first-year college chemistry. Afterwards, on the basis of this assessment, the computer proceeds to teach the material to the student.

Nowadays, this is a common enough situation. Computers have become prominent features in the classroom. How helpful they are in educating the students is still a matter of debate, however, because the success of the enterprise critically depends on the particular software managing the process, and especially, on the accuracy of the assessment component of the software.

A large portion of Part I of this volume is devoted to a number of case studies evaluating the results of one such software system[3], which implements the concepts and methods of *Knowledge Space Theory* (see, for example, Doignon and Falmagne, 1985; Falmagne and Doignon, 1988a; Albert and Lukas, 1999; Doignon and Falmagne, 1999; Falmagne et al., 2006; Falmagne and Doignon, 2011). Part II reviews some recent theoretical developments inspired by such applications.

Knowledge space theory, or *KST* as it is often abbreviated, is unique for its kind in that it allows the representation (in the computer's memory) of an enormously large number of possible *knowledge states* organizing the relevant scholarly subject. According to the theory, a student's competence in the subject at a given instant may be identified with one of these states, which is a collection of the concepts (or skills, facts, problem-solving methods, etc.) that the student has mastered. In a subject such as elementary algebra, the number of feasible knowledge states is on the order of many millions.

[1]Dept. of Cognitive Sciences, University of California, Irvine.

[2]ALEKS Corporation.

[3]This software is the property of the University of California, Irvine, and is licensed to ALEKS Corporation.

J.-C. Falmagne et al. (eds.), *Knowledge Spaces: Applications in Education*,
DOI 10.1007/978-3-642-35329-1_1, © Springer-Verlag Berlin Heidelberg 2013

This large number of states is appropriate. Classroom populations are extremely diverse with respect to proficiency and learning ability. Even though the textbooks may cover the successive topics in more or less the same order, a student's memory of the material follows its own haphazard course. A few months after the end of a course, the nuggets that a student remembers may very well form a disparate ensemble. In any event, an assessment using knowledge space theory strives to assign one of these knowledge states to the student. Note that a knowledge state is not a quantitative estimate of how much a student has learned. Rather, it is a precise description of what the student knows and does not know at a given moment.

In the special case of KST called *Learning Space Theory*, or *LST*, the knowledge state of a student reveals exactly what she is 'ready to learn.' Indeed, it follows from the principles of LST that each knowledge state—except the maximal one—has an 'outer fringe', which is the set of concepts (skills, facts, methods) that the student in that state may start learning[4]. The student can choose to study a concept in the outer fringe of her state. Mastering this concept brings her to another state which, if it is not the maximal state, has its own outer fringe, and the process can continue. The basic principles of LST, along with the precise meaning of 'outer fringe', are described in everyday language in Paragraphs 1.2.4 and 1.2.6 of this chapter. A set-theoretical formulation is given Section 8.1 of Chapter 8, the introductory chapter to Part II (see page 132).

There are similarities between a psychometric test (such as the ACT, SAT[5], or a typical state standards test) and an assessment using KST or its special case LST. In both situations, a student must solve a number of problems selected from some scholarly subject, and receives an overall result that gauges the student's competence in that topic. The differences between the two procedures are multiple and fundamental, however. They concern the result of the test or of the assessment, the underlying theories and, crucially, the mode of selection of the problems.

In contrast to knowledge space theory, psychometric testing—either in the guise of *Classical Test Theory* (CTT) or the more recent *Item Response Theory* (IRT) (c.f. Nunnally and Bernstein, 1994, for example)—aims to produce a numerical score, or a vector of such scores, gauging the proficiency of the student in some broad area. The number of meaningfully different scoring categories in psychometric testing is relatively small—at least compared to the number of possible outcomes of an assessment in the framework of knowledge

[4]By definition, the whole domain, which is the full set of concepts of the subject, does not have an outer fringe.

[5]The ACT (American College Testing) and the SAT are the most prominent scholastic testing procedures used in the U.S. Originally, "SAT" was an acronym for "Scholastic Aptitude Test." In 1990, some doubts regarding exactly what the test was measuring led to changing its name to "Scholastic Assessment Test." Currently, it appears that "SAT" has become a pseudo-acronym, without any associated meaning.

space theory for the same scholarly subject. In the unidimensional versions of CTT or IRT, by far the most widely used, the test is essentially a measurement device placing individuals along a single scoring dimension. The items forming the test are thus required to satisfy a criterion of homogeneity. So, a candidate item may be eliminated because it deviates from the bulk of the other items in the test from the standpoint of some correlation statistics. Such a rejection is made whether or not the item is an integral part of the scholarly subject. To guard against a systematic preparation of the students for a specific test, new items have to be created frequently. In the case of a psychometric test, the issues of the reliability and, especially, the validity of the measurement, are thus legitimate perennial concerns. This is much less of an issue in the applications of KST considered here, which are centered on a particular scholarly curriculum, in a sense soon to be detailed. Note also that a psychometric test is practically always made of multiple choice questions, while an assessment in the framework of KST almost never is.

Knowledge space theory is complex and draws from several mathematical disciplines, such as combinatorics, statistics, and stochastic processes. A comprehensive technical presentation of the theory is contained in the monograph by Falmagne and Doignon (2011)[6]. Particular aspects of the theory have been investigated in depth by various researchers in the U.S. and elsewhere, especially in Austria, Germany, Italy and the Netherlands. We devote the last section of this chapter to a brief historical review of the work in this area, including a short bibliography.

Up to Part II of this volume, however, there will be no need to enter into all the mathematical details underpinning the theory. A solid intuitive understanding of the core concepts will suffice. Fortunately, they are quite commonsensical. The main purpose of this chapter is to provide the reader with such an intuition, which we now endeavor to do.

## 1.2 Basic Concepts: Knowledge States and Spaces

**1.2.1 The items, the instances, and the domain.** We consider a scholarly subject that can be parsed into a set of problem types, or *items*. The set of all these items is called the *domain* for that subject. An example of an item in elementary algebra (Algebra I) is as follows:

*Finding the roots of an equation of the form* $x^2 + bx + c = 0$.

When the computer tests a student on this item, the numbers $b$ and $c$ are selected randomly, but with certain restrictions. For example, one may wish to require that $b$ and $c$ be such that the roots of the equation can be obtained via minimal arithmetic manipulation. The relevant restriction could then be that $b$ and $c$ must be small positive integers (say, smaller than 10) and such that the discriminant $b^2 - 4c$ is a positive integer.

[6]This monograph is a much expanded version of Doignon and Falmagne (1999).

The student would then be given an *instance* of the above item such as

*What are the roots of the equation* $x^2 + 4x + 3 = 0$ ?

The differences among the various instances of an item may sometimes be much more prominent than in the above example, in which they only differ by the number chosen for the parameters $b$ and $c$. This certainly happens if the item is a 'word problem.' Consider, for example, the following instance of a type of word problem combining several decimal operations.

> *Ali works mowing lawns and babysitting. He earns $8.30 an hour for mowing and $9.40 an hour for babysitting. How much will he earn for 5 hours of mowing and 2 hours of babysitting?*

The corresponding item would have, say, seven different story lines involving different activities (selling necklaces, renting DVDs, etc.). In each case, to ensure that the instances are of equivalent difficulty, the dollar amounts could be multiples of $0.10 and chosen in specified intervals (possibly different for each of the story lines) with integer amounts being excluded. In general, all the instances (and story lines, if applicable) of a given item are regarded as equivalent and much care is taken to make them such, so that a student's answer to an instance reveals her mastery of the corresponding item.

Note that in a typical scholarly subject such as elementary algebra, the domain may contain several hundred items[7], and for each item, often several thousand instances.

**1.2.2 The knowledge states.** The *knowledge state* of an individual is represented by the subset of items in the domain that she is capable of answering correctly, barring careless errors or lucky guesses (which frequently occur when the 'multiple choice' format is used[8]). In typical applications of knowledge space theory, much effort is put into avoiding potential lucky guesses. This is consistent with the goal of an assessment, which is to gauge the knowledge state of the student as accurately as possible. Accordingly, multiple choice items are rarely used, and then always with a large number of possible responses.

**1.2.3 Knowledge spaces.** The collection of all the knowledge states forms a *knowledge space* which, by convention, always contains at least the following two special states: (1) the empty state, that is, the state with no items, corresponding to a student knowing nothing in the subject; (2) the entire domain, which is the state of a student knowing everything in the subject. A knowledge space also satisfies another property, which is that for any two states $K$ and $L$ in the space, there is another state in the space, denoted by $K \cup L$, containing

[7] In the current implementation of the ALEKS system, there are 650 items for Beginning Algebra.

[8] We could even say 'systematically occur' in the case of students having taken a test preparation class in which they have been trained to eliminate some of the alternatives and to guess from the remaining ones.

all of the items in $K$ and $L$ combined. To put it concretely: if Kevin is in state $K$ and Lukas is in state $L$, then the theory allows for the possibility that some other student, say Marie, knows everything that Kevin or Lukas knows, that is, everything in $K \cup L$, and nothing more. Except perhaps for very advanced subjects, this property is realistic[9]. It has the important merit of permitting a highly economical summary of the full collection of the states of a knowledge space by a small subcollection of them. Our example with Kevin and Lukas suggests why: starting from $K$ and $L$, we can reconstruct $K \cup L$ whenever we need it without having to store it explicitly. Thus, there is no need to store any union of states in a description of the knowledge space if there is a shortage of computer memory.

In general, we can always represent a knowledge space by a minimal subcollection of states. Such a subcollection is called the *base* of the knowledge space (see Section 8.2 on page 135).

While the concept of a knowledge space is appropriate for assessment purposes, it can be wanting as a core structure for learning. Its chief defect is that its properties do not ensure that the sequencing of the material is adequate: there may be gaps, making it difficult for the system to efficiently channel a student through the material. For example, the four states

$$\varnothing, \{a, b, c\}, \{b, c, d\}, \{a, b, c, d, e\} \tag{1.1}$$

organizing the five items $a$, $b$, $c$, $d$ and $e$, with $\varnothing$ representing the empty state characterizing beginning students, form a knowledge space. However, how can a student progress from the empty state to the next level? It seems that the student must master, at once, either $a$, $b$, and $c$, or $b$, $c$, and $d$. A similar feat is then required at the next step requiring the concurrent mastery of either $d$ and $e$, or $a$ and $e$. Obviously, this is not what would happen: the student would somehow master these items successively, say, first $a$, then $b$. But there is no trace of the possibility of such a structured learning sequence in the knowledge space listed in Line (1.1). Some intermediate knowledge states are clearly missing.

The concept of a 'learning space,' which is a particular kind of knowledge space endowed with a considerable amount of structure, offers a more complete representation, allowing step-by-step mastery of the material by a student.

**1.2.4 Learning spaces.** We illustrate the concept of a learning space by a miniature example[10] in elementary algebra. This example involves 10 items labelled $a, b, \ldots, j$ and is realistic in that the 10 items are part of the domain of an actual learning space used in schools. These 10 items are presented in

**Table 1.1.** The 10 items of the miniature example of Figure 1.1

| Item | One exemplary instance |
|---|---|
| **a.** Quotients of expressions involving exponents | Simplify the following: $\frac{a^4b^5}{5a^6b}$ |
| **b.** Multiplying two binomials | Write the product below in simplest form without parentheses: $(t-6)(t-5)$ |
| **c.** Plotting a point in the coordinate plane using a virtual pencil on a Cartesian graph | Using the pencil, plot the point (-4, 7). |
| **d.** Writing the equation of a line given the slope and a point on the line | A line passes through the point $(x, y) = (-3, 2)$ and has a slope of 6. Write an equation for this line. |
| **e.** Solving a word problem using a system of linear equations (advanced problem) | A scientist has two solutions, which she has labeled Solution A and Solution B. Both A and B contain salt. She knows that A is 45% salt and B is 70% salt. She wants to obtain 70 milliliters of a mixture that is 55% salt. How many milliliters of each solution should she use? |
| **f.** Graphing a line given its equation | Graph the line $5x + 4y = 12$. |
| **g.** Multiplication of a decimal by a whole number | Multiply. $\begin{array}{r} 2.95 \\ \times \quad 5 \\ \hline \end{array}$ |
| **h.** Integer addition (introductory problem) | $-4+4=\square$ $-7+5=\square$ $-3+(-4)=\square$ |
| **i.** Equivalent fractions | Fill in the blank to make the two fractions equivalent: $\frac{3}{5} = \frac{\square}{15}$ |
| **j.** Graphing integer functions | The function h is defined by the rule $h(x) = -x + 1$. Graph $h$ for $x = -1, 0, 1$, and 2. |

Table 1.1. The structure of the learning space organizing these items is shown in Figure 1.1.

[9] At the edge of an advanced field, this property may not hold. If both Kevin and Lukas are experts in an advanced subject, we may not necessarily expect that some other individual has the combined expertise of Kevin and Lukas.

[10] We owe this example, including the graph, to Eric Cosyn.

In this miniature learning space, the 10 items generate 34 possible knowledge states, which are represented by the circles of the figure. The number of dots in each circle indicates the number of items contained in that state. Learning proceeds downward through the figure, following the arrows. The empty state is shown at the top of the figure and the state with all 10 items is shown at the bottom. Clearly, there are many learning paths, that is, many ways to successively acquire the 10 concepts. But, as seen in the graph, there are constraints. For example, it can be checked by close examination that either item $i$ or item $g$ has to be mastered before item $h$. This means that not all learning paths are possible, that is, the corresponding learning space does not contain all possible subsets of the 10-item set. Indeed, there are only 34 states in the graph, while there are $2^{10}$ subsets of a set with 10 elements. This example will be used several times in this volume to illustrate various concepts.

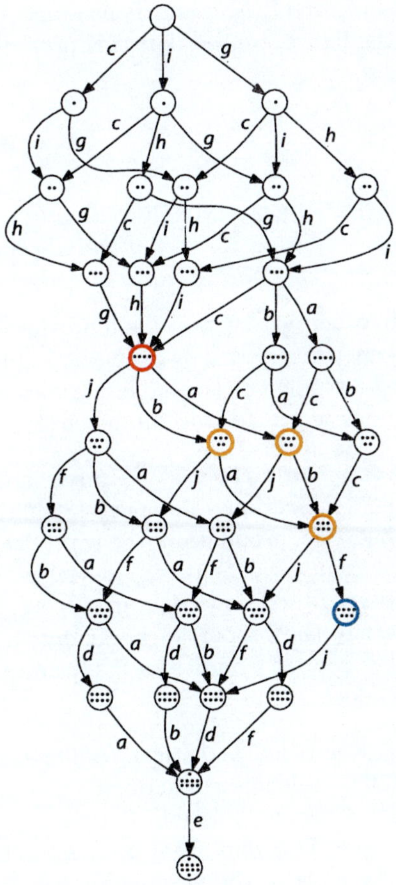


**Figure 1.1.** Graph of the miniature learning space on the items $a, b, \ldots, j$ described in Table 1.1. Each circle stands for a knowledge state. There are 34 states in this example. The top circle is the empty state, and learning proceeds from top to bottom, the items being mastered successively. The number of dots in each circle represents the number of items contained in the corresponding state. For instance, the state represented by the red circle contains the items $c$, $g$, $h$, and $i$, which can be verified by moving down from the empty circle at the top and following the arrows. According to this graph, there are 16 ways to learn these four items and reach the red state: either item $i$ or item $g$ has to precede item $h$. (Ignore the blue circle and the yellow circles for the moment.)

A knowledge space $\mathcal{K}$ is a *learning space* if it satisfies the following two conditions, which can be verified on Figure 1.1:

[L1*] LEARNING SMOOTHNESS. *If the knowledge state $K$ of the student is included in some other knowledge state $L$, then there is a sequence of items $q_1, \ldots, q_n$ that are learnable one at a time, leading the student from state $K$ to state $L$.*

For example, the state represented by the red circle in Figure 1.1 contains the four items $c$, $i$, $h$, and $g$. The state represented by the blue circle contains the same items, plus $b$, $a$, and $f$. There are intermediate states linking these two states. A student can progress from the red state to the blue state by learning successively $b$, $a$, and $f$, passing through some of the yellow states. Examining the figure, we see that this progress can be achieved in exactly two ways:

 $\mapsto a \mapsto b \mapsto f$ and  $\mapsto b \mapsto a \mapsto f$.

[L2*] LEARNING CONSISTENCY. *If item $q$ can be learned by a student in state $K$, then $q$ can also be learned by a student in any state $L$ that includes $K$, or $q$ already belongs to that state $L$.*

In other words, knowing more does not make a student less capable of learning something new. For example, Figure 1.1 indicates that item $j$ can be learned by a student in the red circle state. This item can also be learned by a student in the more advanced blue circle state.

**1.2.5 Remark.** We give a set-theoretic formulation of Conditions [L1*] and [L2*] in Section 8.1 on page 132 in the form of Axioms [L1] and [L2]. In this set-theoretic formulation, the word 'learnable' has disappeared and is replaced by mathematical relations, leaving all pedagogical connotations aside. Learning spaces (and knowledge spaces) are, after all, simply mathematical objects that satisfy the properties in Axioms [L1] and [L2]. We write these properties in the form of Conditions [L1*] and [L2*], though, to provide intuition for the reader.

Many mathematical results can be derived from Conditions [L1*] and [L2*], which are described in detail in Falmagne and Doignon (2011) (see also Chapter 8). One particular consequence of [L1*] and [L2*], called the 'Fringe Theorem', that is essential from an educational standpoint, is recalled informally in the next paragraph.

A knowledge state may sometimes be quite large. In elementary algebra, for example, the knowledge state of a student may contain more than 600 items. Representing a student's mastery of this subject by such a long list of items is not helpful. Fortunately, in a learning space, any knowledge state can be specified by its two 'fringes,' which are almost always much smaller sets.

**1.2.6 The two fringes of a knowledge state.** Examining Figure 1.1 once more, we see that there are three arrows pointing downward from the red state: three larger states can be reached from the red state by learning one of $j$, $b$, or $a$. On the other hand, there are four arrows pointing, from above, directly to the red state: the red state can be reached directly from some smaller state by learning one of $g, h, i$, or $c$. The pair of concepts behind these observations are as follows.

The *outer fringe* of a state $K$ is the set of all the items $q$ such that adding $q$ to $K$ yields another knowledge state. In line with Remark 1.2.5, we can say that the outer fringe of a student's state contains all the items that the student is *ready to learn.*

The *inner fringe* of a state $K$ is the set of all the items $q$ such that removing $q$ from $K$ yields another knowledge state. In other words, the inner fringe contains all the items representing the 'high points' of the student's competence. For example, the inner fringe of the blue state of Figure 1.1 is the set containing just item $f$. The outer fringe of the blue state also contains just one item, namely $j$. Notice that neither of these two sets—the outer fringe and the inner fringe of the blue state—are states.

The crucial role played by the fringes resides in the following result.

FRINGE THEOREM. *In a learning space, the knowledge state of a student is defined by its inner fringe and outer fringe. Thus, if the results of an assessment are summarized in the form of the two fringes of a state, the state is exactly specified* (Falmagne and Doignon, 2011, Theorem 4.1.7, (i) $\Rightarrow$ (v)).

The economy of representation afforded by this result is illustrated by the example of the blue state: in the framework of the learning space of Figure 1.1, this state can be specified by two lists containing just one item each—$f$ for the inner fringe and $j$ for the outer fringe—rather than by the full list of seven items contained in that state. In applications of the theory to a full domain such as elementary algebra, the economy is considerable because a state may contain several dozen to several hundred items, while the outer fringe and the inner fringe may involve fewer than a dozen items.

Tables 1a and 1b contain the two fringes of an exemplary knowledge state in elementary algebra, with each item being represented by one instance. Taken together, the two fringes amount to 14 items. This suffices to specify the 134 items contained in that knowledge state. The economy is notable. Moreover, the summary is more informative to a teacher than a grade or a percentile and certainly more useful. As the type of information contained in Table 1a is available for all the students in a class, it can be used by the teacher to gather students in small, homogeneous groups ready to learn a given item, fostering efficient learning.

**Table 1.2.** A knowledge state in elementary algebra specified by its two fringes

| **Table 1a. Outer fringe** (9 problems): STUDENT IS READY TO LEARN: |
|---|
| *Word problem with linear inequalities:*<br>The sum of two numbers is less than or equal to 13. The second number is 5 less than the first. What are the possible values for the first of the two numbers? |
| *Solving a rational equation that simplifies to a linear equation (Type 1):*<br>Solve for $u$: $-6 = -\frac{8}{u}$. |
| *Word problem on mixed number proportions:*<br>A chocolate chip cookie recipe requires one and one quarter cups of flour for every cup of chocolate chips. If two and one half cups of flour are used, what quantity of chocolate chips will be needed? |
| *Y-intercept of a line:*<br>Find the $y$-intercept of the line whose equation is $y = \frac{17}{15}x - \frac{5}{4}$. |
| *Multiplying polynomials:*<br>Multiply and simplify: $(6z + 6w - 1)(5z + 3w - 3)$. |
| *Word problem on inverse proportions:*<br>Suppose that 8 machines can complete a given task in 5 days. If there were 10 machines, how many days would it take for them to finish the same task? |
| *Word problem on percentage (Type 3):*<br>The price of a gallon of gas has risen to \$2.85 today. Yesterday's price was \$2.79. Find the percentage increase. Round your answer to the nearest tenth of a percent. |
| *Area and perimeter of a rectangle:*<br>The length of a rectangle is twice its width. If the area of the rectangle is 162 $\text{ft}^2$, find its perimeter. |
| *Union and intersection of sets:* The sets $F$ and $A$ are defined by<br>$F = \{x \mid x \text{ is an integer and } -4 < x \leq 0\}$,<br>$A = \{x \mid x \text{ is an integer and } 1 < x \leq 3\}$.<br>Find $F \cup A$ and $F \cap A$. |

| **Table 1b. Inner fringe** (5 problems) : STUDENT IS READY TO LEARN (HIGH POINTS MASTERED): |
|---|
| *Power rule—Positive exponents:*<br>Write without parentheses: $(\frac{y^3}{-2x})^3$. |
| *Squaring a binomial:*<br>Expand the square: $(6x-6)^2$ |
| *Properties of real numbers:*<br>For each equation below, indicate the property of real numbers that justifies the equation:<br>$\frac{3}{4}+(m+b)=(\frac{3}{4}+m)+b$<br>$7\cdot\frac{1}{7}=1$<br>$0=4+(-4)$<br>$m(\frac{3}{5}+7)=m\cdot\frac{3}{5}+m\cdot 7$ |
| *Solving a linear inequality (Type 4):*<br>Solve for $t$: $-\frac{7}{2}t+9>-8t-3$. |
| *Writing a negative number without a negative exponent:*<br>Rewrite without an exponent: $(-3)^{-1}$. |

Some readers of a previous draft of these pages suggested that the outer fringe of a state could be seen as a formal implementation of the concept of *zone of proximal development* or *ZPD* in the sense of Vygotsky (1978) (for a recent reference, see Chaiklin, 2003). This remark is particularly apt because the ZPD explicitly includes the intervention of external agents, such as human teachers or other students (via cooperative problem solving), for the zone of proximal development to be conquered. This is consistent with the possible use of the outer fringe just mentioned, which involves the selection of the part of the class that is prepared for pointed instruction on a particular topic.

## 1.3 Uncovering a Knowledge State

**1.3.1 The assessment engine.** The task of the assessment engine is to uncover, by efficient questioning, the knowledge state of a particular student under examination. Because the number of items in a typical domain is large, only a small subset of items can be used, in practice, in any assessment. So, the successive questions must be cleverly chosen. The situation is similar to that of *adaptive testing*[11], except that the outcome of the assessment here is a knowledge state, rather than a numerical estimate of a student's competence in the subject. The procedure follows a scheme outlined in Figure 1.2. It has two main components: the *questioning rule* and the *updating rule.*

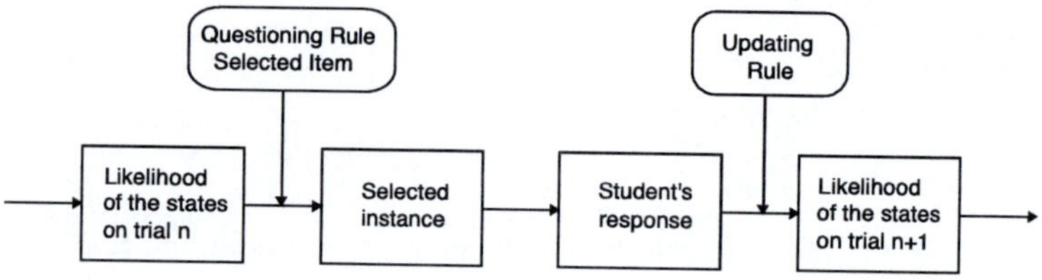


**Figure 1.2.** Diagram of the transitions in the assessment procedure. Four successive operations are involved: the selection of a maximally informative item on the basis of the current probabilities of the states, the choice of an instance, the recording of the student's response and its verification, and the updating of the probabilities.

At the outset of the assessment (at the beginning of trial 1 of the procedure), each of the knowledge states is assigned a certain *a priori* likelihood or probability, which may depend upon the school year of the student if it is known, or some other information. These initial probabilities represent what the assessement engine knows about the student prior to the assessment. The sum of these probabilities is equal to 1. They play no role in the final result of the assessment but may be helpful in shortening it. If no useful information is available, then all the states are assigned the same probability. The first item $q_1$ is chosen so as to be 'maximally informative.' There is more than one way to interpret this term. In most applications of learning space theory, it is construed to mean that, on the basis of the current probabilities of the states, the student has about a 50% chance of knowing how to solve $q_1$. In other words, the sum of the probabilities of all the states containing $q_1$ is as close to .5 as possible[12]. If several items are equally informative in that sense, as may happen at the beginning of an assessment, one of them is chosen at random. The student is then asked to solve an instance of that item, also picked randomly.

The student's response is checked by the system, and the probabilities of all the states are modified according to the following *updating rule.* If the

[11] Such as various K-12 benchmark tests and the computerized forms of some standardized tests, e.g. the GRE and GMAT (c.f. Wainer et al., 2000).

[12] A different interpretation of 'maximally informative' has also been investigated, based on the minimization of the expected 'entropy' of the distribution. (The entropy is a statistical index measuring uncertainty. It is akin to the variance of a distribution of numbers. See the Glossary entry on page 23.) This method did not result in an improvement and was computationally more demanding.

student gave a correct answer to $q_1$, it becomes more likely that the student's knowledge state contains $q_1$. Accordingly, the probabilities of all the states containing $q_1$ are increased and, correspondingly, the probabilities of all the states *not* containing $q_1$ are decreased. The exact updating formula ensures that the overall probability, summed over all the states, remains equal to 1. A false (meaning "incorrect") response given by the student has the opposite effect: the probabilities of all the states *not* containing $q_1$ are increased, and those of the remaining states are decreased. The exact formula of the operator modifying the probability distribution is given in Equation (8.9) on page 142 and Definition 13.44 in Falmagne and Doignon (2011). It is proved in the latter reference that the operator is commutative, in the sense that its cumulative effect in the course of a full assessment does not depend upon the order in which the items have been proposed to the student[13].

In some implementations of LST, a student who does not know how to solve an item is allowed to answer "I don't know"—that is, to click the corresponding button on the screen—instead of guessing. This results in a substantial increase[14] in the probability of the states **not** containing $q_1$, thereby decreasing the total number of questions required to uncover the student's state. Typically, responding "I don't know" only improves the efficiency of the questioning and does not affect the final state uncovered.

Item $q_2$ is then chosen by a mechanism identical to that used for selecting $q_1$, and the probability values are increased or decreased according to the student's answer via the same updating rule. Further problems are dealt with similarly. In the course of the assessment, the probability of some states gradually increases.

The assessment procedure stops when two criteria are fulfilled: (1) the entropy of the probablity distribution, which measures the uncertainty of the assessment system regarding the student's state, reaches a critical low level; and (2) there is no longer any useful question to be asked (all the items have either a very high or a very low probability of being solved correctly). At that moment, a few likely states remain and the system selects the most likely one among them. Note that, because of the probabilistic nature of the assessment procedure, the final state may very well contain an item to which the student gave a false response. Such a response is thus regarded as due to a careless error. Because all the items have either open-ended responses or multiple choice responses with a large number of choices, the probability of lucky guesses is negligible, at least in principle.

[13]This commutativity property is consistent with the fact that, as shown by Mathieu Koppen (see Falmagne and Doignon, 2011, Remark 13.45), this operator is Bayesian.

[14]As compared to the case of an incorrect response, which could be attributed to a careless error.

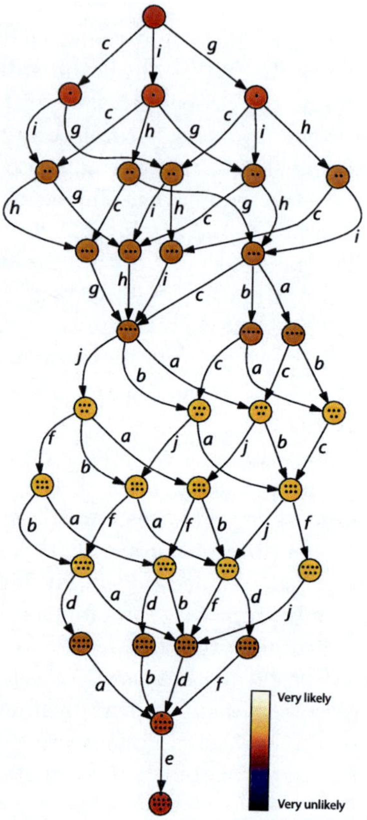


**Figure 1.3.** Initial probabilities of all the states at the beginning of the assessment. The probability is marked by the shading of the circles representing the states. Bluish black means very unlikely, and bright yellow very likely.

**1.3.2 An example.** We illustrate the evolution of an assessment with our example of the 10 items pictured in Figure 1.1. The initial probabilities are pictured on the graph of Figure 1.3 by colors varying from reddish to yellow. Except for these colors, this graph is identical to that of Figure 1.1. So, this graph represents a learning space on the ten items $a$, $b$, ..., $j$ described in Table 1.1. The colors of each circle represents the initial probability of the corresponding state. The full color scale indicated at the bottom goes from bluish black for very unlikely, to bright yellow for very likely. The somewhat lighter colors of some circles shows that the system thinks the student might have mastered about 5 to 7 items.

The system then asks an instance of the first item, which is chosen to be as informative as possible, on the basis of the current probabilities of the states. The adopted rule is to ask the student an item that has about a 50% chance of eliciting a correct response, on the basis of the current probabilities.

Suppose that the first item asked is $a$, and the student's response is correct. The updating rule will then increase the probabilities of all the states containing $a$, and decrease the probabilities of all those states not containing $a$. This is pictured by the leftmost graph of Figure 1.4 in which the colors of the circles are brightened and darkened accordingly. If the next item asked is $f$ and the response is incorrect, the updating rule will then decrease the probabilities of the state containing $f$ and increase the probabilities of the states not containing $f$. The middle graph pictures the result of such an updating. At the end, the situation may be as pictured by the graph on the right. Only one state remains with a very high probability. The system will then chose the state containing $a$, $b$, $g$, $h$, and $i$ as being the current state of the student.

We give a formal, axiomatic description of the stochastic process in Chapter 8. For full details, see Falmagne and Doignon (2011, Chapter 13).

In the case of elementary algebra, which is often called 'beginning algebra', the domain contains about 650 items. A typical assessment terminates after 25 to 30 questions selected in that domain.

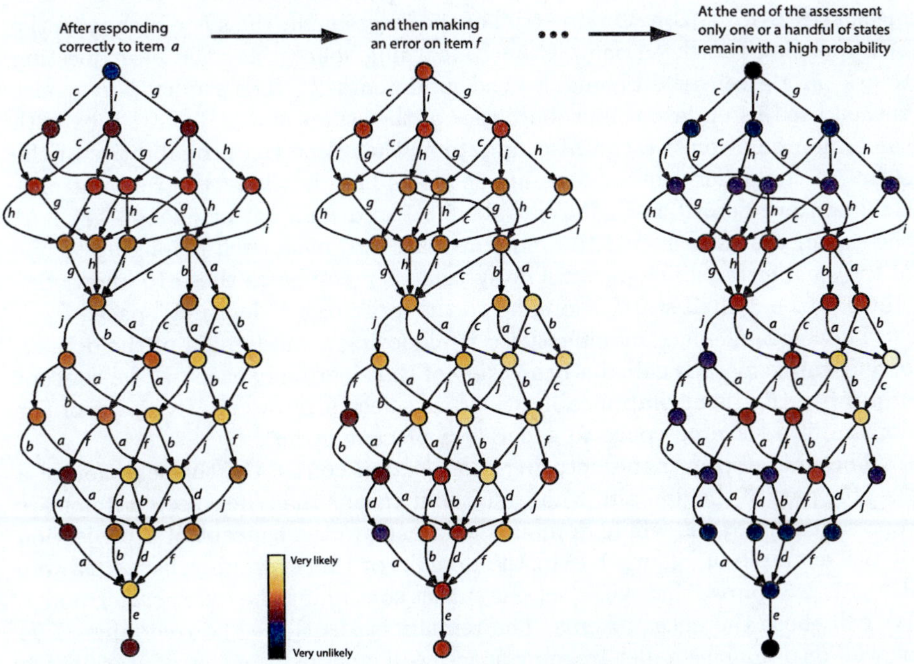


**Figure 1.4.** On the left, the effect of a correct response to item $a$ on the state's probabilities is marked by brightening the yellow shades for states containing $a$, and darkening those shades for those states not containing $a$. The graph in the middle depicts the subsequent effect of an incorrect response to item $f$. The graph on the right represents a possible situation at the end of the assessment. The assessment engine will pick the state containing $a$, $b$, $g$, $h$, and $i$ as being the current state of the student.

## 1.4 The Case of Very Large Knowledge Spaces

**1.4.1 Projections of a learning space.** A problem arises if the learning space is very large, as is almost always the case in practice. Operations on such a learning space cannot be handled by standard servers in the straightforward manner described above. To resolve this problem, the domain of the learning space may be partitioned into subdomains, so that only some part of the learning space has to be dealt with by the computer at any time. We only give here a sketch of the assessment procedure in such a case, the details of which are given Chapter 8 on page 131. Note that this more complex procedure is that routinely used in the ALEKS system, which is practically always dealing with such large learning spaces.

Consider a case in which the domain has been partitioned into 10 subdomains $Q_1, \ldots, Q_{10}$. Each of these 10 subdomains defines a smaller learning space (see Section 1.6 and Theorem 8.4.1 on page 138). We now have 10 learning spaces $\mathcal{L}_1, \ldots, \mathcal{L}_{10}$. On any trial of the assessment, the system chooses the best question to ask among all the 10 learning spaces. Say the best question to ask on a given trial belongs to the subdomain $Q_i$. The student's response is then used to update the probabilities of the states of $\mathcal{L}_i$. Remarkably, with some manipulation, we can also update at that time the probabilities of the states of the other 9 other learning spaces. (The details of the relevant construction are described in Paragraphs 8.8.1 and 8.8.2 on pages 144-145.) At the end of the assessment, 10 final states are obtained, each belonging to one of the smaller learning spaces. The system then combines these 10 knowledge states into one final state belonging to the large parent learning space.

The learning space manufactured by choosing a subdomain of the domain of a learning space is called a *projection* of that learning space[15]. The concept of a projection is also applicable when we select a representative part of the domain of a learning space to generate a placement test.

The first step in manufacturing a projection consists of the selection of a suitable part of the domain, where 'suitable' means that the selected items are representative of the whole domain. We illustrate the concept of a projection by our example of Figure 1.1. In the domain of this example, the items were $a, b, \ldots, j$. Suppose that we select the subset containing the items $c, d, g$, and $j$. We call those the *chosen items*. The remaining items are the *remnants*. Two new kinds of much smaller learning spaces will emerge from this operation, via some construction. Notice that the selection of the chosen items automatically gathers the original 34 knowledge states into classes. The knowledge states in each class differ only by the remnants, while two items in different classes differ by some of the chosen items. The two graphs of Figure 1.5 illustrate this operation. Graph A on the left is identical to that of Figure 1.1, except for the colors which indicate the classes of states defined by the set of chosen

[15] A brief, formal discussion of this concept is given in Section 8.4, p. 137. The relevant paper is Falmagne (2008).

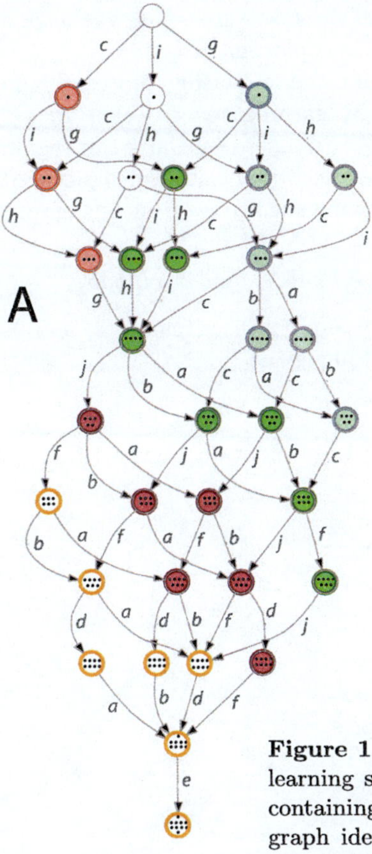


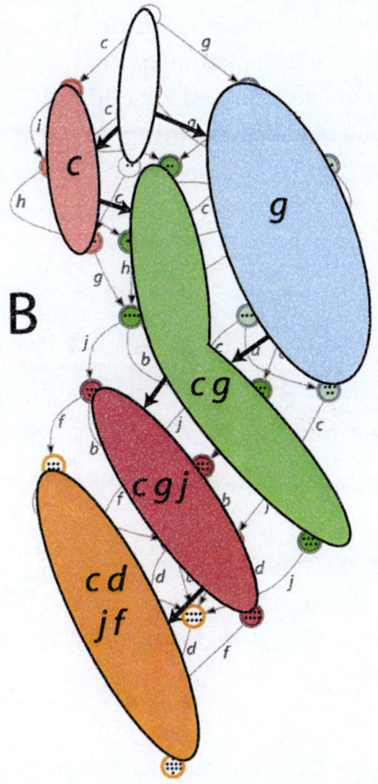


**Figure 1.5.** Two graphs illustrating the projection of the learning space of Figure 1.1 on the subset of the domain containing the chosen items $c, d, g$, and $j$. On the left is a graph identical to that of Figure 1.1 but with colors indicating the classes of states defined by the chosen items. In each class, the states differ only by remnant items. On the right is the projection, which is a learning space. Each class of states corresponds to a state of the projection. The states of the projection only contain chosen items.

items. States of the same color differ only by the remnants, while two states of different color always differ by some of the chosen items. Graph B on the right depicts the new learning space created by the chosen items. Each of the five colored regions is a knowledge state of that learning space, with the white ellipse representing the empty state.

From the empty state at the top the student can progress by learning first $c$ and then $g$, or the other way around. After that, the learning order is determined: $j$ and then $d$. It is worth noting that any subset of items from the domain will generate a learning space of a smaller size. If that subset is well chosen and forms a representative coverage of the whole domain, then an assessment on that smaller learning space will provide a first-pass evaluation of a student's competence in the subject—as in a placement test.

What is remarkable, if less obvious, is that the knowledge states within each of the six classes can also be recast as a learning space. Take, for example, the state containing $c$ and $g$ (which is represented by the green region containing $cg$ in Graph B). This state was generated by the six states pictured by the green circles of Graph A, namely (following the arrows from the empty circle of Graph A)[16]

$$cig, \quad cihg, \quad cihgb, \quad cihga, \quad cihgab, \quad \text{and} \quad cihgabf.$$

Notice that these six sets of items have in common the items $c$, $i$, and $g$. Removing these common items gives the learning space

$$\text{empty state}, \quad h, \quad hb, \quad ha, \quad hab, \quad \text{and} \quad habf\,.$$

Graphing this result in the style of Figure 1.5, we obtain Figure 1.6.

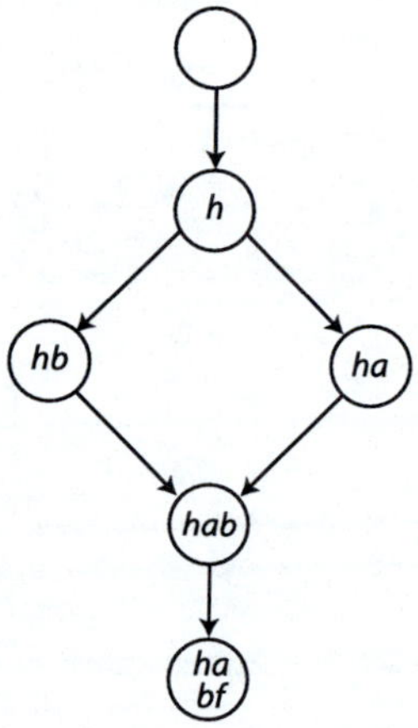


**Figure 1.6.** The learning space represented by the graph on the right is induced by the knowledge state containing $c$ and $g$ in the projection learning space graphed in Figure 1.5.

The concept of a projection can thus be used for two important purposes.One is to allow assessments on very large learning spaces, representing an extensive coverage of a subject. To this end, we partition the domain of the large learning space into a suitable number $N$ of subdomains. We then perform $N$ 'simultaneous', 'parallel' assessments on the $N$ learning spaces generated by such a partition. The final state is then manufactured by combining the final states obtained in these $N$ assessments. For more details, see Section 8.8.

The second purpose is the construction of an efficient placement test. Assessing a student on a full learning space on a subject may be impractical for a placement test because the number of questions would be too large. The solution is then to select a representative subset of the domain, and to perform an assessment on the resulting projection learning space.

A technical discussion of these matters can be found in Section 8.4 of Chapter 8, on p. 137.

[16]Which path is taken is immaterial.

## 1.5 Historical Note

The first paper on the topic of knowledge spaces was Doignon and Falmagne (1985), in which the combinatorial basis of knowledge space theory was presented. A technical follow-up paper was Doignon and Falmagne (1988). The stochastic aspects of the theory were developed in two papers by Falmagne and Doignon (1988a,b). A comprehensive description of the main ideas, intended for non-mathematicians, is contained in Falmagne et al. (1990). Other introductory texts are Doignon and Falmagne (1987), Falmagne (1989a) and Doignon (1994a). During these maturing years other scientists—the second generation—became interested in knowledge space theory, notably Mathieu Koppen in the Netherlands, Düntsch and Gediga, and the group around Cornelia Dowling[17] in Germany, and the teams of Dietrich Albert with Josef Lukas in Germany and with Cord Hockemeyer in Austria. Meanwhile members of these teams became independent and other scientists joined, constituting the third generation dealing with knowledge spaces: Jürgen Heller, Martin Schrepp, Reinhard Suck, Ali Ünlü in Germany, and Luca Stefanutti in Italy—to mention only the most active authors[18].

The literature on the subject grew steadily and now contains several hundred titles. An extensive database of references on knowledge spaces is maintained by Cord Hockemeyer at the University of Graz:

http://liinwww.ira.uka.de/bibliography/Ai/knowledge.spaces.html

(see also Hockemeyer, 2001).

The monograph by Doignon and Falmagne (1999) contains a comprehensive description of knowledge space theory at that time. An educational software grounded on knowledge space theory was developed by a team of software engineers, including in particular Eric Cosyn and Nicolas Thiéry, who are two of the co-authors of Chapter 2, and also Damien Lauly and David Lemoine, under the supervision of Falmagne. This work took place at the University of California, Irvine, with the support of a large NSF grant. Application in schools and universities began around 1999 in the form of the internet-based software `ALEKS`. This provided the impetus for further developments, in the form of statistical analyses of students' data and also of new mathematical results. A source for some of the results is Falmagne et al. (2006), which can be regarded as none-too-technical introduction to the topic. The monograph by Falmagne and Doignon (2011, a much expanded reedition of Doignon and Falmagne, 1999) contains a comprehensive and up-to-date technical presentation of the theory.

The volume edited by Albert and Lukas (1999) contains the results of the KST research group at the University of Heidelberg (Germany). Their aims

[17]Formerly Cornelia Müller; see this name also for references.

[18]See Chapters 9 to 12 in this volume for some recent works by some of these scientists.

have been to clarify the structures and processes underlying KST, as well as to validate and to apply them empirically.

In the course of the above mentioned applications, an important theoretical change of focus took place, which was introduced in Paragraph 1.2.3. Originally, the essential property of a knowledge space resided in the so-called 'closure under union' axiom: if $K$ and $L$ are knowledge states, then there exists a state containing all the items that are either in $K$ or in $L$, or in both. While adoption of such a rule can be convincingly demonstrated to be sound, and pedagogical justifications for it have been given in the past[19], it allows for possible "gaps" in the learning sequence, as can be seen in the example in Paragraph 1.2.3.

These considerations, added to the less than compelling closure under union axiom, led Falmagne to formulate a re-axiomatization of the theory in the guise of a learning space (cf. Paragraph 1.2.4). The new axioms [L1*] and [L2*], stated informally[20] on pages 9 and 10, are considerably more compelling from a pedagogical standpoint and now form the core of the theory. Cosyn and Uzun (2009) have shown that a knowledge structure is a learning space if and only if it is a knowledge space satisfying a condition of 'wellgradedness' (see Theorem 8.1.3 on page 134).

The applications continue to increase in number and diversity, promoting new theoretical advances. This volume is an attempt to outline the current state of development from both an empirical and theoretical standpoint.

## 1.6 A graph of a Larger Structure

Even with a couple of dozen items, the complexity of the structure can be daunting. As an illustration, we give below the graph of the learning space of 37 items in Beginning Algebra. There are 4615 knowledge states in the corresponding learning space. This graph is of the same type as those of Figures 1.1 and 1.3 with the empty state at the top, and learning proceeding downwards. Note that there are 217 different knowledge states containing exactly 25 items. All these knowledge states would be assigned the same psychometric score. This learning space is one of the ten learning spaces discussed in 1.4.1.

[19]The rationale given in Doignon and Falmagne (1999) was that if two students, whatever their respective knowledge states, collaborate intensely, then one of them may end up mastering all the items originally mastered by either or both of them.

[20]For the mathematical formulation of [L1*] and [L2*], see Section 8.1 on page 132.

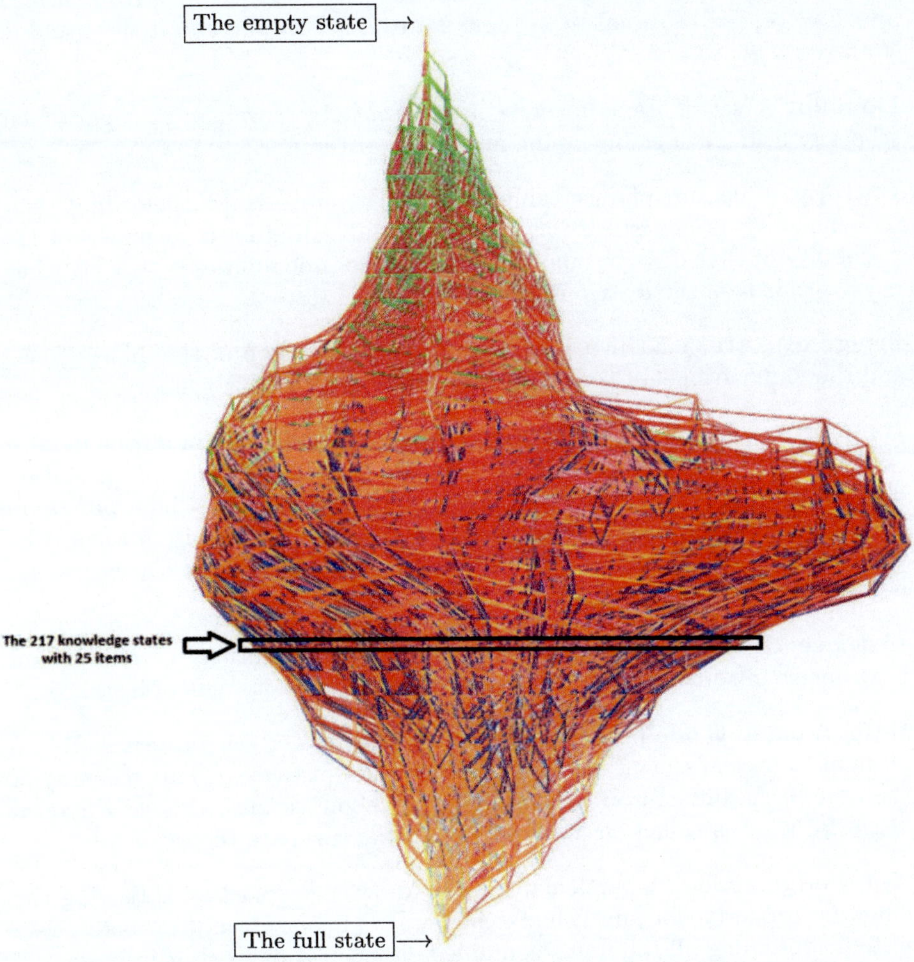


## 1.7 Nontechnical Glossary

The spirit of this glossary is consistent with the rest of this chapter and contains mostly informal, intuitive definitions of the concepts. The reader should consult Chapter 8 for some of the corresponding set-theoretical definitions.

**Base of a knowledge space.** The unique minimal subset of the family of states of the knowledge space; the base can be used to generate any state of the knowledge space by forming the union of some suitable states in the base.

**Distance between two states.** The number of items by which the two states differ; in other words, the number of items contained in one or the other of the two states, but not in both. The distance between the states $K = \{a, b, c\}$ and $L = \{a, c, d\}$ is equal to 2: these states differ by the two items $b$ and $d$. We write $d(K, L) = 2$.

**Domain.** A set of items forming a representative coverage of a particular scholarly field, such as beginning algebra or arithmetic.

**Entropy.** A measure of uncertainty arising in several scientific fields (e.g. thermodynamics); in statistics, the entropy of a distribution is an index of the variability of that distribution: if $n$ events have probabilities $p_1, \ldots, p_n$, then $-\sum_{i=0}^{n} p_i \ln p_i$ is the distribution's entropy.

**Fringe of a state $K$ in a learning space.** The union of the 'inner fringe' and the 'outer fringe' of that state; see these terms in this glossary.

**Inner fringe of a state $K$ in a learning space.** The set of all the items $q$ such that removing $q$ from $K$ forms another state; a common interpretation is that the items in the inner fringe of a state represent the 'high points' in a student's competence. Each one of them could be the last item learned by the student. This interpretation is empirically valid only if the learning space is a faithful representation of the learning process.

**Instance of an item.** A particular way of implementing the item in an assessment; in some cases, an item may have many thousands of instances.

**Item.** A question or a problem expressed in abstract or general terms. To use an item in an assessment, one selects one of its 'instances'. This meaning of the term 'item' differs from that used by psychometricians, who use the term 'item' for what is called an 'instance' in knowledge space theory.

**Knowledge space.** A particular kind of knowledge structure satisfying the following property: for any two states $A$ and $B$ in the structure, there exists a state containing all the items contained either in $A$ or in $B$ or in both.

**Knowledge state.** A set of items representing a possible competence state of a student in a scholarly field: barring careless errors, the student can solve all the items in her knowledge state, and none of the other items.

**Knowledge structure.** The collection of all the feasible knowledge states for a given scholarly domain; thus, a knowledge structure is a set of sets. A knowledge structure always contains at least two special states: the state containing all the items (the student knows everything) and the empty state (the student knows nothing).

**Learning path** or **learning sequence**[21]. A sequence of distinct items $q_1, q_2, \ldots, q_n$ in the domain of a learning space, such that:

[21] Cf. Chapters 13 and 14.

1. $n$ is the number of items in the whole domain; thus, a student knowing everything has mastered all those items;
2. for every sequence of items $q_1, q_2, \ldots, q_j$, with $j < n$, there is a state containing just those items; a student thus can progress from knowing nothing at all to mastering just $q_1$, then mastering $q_2$, etc.

As there are millions of knowledge states in any realistic learning space, there are also certainly trillions of learning paths in such a space.

**Learning space.** A particular knowledge space in which every state has an inner fringe and an outer fringe and, moreover, these two fringes suffice to identify the state; a learning space is specified by the following two principles:

[L1*] If the state $K$ of a student is included in a larger state $K'$, then it is possible for the student to learn, one at a time, in some order, the items of $K'$ that are missing in $K$;

[L2*] If the state $K$ of a student is included in a larger state $K'$ and some item $q$ is learnable by that student, then any student in state $K'$ either has already learned $q$ or can learn that item (without having to learn any other item beforehand).

The intuitive pedagogical language in which these two principles are formulated arise from an interpretation of two set-theoretical axioms given in Section 8.1 on page 132. These axioms imply that a learning space is a knowledge space that also satisfies a 'well-gradedness' condition (see the entry for this term in this glossary).

**Outer fringe of a state $K$ in a learning space.** The set of all the items $q$ such that adding $q$ to $K$ forms another knowledge state; a common interpretation is that a student in state $K$ is 'ready to learn' item $q$. In other words, one could say that the outer fringe of a state $K$ is the set of all the items that a student in state $K$ is ready to learn. As in the case of the inner fringe, this interpretation supposes that the learning space is a faithful representation of the learning process.

**Projection of a learning space.** A learning space obtained by selecting a subset of items from the domain of another learning space $L$, typically of a much larger size. Such a selection automatically gathers the original knowledge states of $L$ into classes. In turn, each of these classes can itself be recast as a learning space. The concept of a projection can be used to generate a placement test. It is also instrumental in the construction of an assessment in the case of very large learning spaces. (Cf. Section 1.4 on page 18 and Section 8.4 on page 137.)

**QUERY.** A type of software, based on an algorithm designed for interviewing an expert, such as a classroom teacher, regarding a knowledge space under construction; the software asks questions such as, "*Suppose that a student is not capable of solving items $q_1, q_2, \ldots, q_j$. Would such a student be capable* *of solving item q?"*. An application of QUERY gives the gradual construction of the particular knowledge space specific to the expert interviewed. This algorithm can also be used to analyze student assessment data.

**Reliability.** *The extent to which a measurement made repeatedly in the same circumstances will yield concordant results.* (O.E.D. 2000 Edition.) In psychometric theory a test is deemed to be reliable if similar versions of the same test administered to the same sample of subjects give highly correlated results. In our terminology, the two versions of the same test would contain different instances of the same items. (An extended discussion of the concept of reliability can be found in Crocker and Algina, 1986, Chapters 6–9.)

**Validity.** A psychometric test is regarded as valid if its result correlates well with a relevant criterion. For example, a standardized test of quantitative abilities, taken before the beginning of a college education, would be regarded as valid if the correlation between the results of the test and the grades obtained in a mathematics course taken during the first semester is sufficiently high. (Chapter 10 of Crocker and Algina, 1986, is devoted to this concept.)

**Well-graded.** A knowledge structure is *well-graded* if for any two states $K$ and $L$ at distance $n$, you can 'travel' from $K$ to $L$ in $n$ steps, hopping from one state to the next, without leaving the structure. It can be proven that the collection of states in a learning space is well-graded. For a more precise definition, see Definition 8.1.2 on page 134.

# 2

# Assessing Mathematical Knowledge in a Learning Space[1]

Eric Cosyn[2]
Christopher Doble[2]
Jean-Claude Falmagne[2,3]
Arnaud Lenoble[2]
Nicolas Thiéry[2]
Hasan Uzun[2]

## 2.1 Introduction

According to Knowledge Space Theory (KST) (cf. Doignon and Falmagne, 1999; Falmagne and Doignon, 2011), a student's competence in a mathematics or science subject, such as elementary school mathematics or first year college chemistry, can be described by the student's 'knowledge state,' which is the set of 'problem types' that the student is capable of solving. (In what follows, we abbreviate 'problem type' as 'problem' or 'item.') As the student masters new problems, she moves to larger and larger states. Some states are closer to the student's state than others, though, based on the material she must learn in order to master the problems in those states. Thus, there is a structure to the collection of states, and this structure gives rise to a 'learning space,' which is a special kind of knowledge space. These concepts have been discussed at length in Chapter 1 of this volume. We recall here that the collection of states forming a *learning space* always contains the 'empty state' (the student knows nothing at all in the scholarly subject considered) and the 'full state' (the student knows everything in the subject). The collection of states must also satisfy two pedagogically cogent principles, which we state below in non-mathematical language.

Consider two hypothetical students S and S', with S' knowing everything that S knows, and more. Then the following hold.

[L1*] Student S can catch up with Student S' by learning the missing concepts one at a time.

---

[1]We are grateful to Brian Junker, Don Laming and two anonymous referees for their useful comments on an early presentation of this material.

[2]ALEKS Corporation.

[3]University of California, Irvine.

J.-C. Falmagne et al. (eds.), *Knowledge Spaces: Applications in Education*, DOI 10.1007/978-3-642-35329-1_2, © Springer-Verlag Berlin Heidelberg 2013

[L2*] Any new concept that Student S is ready to learn either was already mastered by Student S', or S' is also ready to learn it.

Set-theoretical formulations of [L1*] and [L2*] are given in Chapter 8, the introductory chapter to Part II entitled '*Learning Spaces: A Mathematical Compendium*' on page 131. While these two principles may seem rather trite at first blush, we have seen in Chapter 1 that they have strong, non-obvious implications. In particular, the *Fringe Theorem* results from these principles, which states that every state in a learning space is defined by its two fringes (cf. page 11). Thus, we can summarize the result of an assessment by the two fringes of the uncovered knowledge state without any loss of information. The significance of this result from an educational standpoint is that we can interpret the outer fringe as the set of problems that the student is ready to learn. In other words, the assessment gives a precise access to further learning.

Several assessment systems are founded on KST, the most prominent ones being `ALEKS` and `RATH` (see Hockemeyer, 1997a). The focus of this chapter is an examination of the `ALEKS` system, whose assessments are adaptive and taken on-line. We evaluate the validity of these assessments on the basis of student data. Our usage of the term 'validity' in this context requires some comment.

## 2.2 Measuring the Validity/Reliability of an Assessment

An assessment in a learning space contrasts with a psychometric test, whose aim is to obtain a numerical score[4] indicative of the competence of a student in a scholarly subject. In the latter case, the validity of such a measurement is paramount because, *a priori*, there is no immediate, obvious connection between a numerical score of competence in a subject, such as elementary algebra, and the ability to solve a particular problem, such as a quadratic equation by the method of the discriminants, or a word problem on proportions. Such a connection is especially questionable in view of the standard methods used in the construction of such a test, which are based on a criterion of homogeneity of the problems in the test: a problem whose response is poorly correlated with the overall result of the test may be eliminated, even though it may be an integral part of the relevant curriculum. In principle, the situation is quite different in the case of the learning space because the collection of all the items[5] potentially used in any assessment is, by design, a fully comprehensive coverage of a particular curriculum. Arguing that such an assessment, if it is reliable, is also automatically endowed with a corresponding amount of validity is plausible. In other words, granting that the database of problem types is a faithful representation of the curriculum, the measurement of reliability is confounded with that of validity.

[4]Or, in some cases, a numerical vector with a small number of terms.

[5]We recall that we use "item" or "problem" to mean "problem type." The actual question asked is an 'instance' of a problem type.

In any event, we use the following method to evaluate the reliability and validity of the results. In each assessment, an *extra* problem (item) $\boldsymbol{p}$ is randomly selected in a uniform distribution on the set of all problems. Then an instance of $\boldsymbol{p}$, also randomly selected, is given to the student, whose response is not taken into account in assessing the student's state. On the basis of the knowledge state uncovered by the assessment, a prediction can be made regarding the student's response to the extra problem $\boldsymbol{p}$, and the accuracy of the prediction can be evaluated. We shall also investigate the evolution of the accuracy of this prediction in the course of the assessment.

In the rest of this chapter, we describe a large scale study performed to test the validity and reliability of an assessment in elementary school mathematics covering grades 3 to 6. The analysis is based on the particular learning space for this subject used by the `ALEKS` system.

## 2.3 Statistical Methods

**2.3.1 Outline of three methods.** Three different types of data analysis were performed, based on a large number of assessments taken from June 2009 to December 2012.

1) The most obvious method of predicting the student's actual response (correct or false) to the extra problem $\boldsymbol{p}$ is to check whether or not $\boldsymbol{p}$ appears in the student's knowledge state selected by the assessment algorithm at the end of the test. The effectiveness of such predictions can be evaluated using standard measures of correlation between two dichotomous variables, such as the tetrachoric coefficient or the phi-correlation coefficient. This type of analysis is one of the cornerstones of this chapter and is contained in Paragraph 2.4.4. This analysis does not take possible careless errors into account. A variant of the above method, described in Paragraph 2.4.5, uses the same type of data, but corrects the predictions by a factor depending of the probability that the student commits a careless error in responding to a particular problem. The correlation coefficient used is the point biserial. However, limiting our analysis to the data of such $2 \times 2$ correlation matrices does not take full advantage of all the information available.

2) We have seen in Section 1.3 that the core mechanism of the assessment algorithm resides in updating, from one trial to the next and on the basis of the student's response to the question, the likelihood of the knowledge states. A correct response to some item $\boldsymbol{q}$ presented on a trial results in an increase of the probabilities of all the states containing $\boldsymbol{q}$, and an incorrect response in a decrease of all such probabilities. This means that, from the standpoint of the assessment engine, the probability of a correct response to the extra problem $\boldsymbol{p}$ on trial $n$ of the assessment can be obtained by summing the probabilities of all the states containing $\boldsymbol{p}$ on that trial. We use the point biserial coefficient to evaluate the correlation between this continuous variable and the dichotomous 0/1 variable coding the student's response (false/correct) to problem $\boldsymbol{p}$. As this computation can in principle be performed on any trial, we can trace the

evolution of such a prediction in the course of the assessment. The results are the first ones reported in the next section (see Paragraph 2.4.2).

3) The third method for evaluating the validity of the assessment is based on a different idea. At the end of most assessments, the student picks an item in the outer fringe of the state assigned by the assessment engine and begins learning[6]. If the knowledge state assigned to the student is the true one or at least strongly resembles the true one, then the prediction of what the student is capable of learning at that time should be sound. Accordingly, we can gauge the validity of the assessments by the probability that the student successfully masters an item chosen in the outer fringe of the assessed state. We have computed such probabilities for a very large number of assessment/learning trials in both elementary school mathematics and first year college chemistry. Paragraph 2.4.6 contains the results.

**2.3.2 Concepts and notation.** We write $\mathcal{K}$ for the collection of all the knowledge states in the particular subject considered. As recalled above, the likelihoods of the knowledge states are systematically modified in the course of an assessment as a result of a student's response (false or correct). We denote by $\xi_n$ the likelihood distribution on $\mathcal{K}$ on the $n$th trial of the assessment, and by $\xi_n(K)$ the corresponding likelihood of state $K$ on that trial. Thus, the assessment begins with an initial, or a *priori*, distribution $\xi_1$ on $\mathcal{K}$. An item $\boldsymbol{q}_1$ is chosen and given to the student; the student's response $\boldsymbol{r}_1$ is recorded, leading to the transformation of $\xi_1$ into $\xi_2$ by a Bayesian operator, etc. In symbols, we thus have the sequence

$$(\xi_1, \boldsymbol{q}_1, \boldsymbol{r}_1) \longmapsto \ldots \longmapsto (\xi_n, \boldsymbol{q}_n, \boldsymbol{r}_n) \longmapsto \ldots \longmapsto (\xi_{L-1}, \boldsymbol{q}_{L-1}, \boldsymbol{r}_{L-1}) \longmapsto \xi_L,$$

with $L$ denoting the number of the last trial of the assessment. So, the likelihood distribution $\xi_L$ on $\mathcal{K}$ gives the final result of the assessment. The likelihood distribution $\xi_n$ may be regarded as a *probabilistic knowledge state* on trial $n$. If there were no careless errors or lucky guesses, $\xi_n$ would provide, for every item $\boldsymbol{q}$ in the domain, the probability $P_n(\boldsymbol{q})$ of a correct response if an instance of $\boldsymbol{q}$ were presented on trial $n$. Indeed, in our theoretical framework, that probability is the sum of all the probabilities of the states containing $\boldsymbol{q}$. Writing $\mathcal{K}_{\boldsymbol{q}}$ for the subcollection of $\mathcal{K}$ of all the knowledge states containing $\boldsymbol{q}$, we could thus compute $P_n(\boldsymbol{q})$ by the equation

$$P_n(\boldsymbol{q}) = \sum_{K \in \mathcal{K}_{\boldsymbol{q}}} \xi_n(K). \tag{2.1}$$

However, the assumption that there are no careless errors is not warranted and we will shortly enlarge our notation to take care of this aspect of the data. We do, however, suppose that there are no lucky guesses. This assumption is justified because all the problems either have open responses or offer a multiple

[6]Exceptions are cases in which the assessment is a placement test or serves as the final exam of a course.

choice with a very large number of possibilities. We assume that no learning on the part of the student is taking place during the assessment. So, the index $n$ in $\xi_n(K)$ and $P_n(\boldsymbol{q})$ only marks the progress of the assessment.

Note that the last trial number $L$ and the extra problem $\boldsymbol{p}$ depend upon the student. More exactly, they depend on the particular assessment, and so do $\xi_n$ and $P_n(\boldsymbol{p})$ for all trial numbers $n \leq L$. Making this dependence explicit, we refine our notation and define

$$L_{\mathbf{a}} \quad \text{as the number of the last trial in assessment } \mathbf{a}, \tag{2.2}$$

$$\boldsymbol{p}_{\mathbf{a}} \quad \text{as the extra problem in assessment } \mathbf{a}, \tag{2.3}$$

$$\xi_{\mathbf{a},n} \quad \text{as the probabilistic knowledge state on trial } n \leq L_{\mathbf{a}}, \tag{2.4}$$

$$\mathcal{A} \quad \text{as the set of all the assessments.} \tag{2.5}$$

We also define the collection of random variables

$$\mathbf{R}_{\mathbf{a}} = \begin{cases} 0 & \text{if the student's response to problem } \boldsymbol{p}_{\mathbf{a}} \\ & \text{of assessment } \mathbf{a} \text{ is incorrect} \\ 1 & \text{otherwise.} \end{cases}$$

There is no ambiguity in using the abbreviation $\mathbf{R}_{\mathbf{a}} = \mathbf{R}_{\boldsymbol{p}_{\mathbf{a}}}$ since any assessment $\mathbf{a}$ defines a unique extra problem $\boldsymbol{p}_{\mathbf{a}}$.

The possibility of careless errors must be taken into account. We suppose that a careless error probability is attached to each problem and define

$\epsilon_{\boldsymbol{q}}$ as the probability of committing a careless error on problem $\boldsymbol{q}$.

In the framework of the theory, the interpretation of this parameter is straightforward: $\epsilon_{\boldsymbol{q}}$ is the conditional probability that a student whose knowledge state contains $\boldsymbol{q}$ commits a careless error in attempting to solve that problem. We suppose that the parameter $\epsilon_{\boldsymbol{q}}$ only depends on the problem $\boldsymbol{q}$ and does not vary in the course of an assessment. In accordance with the abbreviation convention used above for $\mathbf{R}_{\mathbf{a}}$, from now on we write

$$\epsilon_{\mathbf{a}} = \epsilon_{\boldsymbol{p}_{\mathbf{a}}}, \qquad \text{and} \qquad \mathcal{K}_{\mathbf{a}} = \mathcal{K}_{\boldsymbol{p}_{\mathbf{a}}}$$

for the subcollection of states containing problem $\boldsymbol{p}_{\mathbf{a}}$. As mentioned, we assume that the probability of a lucky guess is zero for each item.

The probability $\mathbf{P}_{\mathbf{a},n}$ that a student correctly solves the extra problem $\boldsymbol{p}_{\mathbf{a}}$, based on the information accumulated by the assessment algorithm up to and including trial $n$ of assessment $\mathbf{a}$, is thus defined by the following equation:

$$\mathbf{P}_{\mathbf{a},n} = (1 - \epsilon_{\mathbf{a}}) \sum_{K \in \mathcal{K}_{\mathbf{a}}} \xi_{\mathbf{a},n}(K) + 0 \times \sum_{K \in \mathcal{K} \setminus \mathcal{K}_{\mathbf{a}}} \xi_{\mathbf{a},n}(K) \tag{2.6}$$

$$= (1 - \epsilon_{\mathbf{a}}) \sum_{K \in \mathcal{K}_{\mathbf{a}}} \xi_{\mathbf{a},n}(K). \tag{2.7}$$

The 0 factor in (2.6) is included as a reminder of the hypothesis that the probability of a lucky guess is zero.

**2.3.3 Estimating the careless error parameters.** Our estimate of $\epsilon_{\boldsymbol{q}}$ is based on the assessments in which $\boldsymbol{q}$ has been presented as the extra problem and also, at least once, as part of the assessment[7]. Coding an error and a correct response as 0 and 1, respectively, our sample space for estimating the careless error probability $\epsilon_{\boldsymbol{q}}$ of item $\boldsymbol{q}$ is thus the set $\{(0,0),(0,1),(1,0),(1,1)\}$, in which, by convention, the first term of every pair codes the response to $\boldsymbol{q}$ as the extra problem, and the second term the response to the other relevant presentation of $\boldsymbol{q}$ in that assessment. Denoting by $p_{\boldsymbol{q}}(i,j)$ the probability of sampling the point $(i,j)$ and by $\gamma_{\boldsymbol{q}}$ the probability that the knowledge state of the student belongs to $\mathcal{K}_{\boldsymbol{q}}$, we have the following (compare with Eq. (2.7)):

$$p_{\boldsymbol{q}}(0,0) = \epsilon_{\boldsymbol{q}}^2\gamma_{\boldsymbol{q}} + (1-\gamma_{\boldsymbol{q}}) \tag{2.8}$$
$$p_{\boldsymbol{q}}(0,1) = \epsilon_{\boldsymbol{q}}(1-\epsilon_{\boldsymbol{q}})\gamma_{\boldsymbol{q}} \tag{2.9}$$
$$p_{\boldsymbol{q}}(1,0) = (1-\epsilon_{\boldsymbol{q}})\epsilon_{\boldsymbol{q}}\gamma_{\boldsymbol{q}} \tag{2.10}$$
$$p_{\boldsymbol{q}}(1,1) = (1-\epsilon_{\boldsymbol{q}})^2\gamma_{\boldsymbol{q}}. \tag{2.11}$$

Writing $N_{\boldsymbol{q}}$ for the number of assessments having at least two presentations of item $\boldsymbol{q}$, with one of them as the extra problem, and $N_{\boldsymbol{q}}(i,j)$ for the number of times $(i,j)$ is realized among the $N_{\boldsymbol{q}}$ assessments, we obtain the statistic

$$\mathtt{Chi}_{\boldsymbol{q}}(\epsilon_{\boldsymbol{q}},\gamma_{\boldsymbol{q}}) = \sum_{i,j}\frac{\left(N_{\boldsymbol{q}}(i,j) - N_{\boldsymbol{q}}\cdot p_{\boldsymbol{q}}(i,j)\right)^2}{N_{\boldsymbol{q}}\cdot p_{\boldsymbol{q}}(i,j)}, \tag{2.12}$$

in which the $p_{\boldsymbol{q}}(i,j)$'s are defined by (2.8)-(2.11). The parameters $\epsilon_{\boldsymbol{q}}$ and $\gamma_{\boldsymbol{q}}$ are estimated by minimizing the Chi-square statistic (2.12). The details are relegated to the Appendix 2.6 on page 49. The estimators for $\epsilon_{\boldsymbol{q}}$ and $\gamma_{\boldsymbol{q}}$ are

$$\hat{\epsilon}_{\boldsymbol{q}} = \frac{N_{\boldsymbol{q}}(0,1)+N_{\boldsymbol{q}}(1,0)}{N_{\boldsymbol{q}}(0,1)+N_{\boldsymbol{q}}(1,0)+2N_{\boldsymbol{q}}(1,1)}, \tag{2.13}$$
$$\hat{\gamma}_{\boldsymbol{q}} = \frac{(N_{\boldsymbol{q}}(0,1)+N_{\boldsymbol{q}}(1,0)+2N_{\boldsymbol{q}}(1,1))^2}{4N_{\boldsymbol{q}}(1,1)\cdot N_{\boldsymbol{q}}}. \tag{2.14}$$

We may regard $\mathtt{Chi}_{\boldsymbol{q}}(\hat{\epsilon}_{\boldsymbol{q}},\hat{\gamma}_{\boldsymbol{q}})$ as a $\chi_1^2$ random variable[8] with $3-2=1$ degree of freedom (three degrees of freedom in the $2\times 2$ table of the $p_{\boldsymbol{q}}(i,j)$'s minus two estimated parameters). As such, $\mathtt{Chi}_{\boldsymbol{q}}(\hat{\epsilon}_{\boldsymbol{q}},\hat{\gamma}_{\boldsymbol{q}})$ may serve as an additional indicator of the adequacy of a model based on an "all-or-none" assumption regarding the mastery of an item, and a zero probability of a lucky guess.

[7] We only consider the first of these non-extra-problem presentations of $\boldsymbol{q}$.
[8] However, see Remark 2.3.4 (b).

### 2.3.4 Remarks.

(a) Another possibility for estimating the parameter $\epsilon_{\boldsymbol{q}}$ would rely on those assessments in which the final state chosen by the assessment algorithm contains item $\boldsymbol{q}$, with this item being presented at least once during the assessment (either as the extra problem or otherwise) and answered incorrectly by the student (in the first presentation).

The objection to this method is that the choice of a knowledge state for a student at the end of some assessment $\mathbf{a}$ is based on choosing the most likely state in the final probabilistic state $\xi_{\mathbf{a},L_{\mathbf{a}}}$ (cf. (2.2), (2.4) for this notation). At that time, there may still be several states with a maximally high likelihood. The algorithm then chooses randomly among them. This choice is not very critical from an assessment or even learning standpoint because these states very much resemble each other.[9] Nevertheless, the remaining uncertainty regarding the exact state of the student makes this method for estimating careless errors questionable. Note that the method actually used, i.e., the one described in Paragraph 2.3.3, does not suffer from this shortcoming. In particular, the method used does not rely on the final assessed state.

(b) The phrase "*probability that the knowledge state of the student contains item* $\boldsymbol{q}$" is ambiguous. Its interpretation as the parameter $\gamma_{\boldsymbol{q}}$ in the chi-square statistic of Eq. (2.12) is a device allowing us to estimate $\epsilon_{\boldsymbol{q}}$, which is our primary concern. However, while $\epsilon_{\boldsymbol{q}}$ has a legitimate place in our theory, the parameter $\gamma_{\boldsymbol{q}}$ that we estimate by minimizing $\mathtt{Chi}_{\boldsymbol{q}}(\epsilon_{\boldsymbol{q}}, \gamma_{\boldsymbol{q}})$ lies outside. We could have estimated $\gamma_{\boldsymbol{q}}$ differently, for example by averaging the final $\xi_{\mathbf{a},L_{\mathbf{a}}}$ values appropriately. However, this method suffers from the same objection spelled out in (a) above.

### 2.3.5 Temporal course of the assessments. The Vincent curves.

For every assessment $\mathbf{a}$ and every trial number $n$ of that assessment, we have a pair $(\mathbf{P}_{\mathbf{a},n}, \mathbf{R}_{\mathbf{a}})$ of numbers[10]. The first one $\mathbf{P}_{\mathbf{a},n}$ is computed from Eq. (2.7) and is the prediction of the algorithm for the probability of a correct response to the extra problem $\boldsymbol{p}_{\mathbf{a}}$ on trial $n$ of assessment $\mathbf{a}$, taking into account possible careless errors. The second number $\mathbf{R}_{\mathbf{a}}$ is a dichotomous (0/1) variable coding the student's (false/correct) response to $\boldsymbol{p}_{\mathbf{a}}$. This number is constant for a given assessment. If we align all the assessments on their last trial, we can compute such a correlation between the $\mathbf{P}_{\mathbf{a},L_{\mathbf{a}}}$ values, which vary continuously between 0 and 1, and the dichotomous values $\mathbf{R}_{\mathbf{a}}$. This correlation, computed

[9] Continuing the assessment and asking more questions might help in discriminating among these highly likely states, but the benefit would be slight and the cost to the student heavy.

[10] To avoid complicating our notation, we do not distinguish between the random variables $\mathbf{P}$ and $\mathbf{R}$ on the one hand, and their realizations on the other hand.

for all pairs $(\mathbf{P}_{\mathbf{a},L_{\mathbf{a}}}, \mathbf{R}_{\mathbf{a}})$ for $\mathbf{a} \in \mathcal{A}$, is a measure of the success of the assessment algorithm in uncovering the student's knowledge state. We have computed such correlations as part of a more extensive analysis of the evolution of the correlation in the course of the assessment. This analysis requires aligning all the assessments, which are typically of different lengths, from the first to the last trial. The method of choice to perform such an alignment is the *Vincent curve*, which is a standard tool for the analysis of learning data[11]. In our case, this consists of splitting all the assessments into the same number of parts—we have chosen 10—and gathering data pertaining to the same parts in all the assessments. Specifically, for each of the ten deciles of each assessment, we combine the data of the last trial for that decile for all the assessments.

An example involving three hypothetical assessments **a**, **b**, and **c** is displayed on Table 2.1. The first line in the body of the table concerns the assessment **a** which takes 40 trials. Each of the 10 deciles for that assessment consists of four trials, the last one of which is retained for the construction of the Vincent curve. We also include the initial trial in this analysis. In other words, only the probabilistic knowledge states $\mathbf{P}_{\mathbf{a},1}$, $\mathbf{P}_{\mathbf{a},4}$, $\mathbf{P}_{\mathbf{a},8}$, ..., $\mathbf{P}_{\mathbf{a},36}$, $\mathbf{P}_{\mathbf{a},40}$ are taken into account for the computation of the correlation coefficients. The second and third line concern the assessments **b** and **c**, which take 30 and 20 trials, respectively. The last trial in each of their deciles is aligned with the corresponding trial of assessment **a**. We also include the very first trial in our analysis.

The value of the correlation coefficient for the first trial measures how much the learning space $\mathcal{K}$, equipped with the *a priori* probability distribution on $\mathcal{K}$, already knows about the student population before the beginning of the assessment. When the length $L_{\mathbf{d}}$ of some assessment is not a multiple of 10, the rule for specifying the decile is a generalization of the above rule, namely, the trial number retained for the computation of the correlation coefficient in decile $i$ is the smallest integer greater than or equal to $i \times L_{\mathbf{d}}/10$. For example, for $N_{\mathbf{d}} = 17$ the trial numbers retained (including the initial trial) are 1, 2, 4, 6, 7, 9, 11, 12, 14, 16, and 17.

The columns of the table that are relevant to the analysis in terms of Vincent curves are printed in red, and so is the initial column, corresponding to the probabilities at the outset of the assessment. For each of these 11 columns, we have computed the correlations between the $\mathbf{P}$ values measuring the probability that the student responds correctly to the extra question, and the actual response $\mathbf{R}$ coded as 0 or 1 for false or correct, respectively.

[11] From Stella B. Vincent, who is the first researcher on record to have used this type of analysis (Vincent, 1912).

**Table 2.1.** The initial trial and the ten deciles in the Vincent curve analysis. The first column lists the assessments. The data of the 11 columns in red are the only ones retained for the correlation analysis.

| Assessment | Deciles 1 | 2 | ... | 10 |
|---|---|---|---|---|
| **a** | $\mathbf{P}_{\mathbf{a},1}$ $\mathbf{P}_{\mathbf{a},2}$ $\mathbf{P}_{\mathbf{a},3}$ $\mathbf{P}_{\mathbf{a},4}$ | $\mathbf{P}_{\mathbf{a},5}$ $\mathbf{P}_{\mathbf{a},6}$ $\mathbf{P}_{\mathbf{a},7}$ $\mathbf{P}_{\mathbf{a},8}$ | ... | $\mathbf{P}_{\mathbf{a},37}$ $\mathbf{P}_{\mathbf{a},38}$ $\mathbf{P}_{\mathbf{a},39}$ $\mathbf{P}_{\mathbf{a},40}$ |
| **b** | $\mathbf{P}_{\mathbf{b},1}$ $\mathbf{P}_{\mathbf{b},2}$ $\mathbf{P}_{\mathbf{b},3}$ | $\mathbf{P}_{\mathbf{b},4}$ $\mathbf{P}_{\mathbf{b},5}$ $\mathbf{P}_{\mathbf{b},6}$ | ... | $\mathbf{P}_{\mathbf{b},28}$ $\mathbf{P}_{\mathbf{b},29}$ $\mathbf{P}_{\mathbf{b},30}$ |
| **c** | $\mathbf{P}_{\mathbf{c},1}$ $\mathbf{P}_{\mathbf{c},2}$ | $\mathbf{P}_{\mathbf{c},3}$ $\mathbf{P}_{\mathbf{c},4}$ | ... | $\mathbf{P}_{\mathbf{c},19}$ $\mathbf{P}_{\mathbf{c},20}$ |
| ⋮ | ⋮ ⋮ ⋮ | ⋮ ⋮ ⋮ | ⋮ | ⋮ ⋮ ⋮ |

**2.3.6 The correlation coefficients.** We computed the correlations between the variables **P** and **R** by the point biserial coefficient

$$r_{pbis} = \frac{M_1 - M_0}{s_n} \sqrt{\frac{n_1 n_0}{n^2}}$$

where

| | |
|---|---|
| $n$ | is the number of pairs $(\mathbf{P}, \mathbf{R})$, |
| $s_n$ | is the standard deviation of the continuous variable $\mathbf{P}$, |
| $n_1, n_0$ | are the numbers of $\mathbf{R} = 1$, $\mathbf{R} = 0$ cases, respectively, |
| $M_1, M_0$ | are the conditional means of $\mathbf{P}$ given $\mathbf{R} = 1$ and $\mathbf{R} = 0$ |

(see e.g. Tate, 1954). This correlation coefficient provides an estimate of a Pearson correlation coefficient $\rho$ under some hypotheses regarding the joint distribution of the two random variables involved. Applied to our situation, these hypotheses are as follows:

(1) the variable **R** is obtained by dichotomizing some underlying continuous random variable **X**;

(2) the joint distribution of **P** and **X** is Gaussian;

(3) the marginal distributions have equal variances;

(4) the conditional variances of **P** given $\mathbf{R} = 0$ and $\mathbf{R} = 1$ are equal.

These hypotheses are not satisfied in our case. For one thing, **P** is bounded, taking its values in the interval $[0, 1]$. More critically, experimental plots of the distribution of **P** indicate that the underlying random variable is bimodal, which is at odds with Hypothesis (2). Our data also point to a contradiction of Hypothesis (3), showing that the conditional variance of **P** given $\mathbf{R} = 1$ is substantially larger than that given $\mathbf{R} = 0$. Thus, the values obtained for $r_{pbis}$ should not generally be regarded as estimates of a Pearson correlation coefficient[12]. Nevertheless, we adopted the point biserial coefficient in view of its frequent use in psychometrics to compute the item-test correlation. Even though our variable $\mathbf{P_a}$ is different from the overall result of a standardized test, it quantifies the final result of an assessment, making a comparison worthwhile.

Two other correlation coefficients have also been used for cases in which the data take the form of double dichotomies, namely the tetrachoric coefficient and the phi coefficient. The tetrachoric coefficient tends to give higher values than the phi coefficient, and that is what we shall observe in Paragraph 2.4.4 where we show that these coefficients give substantially different numbers for the same data. Their values should be regarded as complementary.

## 2.4 Data Analysis

We recall that the data pertains to elementary school mathematics.

**2.4.1 The participants.** The participants were elementary school students in grades 3 to 6. The students took assessments via the internet, either in school or at home. In most cases, such assessments were carried out in the framework of a computerized course on the subject. The data only concern the initial assessments taken by the students. This ensures that performance on the assessment is independent from any knowledge acquired on the ALEKS computerized courses[13]. The students used an interface that made it clear what form the answer to a given problem should take (a number, an algebraic expression, a graph, etc.). Students were given a short, 15-minute tutorial on using the computer system's answer input tools and were offered (pre-packaged) online help with the tools during the assessment. Otherwise, no help or feedback was given.

**2.4.2 The Vincent curve analysis.** The data used here are based on the assessment algorithm outlined in Section 8.8 of Chapter 8 of this volume. This algorithm allowed us to conduct assessments on the very large knowledge structure associated to the domain. The algorithm partitions the domain into several parts on which the structure is 'projected' (see Section 1.4)[14].

[12] There are indications in the literature (cf. Kraemer, 2006; Karabinus, 1975) that $r_{pbis}$ is a robust statistic in some situations, such as for testing the $\rho = 0$ hypothesis. This robustness of $r_{pbis}$ does not seem to extend to cases where $\rho$ is far from 0.

[13] As discussed in Section 2.2, this precaution is somehow superfluous since the domain of knowledge is confounded with the curriculum by design.

[14] Cf. the Glossary on page 25. We discussed in Section 1.4 how and why a large learning space may be split into several parts for the purpose of performing assessments. The concept of a 'projection' was developed for this purpose. In particular, for each item **a**, Eq. (2.15) is defined on the projection of the knowledge structure on the sub-domain that contains **a**.

As mentioned earlier, we use the point biserial coefficient to compute the correlation between the probability of a correct response to the extra problem on trial $n$, specified by the equation

$$\mathbf{P}_{\mathbf{a},n} = (1 - \epsilon_{\mathbf{a}}) \sum_{K \in \mathcal{K}_{\mathbf{a}}} \xi_{\mathbf{a},n}(K), \tag{2.15}$$

and the actual response to that problem coded as 0 or 1. Note that estimates of the careless error rate $\epsilon_{\mathbf{a}}$ were not available for all items. However, since this rate is defined as constant for a given item, it does not affect the point biserial correlation for that item.

Figure 2.1 traces the evolution of the medians of the distributions of the point biserial coefficient in the course of the assessment. These data concern 300 items and are based on 125,786 assessments. (Seventy items were discarded because the relevant data were too sparse.) All the assessments available were used except those in which the extra problem had also been presented as part of the assessment, which occasionally happened since the extra problem was selected randomly among the available problems.

The initial values (at the zero abscissa) of the correlation are obtained from the *a priori* distribution on the set of knowledge states, before the assessment begins. The fact that the median value, about .19, is substantially above zero shows that something is already known by the assessment engine about the population of students: they are more or less ready to take an assessment in elementary school mathematics, or are already engaged in a course on that subject. Then, the curve reaches what appears to be an asymptote at .46. The upper quartile reaches the value .55 and the lower quartile the value .35.

**2.4.3 Vincent curves for different categories of problems.** We first investigated how sensitive the Vincent curves were to the careless error rate of the items. The analysis was restricted to the items with sufficient data to estimate their careless error rates as defined by Eq. 2.13. As explained earlier in Section 2.3.3, these estimates rest on assessments where an item as been presented twice: once as the extra question and once as part of the assessment. To reduce the selection bias arising from one presentation being part of the adaptive assessment, we only kept those assessments that presented the question within the first five questions. We obtained estimates for 143 items, with a median careless error rate value of 19%. We divided the items in two groups, the first group comprising the 67 items with a careless error rate less than 19% and the second group comprising the remaining 76 items. The Vincent curves for the median value of each group are displayed in Figure 2.2.

Items with a lower careless error rate tend to fare better than items with a higher careless error rate. The median $r_{pbis}$ reaches the value of .51 for the former and .48 for the latter. The difference between the two Vincent curves however is quite small in view of the difference between quartiles in Fig.2.1.

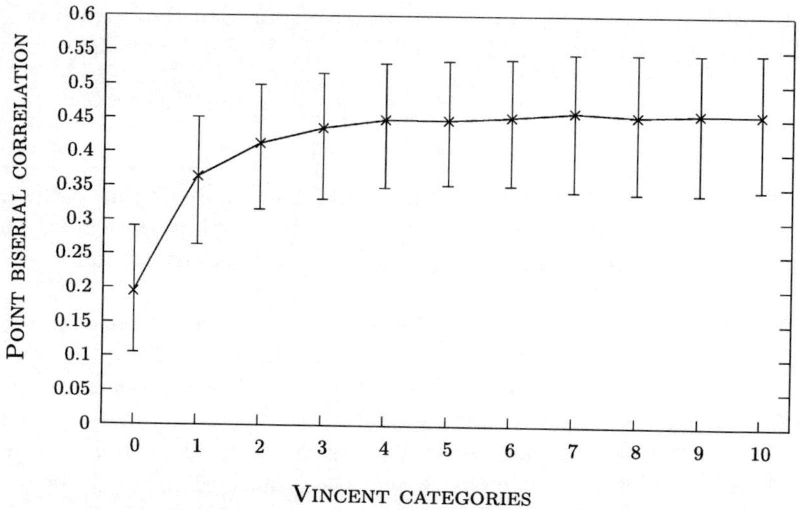


**Figure 2.1.** Evolution, during the initial part of the assessment, of the point biserial correlation between the probability of a correct response to the extra problem predicted by the assessment engine via Eq. (2.15) and the actual response to that problem. The midpoint of each vertical segment represents the median value of the correlation, the bottom end the 25th percentile value and the top end the 75th percentile value. The abscissae 1, ..., 10 correspond to the ten Vincent categories (cf. Table 2.1). The zero abscissa indicates the initial correlation, before the first trial of the assessment.

We also notice that the median value for both groups of items is greater than the overall median value of .46 from Fig.2.1. The 143 items used for the two Vincent curves in Fig. 2.2 were the most popular items in the early part of the assessment because of the way their careless error rates were computed. Such items tend to be of a middle level of difficulty. The next paragraph provides a closer look at this aspect.

Second, we examined how sensitive the Vincent curves were to the overall difficulty of the items. A simple difficulty index was defined as the proportion of incorrect answers when the item was presented as the extra question. Such a ratio can be used as a rough and straightforward measure of the item difficulty. In particular, it does not attempt to correct for careless errors. We divided the 300 items with sufficient data to compute their point biserial correlations in three groups. The first group has 22 items with a difficulty index lower than 33%, the second group has 84 items with a difficulty index between 33% and 66%, and the third group has 194 items with a difficulty index greater than

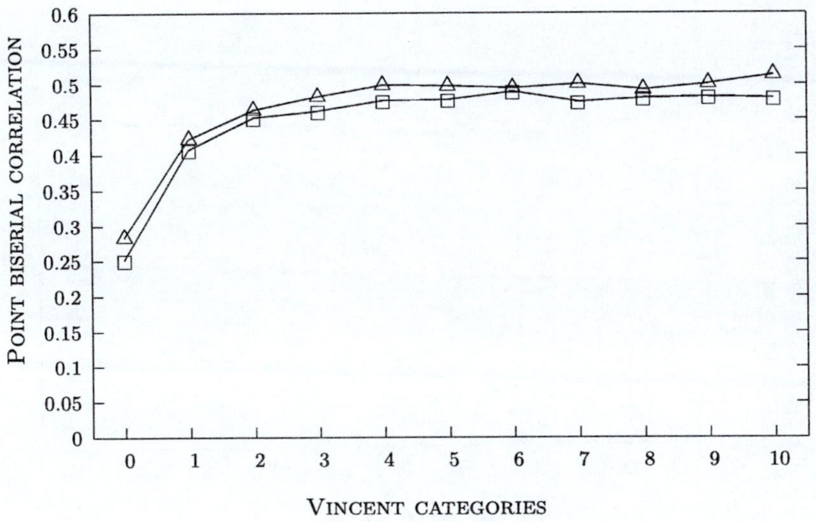


**Figure 2.2.** Vincent curves similar to the one of Figure 2.1. The curve connecting triangular dots concerns problems with a careless error rate less than 19%. The curve connecting square dots concerns problems with a careless error rate greater than or equal to 19%.

66%. Figure 2.3 traces the Vincent curves for the median values of the point biserial correlation of each group.

The figure makes it clear that items of medium difficulty yield better correlations than items that are either easier or more difficult, with ending correlation values of .51, .44, and .44, respectively. Let us recall that the assessments under examination are initial assessments that took place before any learning. In other words, the population who took them had a limited knowledge of the curriculum as evidenced by most items being classified in the higher difficulty group.

**2.4.4 The final state predictions.** The Vincent curve analysis that we just discussed does not tell us how successful the assessment ultimately is. We now consider the final result of the assessment, that is, the knowledge state chosen by the assessment engine, and investigate how predictive of the competence of a student that final state is. We still rely on the extra problem methodology. The relevant data for a particular extra problem **a** is a $2 \times 2$ matrix of the form of Table 2.2:

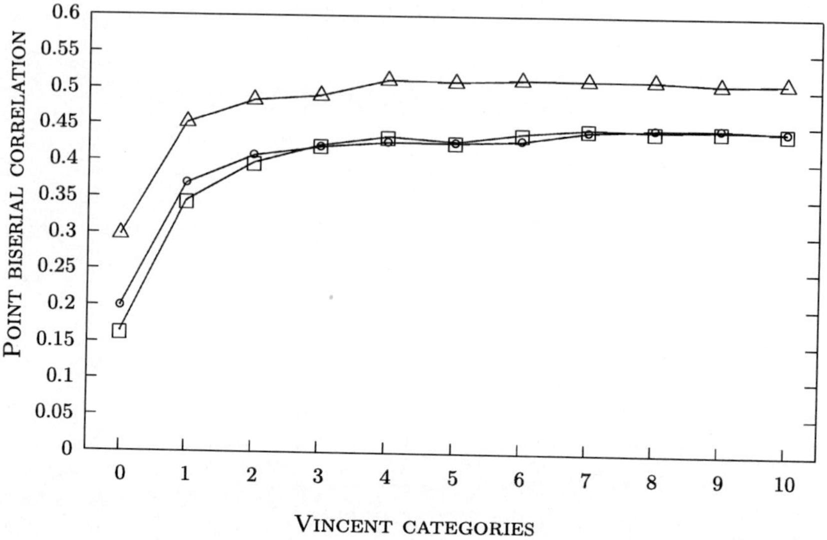


**Figure 2.3.** Vincent curves similar to the one of Figure 2.1. The curves connecting circular, triangular, and square dots concern problems of lower, medium, and higher difficulty, respectively.

**Table 2.2.** Basic data matrix for the computation of the correlation between the cases 'in or out of the final state' and the student's response coded as 0/1 for false/correct by the variable $\mathbf{R_a}$.

| | | $\mathbf{R_a}$ | |
|---|---|---|---|
| | | 0 | 1 |
| State | **a** in | $x$ | $y$ |
| | **a** out | $z$ | $w$ |

The letter $x$ in this matrix represents the number of cases in which the extra problem **a** was in the final state assigned to the student and the response was incorrect; $y$, $z$, and $w$ have similar interpretations. From the standpoint of the assessment engine, $x$ can thus be regarded as the number of careless errors committed in the $x + y + z + w$ cases in which **a** was presented.

As no correlation coefficient is available that would be completely adequate for our situation and without bias of some sort, we have used three of them. The first two, which are the tetrachoric and the phi coefficients, act directly on matrices of the type displayed in Table 2.2 and do not take the careless errors into account. The third coefficient is the point biserial. It is applied to the same data matrices but involves a correction for careless errors. The results are described in Paragraph 2.4.5.

The data analyzed in this section are based on the same 125,786 assessments used for the Vincent curve analysis. For the purpose of this analysis, the assessments were replayed with a simple extracting rule that ascribes to the student's knowledge state any item $\boldsymbol{q}$ such that

$$\sum_{K \in \mathcal{K}_q} \xi_n(K) > 0.5,$$

where $\xi_n$ is the final probability distribution on the knowledge structure over the sub-domain containing $\boldsymbol{q}$. (For practical reasons, assessments sometimes stopped while there were still items for which the left hand side of the above inequality was around .5, making them potential questions to ask. Such items would typically not be ascribed to the student's actual knowledge state.)

Figure 2.4 displays the distribution of the tetrachoric coefficient values pertaining to 324 problems out of the 370 problems forming the elementary school mathematics domain in ALEKS. Forty-six problems were discarded because the relevant data were too sparse for reliably estimating the coefficient. The tetrachoric coefficient is based on the hypothesis of an underlying 2-dimensional Gaussian random variable. Thus, the double dichotomy matrix arises from splitting each of the two Gaussian marginals into two categories[15]. The median of the distribution is around .68, which is quite high in such a context. The grouped data, obtained from gathering the 324 individual $2 \times 2$ matrices into one, yields a still higher correlation of about .80. For high values, the tetrachoric coefficient is sometimes regarded as biased upward (however, see Greer et al., 2003).

Figure 2.5 contains a similar analysis using the phi coefficient. All the values are noticeably lower, yielding a median of .43 (in contrast to the .68 value obtained for the tetrachoric coefficient) and a grouped data value of .58 (instead of .80). Of particular interest are the very low correlation values obtained for some problems. In our view, any problem with a correlation below .2 for this coefficient deserves some examination. A low correlation value could be due, for example, to a high careless error rate or a misplacement in the structure. We go back to this issue in Paragraphs 2.4.5 and 2.4.6.

---

[15] Admittedly, this hypothesis does not fit the situation very well. See Drasgow (1988) or Nunnally and Bernstein (1994) for a discussion of the hypotheses underlying the use of the tetrachoric and the phi coefficients.

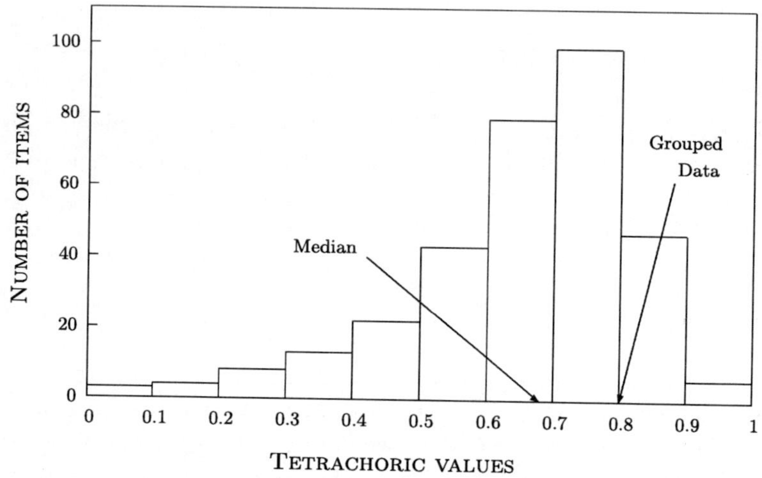


**Figure 2.4.** Correlation between the in/out cases for the final knowledge state and the actual false/correct response of the student. The figure displays the distribution of the tetrachoric coefficient values for 324 of the 370 problems forming the domain of the learning space for elementary school mathematics.

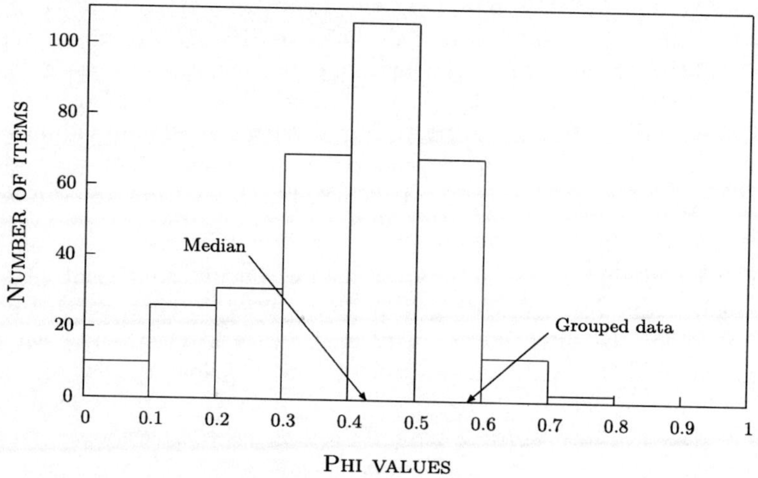


**Figure 2.5.** The similar distribution, based on the same data, for the phi coefficient.

**2.4.5 Correcting for careless errors.** The analyses in terms of the tetrachoric and phi coefficients that we just described did not take the careless errors into consideration. However, the basic data matrix (2.2) can be amended to include the effect of careless errors. To this end, we define the new variable

$$\mathbf{S_a} = \begin{cases} (1 - \epsilon_{\mathbf{a}}) & \text{if the final state contains the extra question } \mathbf{a} \\ 0 & \text{otherwise.} \end{cases}$$

Thus, $\mathbf{S_a}$ is the probability of not committing a careless error on the extra problem **a** computed based on the final state at the end of the assessment. The variable $\mathbf{S_a}$ is neither exactly continuous nor exactly discrete[16]. Nevertheless, again for the purpose of comparison with similar analyses performed in psychometric situations, we have used the point biserial coefficient $r_{pbis}$ to compute the correlation between the variables $\mathbf{S_a}$ and $\mathbf{R_a}$ for the grouped data over the same 143 items used in the first paragraph of Section 2.4.3. The value obtained for $r_{pbis}$ was .67, sensitively higher than the .57 obtained for the phi coefficient for the same grouped data.

**2.4.6 Validity and learning readiness.** Finally, we discuss an indirect but nevertheless revealing way of gauging the validity of an assessment. We recall from Chapter 1 that the outer fringe of a knowledge state is the set of problems that the student in that state is ready to learn[17]. Consider a situation in which an assessment is a prelude to learning and the system routinely prompts the student to start learning how to solve the problems in the outer fringe of her state. The capability of the student to learn ("master") such problems should be revealing of the validity of the assessment. This may be evaluated by the conditional probability that a student is capable of mastering a problem, given that it is located in the outer fringe of her knowledge state (and so accessible for learning). These probabilities can be estimated from our learning data. For elementary school mathematics, the median of the distribution of the estimated (conditional) success probabilities is .93. To understand the import of this number, some details about the learning process in the system examined here must be given.

When a problem is located in the outer fringe of a student's state, the student may select that problem as the next one to learn. This choice initiates a random walk keeping track of the learning stages for that problem. The random walk takes place on an interval of the integers and has two absorbing barriers (see Figure 2.6). At each step of that random walk, an instance of the problem is proposed and the student is asked to solve it. In case of failure, an explanation of the solution that is centered on that instance is offered. The problem enters the random walk at the point 0 and moves left or right depending on the student's response to the instance presented. The general principle is that a success in the solution of an instance provokes a move to the right, and an error a move to the left. The problem is considered to be learned when the random walk hits the right barrier. Hitting the left barrier means

[16] For example, the distribution of $\mathbf{S_a}$ vanishes in a positive neighborhood of 0, but is positive at the point 0 itself.

[17] See Paragraph 1.2.1 on page 11 for an introduction to this concept, and Definition 8.3.1 on page 136 for a formal definition.

that the student is not capable of learning the problem type at that time. The learning of this problem is then postponed and the student's knowledge state may be readjusted.

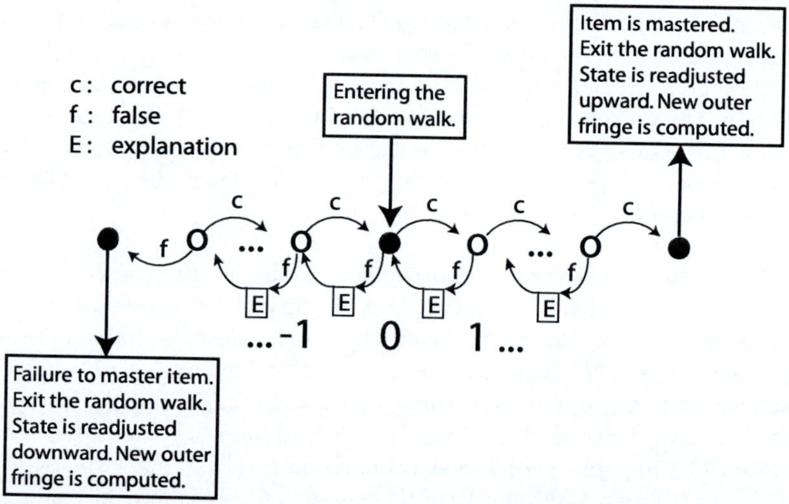


**Figure 2.6.** Illustration of the random walk on the integers with two absorbing barriers. The left barrier corresponds to the failure to master the problem at this time, and the right barrier to the success. The random walk keeps track of the intermediate learning stages for a problem. The problem enters the random walk at the point 0 and moves left or right depending on the student's response to the instance presented. In case of a false response, an explanation is given to the student, and the random walk moves one unit to the left. A correct response initiates a move of one unit to the right. The reader should keep in mind that the successive instances proposed to the student may be quite different.

Figure 2.7 displays the distribution of the conditional probabilities that a problem entering the random walk ends up at the right bound, and so is regarded as mastered, at least for the present. (A later assessment would verify the fact.) An examination of the graph shows that many items are satisfactorily handled: 90% of them have a probability of success of at least .83, with the median of the distribution at .93. Nevertheless, the left tail of the distribution indicates that a few problems are not learned easily and that adjustments deserve to be made. For example, some intermediate problems may be missing and should then be added to the domain. Also, a given problem may be misplaced in the structure, or its explanation may be defective and should be rewritten. Care is taken in the design of the various instances of a problem so that a correct response would not be due to a trivial device, unrelated to the understanding of the problem. The data analyzed are based on 1,940,473 such random walks.

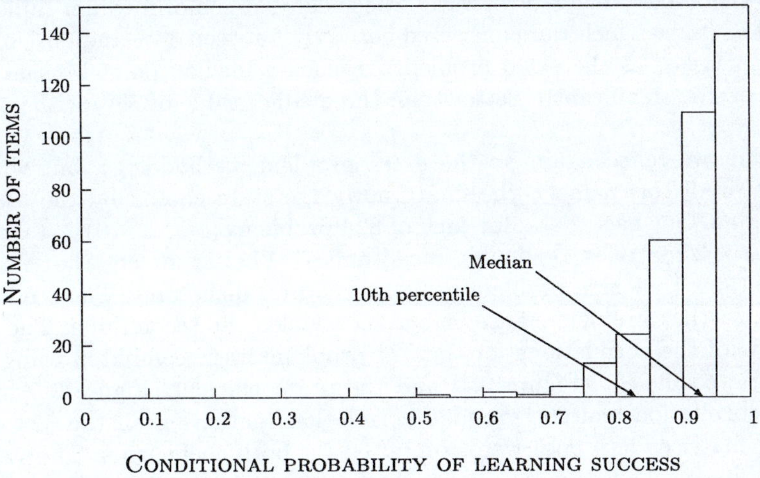


**Figure 2.7.** For the 353 items in elementary school mathematics (out of 370), the distribution of the estimated values of the conditional probabilities that a student entering the random walk reaches its right barrier. The problem is then regarded as having been mastered. The median probability is .93, with the 10th percentile at .83. These data are based on 1,940,473 random walks.

## 2.5 Summary and Discussion

The aim of this work was to evaluate the extent to which an assessment in a learning space, performed by the ALEKS system, is predictive of a student's mastery of a scholarly subject. In other words, we wanted to appraise the validity of such an assessment. We took elementary school mathematics as the exemplary subject. The method used for the appraisal was systematically to ask the student an extra problem $\boldsymbol{p}$, randomly selected, and predict the student's response to that problem on the basis of the rest of the information provided by the assessment. The accuracy of the prediction was measured by correlation coefficients (three of them were used in different parts of the study). Two aspects of the data were taken into account for this correlation analysis.

First, we computed Vincent curves tracing the evolution of the median point biserial correlations during the assessment. The graph of Figure 2.1

shows that, beginning with a median correlation of about .19, the median grows steadily to an asymptote that appears to be around .46. We also computed Vincent curves comparing problems with respect to their careless error rate (Figure 2.2) and with respect to their overall difficulty (Figure 2.3). The overall difficulty turned out to account for a greater variability of the correlation than the careless error rate. Specifically, items of medium difficulty (defined as items which were answered correctly between 33% and 66% of the time when asked as the extra problem) reached a median point biserial correlation of .51, significantly better than the median value of .44 for the other items.

Second, we relied again on the extra problem methodology, but with a different predictor, namely, the final knowledge state chosen by the assessment engine. Our basic data, for each of 324 problems (out of 370)[18], take the form of $2 \times 2$ matrices, or double dichotomies. Whether or not the selected state contains the extra problem provides the first dichotomy. The student's response to that problem—false or correct—yields the second one. The distributions of the correlations for the 324 problems were computed using the tetrachoric coefficient (Figure 2.4) and the phi coefficient (Figure 2.5). The median correlation and the grouped correlation (gathering all the 324 contingency matrices into one) were computed for both coefficients. The values obtained are recalled below:

| | tetrachoric | phi |
|---|---|---|
| median | .68 | .43 |
| grouped data | .80 | .58 |

These correlations do not take into account the careless errors. So, we also computed the correlation between the response to the extra problem and a different prediction variable, which also depends on the final state of the student. This variable has value 0 when the final state of assessment $\mathbf{a}$ does not contain the extra problem $p_{\mathbf{a}}$, and value $1 - \epsilon_{\mathbf{a}}$ (the probability that no careless error occurs) when the final state contains $p_{\mathbf{a}}$. We used the point biserial coefficient to evaluate this correlation and obtained $r_{pbis} = .67$, which is a high number in a psychometric context.

Our last analysis was of a different type, and its bearing on the validity of the assessment, while indirect, is important in practice. If the final state selected by the assessment is a valid representation of a student's competence, then the outer fringe of that state should contain problems that the student is ready to learn. Thus, if a student chooses to work on one of these problems, the probability of success should be high. We computed these probabilities for 1,940,473 learning experiences. The median probability of learning success was .93. The details are given by Figure 2.7.

[18] As indicated, 46 problems were discarded because the correlation coefficients could not be reliably computed.

While overall the validity results presented in this chapter should be regarded as satisfactory, an examination of Figures 2.1, 2.5, and 2.7 reveals the weaknesses of some items: the correlations measuring the validity are too low, and the probability of learning success is also too low. As argued earlier in this chapter, eliminating low performing items is not part of a solution: all items are included in the domain because they are considered an integral part of the curriculum. In particular, Figure 2.3 tells us that items of medium difficulty with respect to the population yielded better point biserial correlations. We only considered in this analysis initial assessments and a large majority of items were of high difficulty with respect to the current knowledge of the students. We may expect that, tested against the same students at a later stage after they have learned and increased their knowledge, the point biserial correlation of some of these items would show improvement. For general improvements, we discuss possible remedies below.

**On improving the items and the structure.** Let us first consider the careless errors. One might posit that a high value of $\epsilon_{\mathbf{a}}$ might be due to a mistake in the placement of item **a** in the structure of the learning space. Actually, this is unlikely in view of the procedure used to estimate the careless error probabilities, which does not involve the learning space. Indeed, the four equations (2.8)-(2.11) of Paragraph 2.3.3 rely on a dummy parameter $\gamma_{\boldsymbol{q}}$ measuring the probability that the knowledge state contains $\boldsymbol{q}$. The estimated value of this parameter resulting from minimizing the chi-square expression (2.12) has no formal relation with the structure. The most plausible hypothesis is that the high careless error of an item is intrinsic. Assuming this, there are two major reasons why an item could have a high careless error rate.

1. THE INSTANCES OF AN ITEM ARE NOT HOMOGENEOUS. This could arise, for example, if the instances are not of equal difficulty, or if some of them are ambiguously phrased. So, an item may be identified largely by its "good" instances, but a "bad" instance will occasionally be asked as the extra problem. The cure is straightforward, if tedious: all the instances of items having a high careless error rates must be carefully examined, and adjustments made if need be.

2. THE NATURE OF AN ITEM INDUCES POSSIBLE CARELESS ERRORS. This could arise, for example, when an item involves several numerical operations, each of which can elicit a careless error. This is unavoidable but there are ways to mitigate the high careless error rates of such items. `ALEKS` already uses specialized feedbacks, in some cases, to induce the student to check for careless errors. Such feedbacks give no hint of what the correct response is. A more thorough use of such feedbacks could probably palliate some of these high careless error rates.

There are also a couple of reasons for an item to have a low correlation value (whether in the form of the point biserial, tetrachoric, or phi) and/or a poor learning success rate (say, below the 10th percentile in Figure 2.7). First, a lack of instance homogeneity could contribute to either situation. Second, and in contrast to high careless error rates, low correlation or learning success values could certainly be due to a misplacement of an item in the structure. This means that the item must be removed from some states, and possibly added to some other states. While such a restructuring is a substantial enterprise, the tools are available to achieve it. We only sketch the main steps here. Suppose that some item $\boldsymbol{q}$ is distinguished as misplaced in a learning space $\mathcal{K}$. Removing this item from the domain of $\mathcal{K}$ creates a structure $\mathcal{K}^{-\boldsymbol{q}}$ that is still a learning space; that is, Axioms [L1*] and [L2*] are satisfied. We can now regard $\mathcal{K}^{-\boldsymbol{q}}$ as if it were the result of a partial construction of $\mathcal{K}$ and use the standard tools to complete the construction; that is, add the item $\boldsymbol{q}$ and give it a better placement in the structure. One of these tools is the QUERY algorithm[19], which can be used either with human experts, for whom it was originally conceived, or with statistics of conditional probabilities of solving or failing to solve problems, estimated from the data. The original implementation of the QUERY algorithm resulted in a structure that was a knowledge space but not necessarily a learning space. However, the current version of the algorithm corrects this shortcoming and substantially eases the completion process. Alternatively, how to complete a knowledge space into a learning space by a minimal addition of states formed from items in the same domain is the topic of Eppstein et al. (2009).

Finally, we must consider the learning success data reported in Figure 2.7, which shows that not all of the items are successfully learned when selected in the outer fringe of the student. Any item with a success rate below .8 may be regarded as flawed in some fashion. There are about 6% of such items. It is possible that these items are also misplaced in the structure and that the defect would be corrected by restructuring the learning space. A second possibility is that the explanation of these items is not clear enough and should be rewritten. A third possibility is that the domain lacks 'granularity' with respect to the skills tested by theses items and that the addition of 'step items' would benefit their learning.

From the analysis presented in this chapter, the `ALEKS` system appears to provide a valid assessment of what a student knows, does not know, and is ready to learn. Especially noteworthy are the high probabilities of learning success revealed by Figure 2.7 for all but a few items in elementary school mathematics. The system is not perfect: we have remarked that some items deserve improvements, whether in their placement in the structure, their instance formulation or their explanation. This discussion indicates that the means are available to achieve the necessary developments.

---

[19] Cf. page 25 in the Glossary.

## 2.6 Appendix

We recall that $N_q$ stands for the number of assessments having at least two presentations of item $\boldsymbol{q}$, with one of them as the extra problem, and $N_q(i,j)$, $i,j \in \{0,1\}$, stands for the number of times $(i,j)$ is realized among the $N_q$ assessments[20]. We estimate for each item $\boldsymbol{q}$ the careless error probability $\epsilon_q$ and the probability $\gamma_q$ that the students has mastered item $\boldsymbol{q}$ by minimizing the Chi-square statistic

$$\begin{aligned}
\mathtt{Chi}_q(\epsilon_q,\gamma_q) &= \sum_{i,j} \frac{(N_q(i,j) - N_q p_q(i,j))^2}{N_q p_q(i,j)} \qquad (2.16)\\
&= \frac{(N_q(0,0) - N_q(\epsilon_q^2\gamma_q + 1 - \gamma_q))^2}{N_q(\epsilon_q^2\gamma_q + 1 - \gamma_q)} + \frac{(N_q(0,1) - N_q\epsilon_q(1-\epsilon_q)\gamma_q)^2}{N_q\epsilon_q(1-\epsilon_q)\gamma_q}\\
&+ \frac{(N_q(1,0) - N_q(1-\epsilon_q)\epsilon_q\gamma_q)^2}{N_q(1-\epsilon_q)\epsilon_q\gamma_q} + \frac{(N_q(1,1) - N_q(1-\epsilon_q)^2\gamma_q)^2}{N_q(1-\epsilon_q)^2\gamma_q}.
\end{aligned}$$

Thus, the probabilities $p_q(i,j)$ in (2.16) are defined by (2.8)-(2.11).

We obtain the minimum of $\mathtt{Chi}_q(\epsilon_q,\gamma_q)$ by the Lagrange multipliers method. We simplify the notation and set

$$x = N_q(\epsilon_q^2\gamma_q + 1 - \gamma_q) \qquad (2.17)$$

$$y = N_q\epsilon_q(1-\epsilon_q)\gamma_q \qquad (2.18)$$

$$z = N_q(1-\epsilon_q)^2\gamma_q. \qquad (2.19)$$

Notice that the ratio of the last two equations only depends upon $\epsilon_q$ since

$$\frac{z}{y} = \frac{1-\epsilon_q}{\epsilon_q},$$

and so

$$\hat{\epsilon}_q = \frac{y}{z+y}. \qquad (2.20)$$

Replacing $\epsilon_q$ in (2.18) by its expression in (2.20) yields

$$y = N_q \frac{yz}{(z+y)^2}\gamma_q.$$

Canceling the $y$'s and rearranging, we obtain

$$\hat{\gamma}_q = \frac{(z+y)^2}{zN_q}. \qquad (2.21)$$

[20]Thus, for example, $(1,0)$ stands for the event that the student responds correctly to the first presentation of $\boldsymbol{q}$ in the assessment, and incorrectly to the second one.

Thus, $\epsilon_{\boldsymbol{q}}$ and $\gamma_{\boldsymbol{q}}$ are defined by $y$ and $z$. To obtain the values of $y$ and $z$, we minimize the function

$$(x,y,z) \mapsto \frac{(N_{\boldsymbol{q}}(0,0)-x)^2}{x}+\frac{(N_{\boldsymbol{q}}(0,1)-y)^2}{y}+\frac{(N_{\boldsymbol{q}}(1,0)-y)^2}{y}+\frac{(N_{\boldsymbol{q}}(1,1)-z)^2}{z}$$

with respect to $x$, $y$, and $z$, subject to the constraint

$$x+2y+z=N_{\boldsymbol{q}}.$$

To this effect, we define

$$\begin{aligned}\Lambda(x,y,z,\lambda) &= \frac{(N_{\boldsymbol{q}}(0,0)-x)^2}{x}+\frac{(N_{\boldsymbol{q}}(0,1)-y)^2+(N_{\boldsymbol{q}}(1,0)-y)^2}{y}\\ &\quad+\frac{(N_{\boldsymbol{q}}(1,1)-z)^2}{z}+\lambda(x+2y+z-N_{\boldsymbol{q}}).\end{aligned} \tag{2.22}$$

We then compute the derivatives of $\Lambda$ with respect to its four variables and solve, with respect to $x$, $y$, $z$, and $\lambda$, the system of the four equations

$$\frac{d\Lambda}{x}=\frac{d\Lambda}{y}=\frac{d\Lambda}{z}=\frac{d\Lambda}{\lambda}=0.$$

This gives

$$\frac{N_{\boldsymbol{q}}(0,0)^2}{x^2}=\frac{N_{\boldsymbol{q}}(0,1)^2+N_{\boldsymbol{q}}(1,0)^2}{2y^2}=\frac{N_{\boldsymbol{q}}(1,1)^2}{z^2}=\lambda+1, \tag{2.23}$$

leading to

$$\frac{z}{y}=\sqrt{\frac{2N_{\boldsymbol{q}}(1,1)^2}{N_{\boldsymbol{q}}(0,1)^2+N_{\boldsymbol{q}}(1,0)^2}}. \tag{2.24}$$

Replacing $\frac{z}{y}$ in (2.20) by its expression in (2.24) gives

$$\hat{\epsilon}_{\boldsymbol{q}}=\frac{\sqrt{N(0,1)^2+N(1,0)^2}}{\sqrt{2N(1,1)^2}+\sqrt{N(0,1)^2+N(1,0)^2}}. \tag{2.25}$$

The derivations giving an estimation formula for $\gamma_{\boldsymbol{q}}$ are similar. From (2.17), (2.19) and (2.23), we get, after canceling the $N_{\boldsymbol{q}}$'s,

$$\frac{x}{z}=\frac{N_{\boldsymbol{q}}(0,0)}{N_{\boldsymbol{q}}(1,1)}=\frac{\epsilon_{\boldsymbol{q}}^2\gamma_{\boldsymbol{q}}+1-\gamma_{\boldsymbol{q}}}{(1-\epsilon_{\boldsymbol{q}})^2\gamma_{\boldsymbol{q}}}.$$

The last equation is linear in $\gamma_{\boldsymbol{q}}$. Solving this equation for $\gamma_{\boldsymbol{q}}$ and replacing $\epsilon_{\boldsymbol{q}}$ by its expression in (2.25) gives finally, after rearranging,

$$\hat{\gamma}_{\boldsymbol{q}}=\frac{\left(\sqrt{2N_{\boldsymbol{q}}(1,1)^2}+\sqrt{N_{\boldsymbol{q}}(0,1)^2+N_{\boldsymbol{q}}(1,0)^2}\right)^2}{2N_{\boldsymbol{q}}(1,1)\left(N_{\boldsymbol{q}}(0,0)+N_{\boldsymbol{q}}(1,1)+\sqrt{2(N_{\boldsymbol{q}}(0,1)^2+N_{\boldsymbol{q}}(1,0)^2)}\right)}. \tag{2.26}$$

# 3

# ALEKS-based Placement at the University of Illinois

Alison Ahlgren Reddy[1]

Marc Harper[2]

## 3.1 Introduction

ALEKS is an adaptive assessment and learning mechanism used as a course companion and assessment tool. The use of ALEKS as an assessment mechanism in higher education at the level between college algebra and calculus began recently, following the work of Carpenter and Hanna which showed that ALEKS could serve as a preparedness measure (Carpenter and Hanna, 2006). This chapter investigates the use of ALEKS as part of a placement mechanism at the University of Illinois, comparing the results to the previously used mechanism based on ACT scores. The effectiveness of standardized college entrance exams as predictors of student performance has been previously investigated, particularly for the SAT (Baron and Norman, 1992). An ALEKS-based mechanism has been implemented at Boise State University (Bullock et al., 2009), using ALEKS as a course companion and as an assessment mechanism. Emulating the implementation at the University of Illinois, the Boise State Mechanism has had similar results regarding the decline of failure proportions in placement courses and enrollment changes. Other studies investigate the use of ALEKS as a course companion (Hagerty and Smith, 2005; Hampikian et al., 2006; Hampikian, 2007; Callahan et al., 2008; DeLucia, 2008, see also Chapters 4 and 5 in this volume). For more information regarding the implementation details of ALEKS see Falmagne et al. (2006) or Falmagne and Doignon (2011).

Ineffective course placement has many negative effects. Students may face significant consequences for failure or withdrawal from a course, in addition to increasing time towards degree completion, because they do not recognize the risk of failure until the first midterm examination (generally four weeks into a semester and two weeks past the add-deadline). At the University of Illinois, the standard introductory calculus course (Calc: Calculus I) is a five credit course, the withdrawal from which beyond the add-deadline may reduce students to a credit total below full-time status, resulting in the loss of tuition benefits, health benefits, scholarships, and athletic eligibility.

[1]Dept. of Mathematics, University of Illinois, Urbana-Champaign, IL

[2]Dept. of Genomics and Proteomics, University of California, Los Angeles, CA

J.-C. Falmagne et al. (eds.), *Knowledge Spaces: Applications in Education*,
DOI 10.1007/978-3-642-35329-1_3, 

Improper course placement also results in students being taught material they already know or are unprepared for and introduces challenges for advisors and faculty, including course planning. Poor course performance from overplacement may lead to forced change of major and academic probation. A loss of time from underplacement may force the student to stay an additional semester, incurring substantial additional financial costs.

## 3.2 Placement at the University of Illinois

The Department of Mathematics at the University of Illinois at Urbana-Champaign, at the impetus of the first author, began searching for a cross-platform web-based assessment and placement instrument in October of 2005. Placement at the University of Illinois at that time was based on ACT scores. The Mathematics department desired a new placement mechanism that would reduce the high failure and withdrawal rates in many of the introductory mathematics courses.

Several reasons were proposed for the high failure rates. Students at the University of Illinois come from a variety of geographic locations and educational backgrounds. High school students in the state of Illinois are only required to take three years of mathematics courses. Many students elected not to take mathematics in their final year of high school and may have forgotten significant amounts of mathematical knowledge in the more than a year that passed before they arrived at the university. Moreover, the indicators of a student's mathematical knowledge and skill – their high school transcript and standardized test scores – were captured at the height of their knowledge rather than at the time of course enrollment.

Many examinations and systems were evaluated and piloted. ALEKS was chosen because of its ability to measure students' knowledge and the facts that it is cross-platform (requiring only a web-browser), non-multiple choice, and adaptive. The assessments provided by ALEKS were then used as a basis for a placement mechanism under the assumption that the initial knowledge of a student, measured shortly before entering a course, would be predictive of student success.

## 3.3 The University of Illinois Math Placement Program

In the summer of 2007 the University of Illinois Department of Mathematics began using ALEKS to assess students for course readiness. The placement program focuses on four courses: Preparation for Calculus (Math 115: **Pre-Calc**), Calculus I (Math 220: **Calc**), Calculus I for students with experience (Math 221: **CalcExp**), and Business Calculus (Math 234: **BusCalc**)[3].

[3]For a description of these courses, please see:
http://courses.uiuc.edu/cis/catalog/urbana/2007/Fall/MATH/index.html.

The placement program was required for course placement by the mathematics department for all students enrolling in the focus courses and became a university requirement for all incoming students in 2008.

**3.3.1 Assessment Procedure.** The placement exam is an ALEKS assessment composed of items from the "Prep for Calculus" course product, slightly customized to remove some items that do not appear in the syllabi of the focus courses at the University of Illinois. Students access the assessment in a non-proctored setting such as their homes or a campus computer lab. Students who fail to reach the required score for placement for a particular course have the option of retaking the placement exam or self-remediating using the ALEKS learning module (and other methods of their volition). Students are allowed to repeat assessments during the five months prior to the course add-deadline, which is always the fifth day of enrollment. Students that ultimately take another course in the placement pool must take another assessment regardless of the grade obtained in a prerequisite course.

**3.3.2 Determination of Placement Cutscores.** Data were also collected the previous semester by offering students a small grade incentive (to ensure proper effort) in the focus classes, providing initial cutscore choices. These were lowered slightly for the first year of implementation to 40% for BusCalc and PreCalc and 60% for Calc and CalcExp to account for any bias in the initial testing procedure. After the first year of data collection the cutscores were raised to 50% and 70%, scores indicated by the data to be more effective in reducing overplacement.

**3.3.3 Policy Enforcement.** In the first year, the placement policy was enforced by making ten percent of the student's final grade contingent on completing the ALEKS requirement, all or nothing. The remaining 90% of the grade was calculated based on homework, quizzes, and exams (depending on the teaching style of the instructor). In the subsequent year is was possible to make the placement exam a prerequisite (in fact the only prerequisite) for enrolling in a course.

The policy change emphasized the positive role of assessment for the students, by providing an accurate and current assessment of their skills to them and to the mathematics department. For many students, the ALEKS assessment may have been the first objective evaluation of their mathematical skills that they had received in years (or ever).

## 3.4 Description of the Data

The data consist of the following information for each student in any of the focus courses for which an ALEKS assessment is required, for each semester in which the policy affects (Fall 2007, Spring 2008, Fall 2008):

1. enrollment and grades (including withdrawals) reported as
   A+, A, A−, B+, B, B−, C+, C, C−, D+, D, D−, F, or W;
2. ALEKS assessment reports for all assessments completed, including subscores, with scores as percentages;
3. ACT Math scores.

Some students enrolled in more than one course during the three semesters. The total number of assessments exceeds 15,000 for approximately 10,000 students. Approximately 20% of the students took more than one assessment (see Figure 3.1).

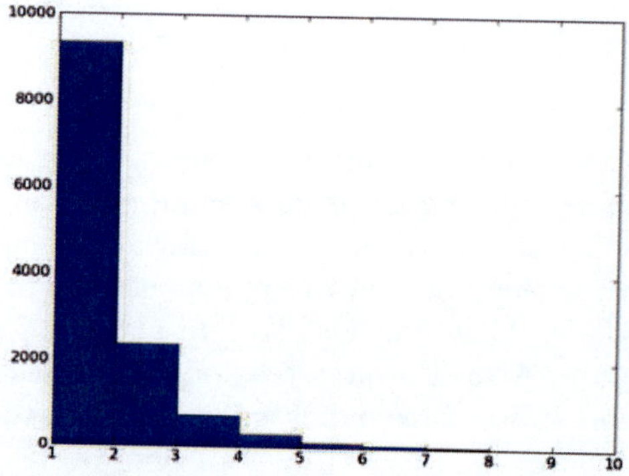


**Figure 3.1.** Histogram of ALEKS Assessments per student: Most students took a single assessment from which their placement was determined. Around 20% took a second assessment, and relatively few students took three or more assessments.

**3.4.1 Histograms of ACT and ALEKS scores.** Visualizations and basic statistics of these data are given in the figures below. The legends contain the mean and standard deviation for each course in Figures 3.2 and 3.3, which are normalized for direct comparison. Notice the bimodalities that emerge in the ALEKS histograms versus the single modalities of the ACT histograms. The authors believe the bimodality to result from the state policy of only requiring three years of mathematics in high school, but this hypothesis was not confirmed analytically.

In the first year of implementation, the cutoff scores for PreCalc and BusCalc was 40% and for Calc and CalcExp was 60%. In the second year of implementation, these scores were changed to 50% and 70%, respectively, based on analysis of the data from the first year. This is clearly evident in the distributions in Figure 3.3 when comparing the Fall 2008 scores to the prior two semesters.

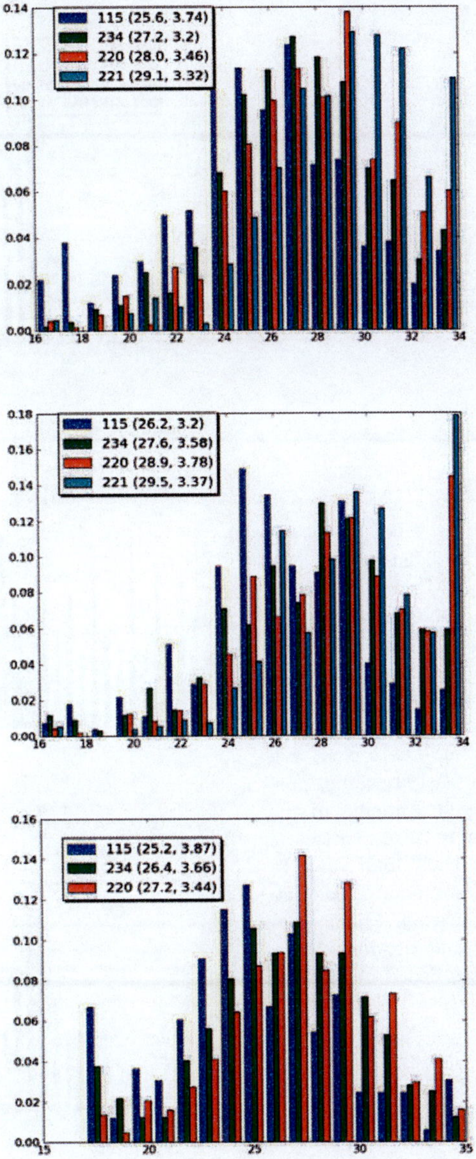


**Figure 3.2.** ACT Math Histograms, normalized. Top: Fall 2007. Middle: Fall 2008. Bottom: Spring 2008. Each bar shows, for the corresponding course (Math 115: PreCalc, Math 234: BusCalc, and Math 220: Calc), the proportion of students with a particular ACT Math score. In the more difficult courses, the means of the distributions are slightly larger, while the variances are largely similar.

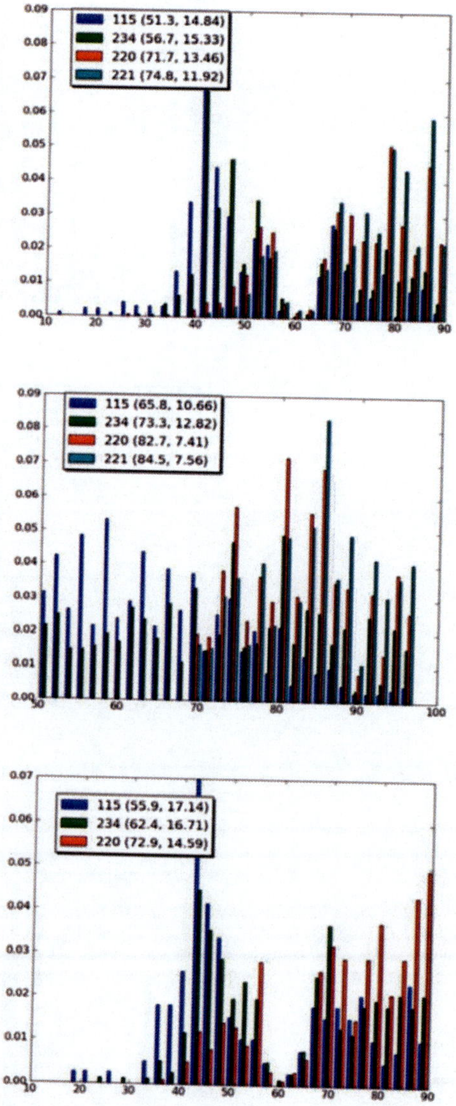


**Figure 3.3.** ALEKS Score Histograms, normalized. Top: Fall 2007. Middle: Fall 2008. Bottom: Spring 2008. Each bar shows, for the corresponding course (Math 115: PreCalc, Math 234: BusCalc, Math 220: Calc, and Math 221: CalcExp), the proportion of students with a particular ALEKS score in the corresponding course. There is an evident bimodality to the distributions, possibly because high school students in the state of Illinois are not required to take a mathematics course in their final year. In contrast to the ACT Math scores, the distributions are more separated. This is partially due to the fact that students were placed into these courses based on achieving a minimum score.

## 3.5 Analysis of Data

The focus courses had many instructors and teaching assistants each semester. Most of the students were in large lecture courses with weekly discussions, but some were in small discussion oriented courses or small lectures. No attempt was made to control for these factors in the grade data because the decision to enroll in a particular style of course is not determined by any placement mechanism, rather by student preference in style or time of course meetings.

### 3.5.1 Effect of placement on DFW rates and enrollment.

A change in assessment policy affects the enrollments of the focus courses and those courses immediately in the sequence before or after focus courses. The relative enrollment changes (relative to Fall 2006 enrollments which are before the use of ALEKS ) are given below. Note that Preparation for Calculus (PreCalc) was first offered in Spring 2007, and so Fall 2006 comparisons are not possible.

Lowering DFW (D+, D, D−, F, or withdrawal) percentages was a principal motivation for the placement program. In addition to providing information for advisors and department members to help students land in the appropriate course, ALEKS also provides students with a measure of their current mathematical skills. Many students may have been unaware of their relative lack of preparedness before completing an ALEKS assessment and subsequently altered their enrollments appropriately, but it is not possible from the data to determine how many students failed to meet their initial goal and enrolled in a prerequisite course.

The relative changes in fail (D+, D, D−, and F) percentages and withdrawal percentages are given in Tables 3.1 and 3.2. These tables reveal several changes to the distribution of students and final grades. In particular, withdrawals were dramatically reduced and enrollments shifted toward the more advanced courses. These percentages are compiled over all sections and instructors of the courses which were not consistent in the three relevant years. The proportion of students failing BusCalc was greatly reduced. In all four placement courses, the number of DFW students averaged in 2007 and 2008 was lower than in 2006.

| Course | % Decrease W | % Decrease (D±, F) | % Increase Enrollment |
|---|---|---|---|
| BusCalc | 57% | 40% | -7% |
| Calc | 49% | -12% | 3% |
| CalcExp | 67% | 62% | 21% |

**Table 3.1.** Changes in Withdraw, Failure, and Enrollment for Fall 2007 relative to Fall 2006

| Course | % Decrease W | % Decrease (D±, F) | % Increase Enrollment |
|---|---|---|---|
| BusCalc | 19% | 33% | -18% |
| Calc | 81% | -16% | 8% |
| CalcExp | 42% | -0.7% | 38% |

**Table 3.2.** Changes in Withdraw, Failure, and Enrollment for Fall 2008 relative to Fall 2006

The withdrawal percentages for all courses dropped substantially relative to the fall semester of 2006, which is the last semester in which ALEKS was not used as a placement mechanism. This indicates that the placement mechanism reduced severe cases of overplacement, although there is a increase in the number of students receiving a score of D+ or below in Calc. Notice that enrollments changed significantly as more students placed into the calculus courses Calc and CalcExp or chose not to take BusCalc, a calculus course of less rigor, which is often not accepted in graduate programs in business as fulfilling the undergraduate calculus requirement.

**3.5.2 Relationship of Placement Scores and Grades.** National standardized collegiate entrance examinations, such as the ACT and SAT, are generally taken by students in their junior year, providing a snapshot of student abilities at a time that is significantly prior to enrollment. A mechanism relying on an assessment with a temporal delay is a potential source of underplacement, in the case that the student completed additional mathematics courses after the examination date, and a potential source of overplacement, in the case that the student's skill level decreased from lack of use.

The ALEKS distributions provide more granular information on the mathematical abilities of the students in the placement population than standardized tests. These scores can be interpreted as the percentage of concepts of prerequisite material known by the students rather than a renormalized national standardized test score, of which the meaning is less substantial. The placement program uses the total percentage of concepts demonstrated and does not incorporate subscores for specific content areas.

The distribution on the ALEKS plot in Figure 3.4 is increasing with grade in both the medians (the horizontal line within the box) and the interquartile range. Figure 3.5 shows the correlation between ACT scores and ALEKS scores with final grade. In the ALEKS plot, scores are aggregated over 5 point intervals (as in a histogram). The numbers near each point indicate how many scores contributed to the mean computation.

**3.5.3 ALEKS Subscores.** The ALEKS exam content is broken into several subcategories, listed in Table 3.3 on page 61.

Each subscore contributes to the overall correspondence between the ALEKS score and student performance, though there are dependencies among the subscores. For every course and every semester, the subcategories for functions and for trigonometry had the highest mutual information with the grade distribution. See Table 3.4 (also on page 61). The categories for rational expressions and radical expressions are also relatively large in proportion. The numbers category is relatively poor, which is explained by the fact that most students obtain complete or nearly complete mastery in this category. The

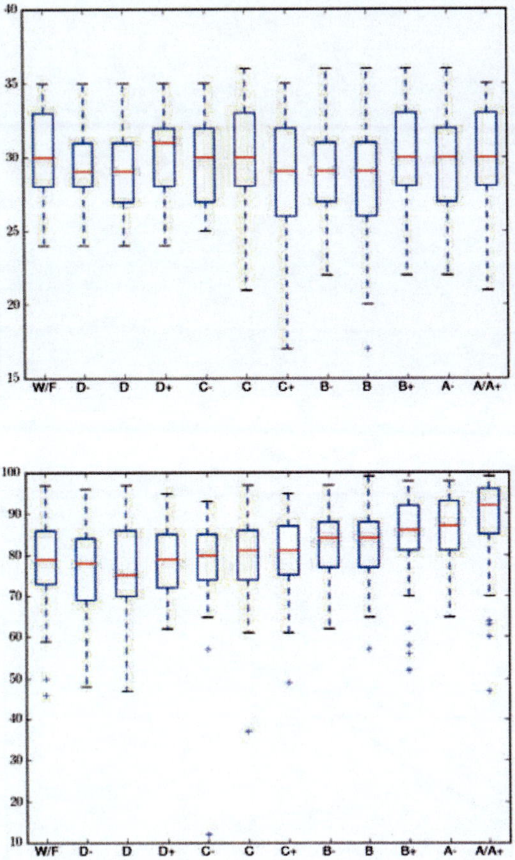


**Figure 3.4.** CalcExp Fall 2008: Box plots of grade distributions for ACT Math score (top) and ALEKS score (bottom). Notice that the medians (indicated by the central red lines) are relatively flat for the ACT Math score yet are increasing for the ALEKS score as grades increase. Similarly for the interquartile ranges (blue boxes).

logarithms and exponentials category is also relatively poor in proportion, for the opposite reason as most students performed weakly in this category.

That the information from the function subscore correlates highly with the grade distribution is not surprising since the content of the focal courses is heavily dependent on shifting and plotting functions, modeling using functions, and operations on functions, such as limits and derivatives. Interestingly, trigonometry has a strong subcategory correlation for BusCalc in the Fall 2008 semester despite the fact that trigonometry is not used in the course.

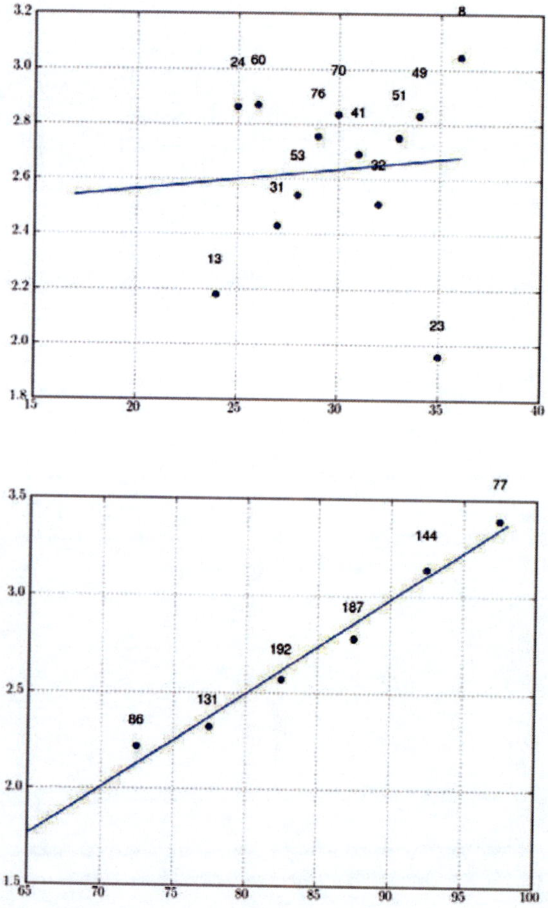


**Figure 3.5.** CalcExp Fall 2008: Mean Grade versus Exam Scores for ACT Math (top) and ALEKS (bottom). This plot reflects the behavior of the distributions in Figure 3.4. The ALEKS scores are grouped into buckets (width 5) and the points have a correlation coefficient of $r^2 = 0.97906$. Grouping scores did not substantially increase correlation coefficient of the ACT Math score $r^2 = 0.00895$.

The higher correlation may result from a relationship between trigonometric knowledge and mathematical maturity. See Figure 3.6.

### 3.5.4 Effect of Placement Policy on ALEKS Subscores.

Figure 3.7 (page 63) shows the subscore distributions for PreCalc in the fall of 2007, in which a soft requirement of 40% was used, compared to the fall of 2008, in which a hard requirement of 50% was used. Figure 3.8 is the same

| Subcategory | Content | Items |
|---|---|---|
| Numbers | Real Numbers | 23 |
| Equations | Equations and Inequalities | 29 |
| Functions | Linear and Quadratic Functions | 40 |
| Polynomials | Exponents and Polynomials | 26 |
| Rationals | Rational Expressions | 23 |
| Radicals | Radical Expressions | 21 |
| Logarithms | Exponentials and Logarithms | 17 |
| Trigonometry | Geometry and Trigonometry | 36 |

**Table 3.3.** ALEKS Assessment Subcategories. The total set of items in the assessment is partitioned into these subcategories.

| Course | Semester | Total ALEKS | Functions | Trig |
|---|---|---|---|---|
| PreCalc | Fall 2007 | 0.461 | 0.239 | 0.240 |
| PreCalc | Spring 2008 | 0.574 | 0.306 | 0.430 |
| PreCalc | Fall 2008 | 0.310 | 0.173 | 0.142 |
| BusCalc | Fall 2007 | 0.634 | 0.325 | 0.370 |
| BusCalc | Spring 2008 | 0.472 | 0.222 | 0.231 |
| BusCalc | Fall 2008 | 0.317 | 0.203 | 0.176 |
| Calc | Fall 2007 | 0.660 | 0.433 | 0.341 |
| Calc | Spring 2008 | 0.361 | 0.196 | 0.202 |
| Calc | Fall 2008 | 0.166 | 0.102 | 0.097 |
| CalcExp | Fall 2007 | 0.719 | 0.390 | 0.415 |
| CalcExp | Fall 2008 | 0.156 | 0.081 | 0.096 |

**Table 3.4.** Proportion of grade distribution entropy captured by ALEKS score and subscores. Mutual information is a measure of dependence between two variables (Cover and Thomas, 2006). The subscores capture a significant portion of the total information captured by the total ALEKS score.

comparison for Calc, with cutoffs of 60% and 70%. The higher cutoff led to better prepared students, evident from the larger medians in each subscore.

**3.5.5 ALEKS as a placement measurement tool.** Students fail courses for many reasons that are difficult to determine simply from their placement scores and unrelated to their academic preparedness, such as negative housing situations (for instance due to randomly assigned roommates in dormitories), financial stresses, and poor time management. For incoming students the number of challenges is even greater. Accordingly, no assessment process is expected to predict student performance completely. Nevertheless, proper preparation is expected to be a factor in student performance and so correlation of the initial preparation measure and final grade performance is an indication of effectiveness. The data indicate that the ALEKS assessment re-

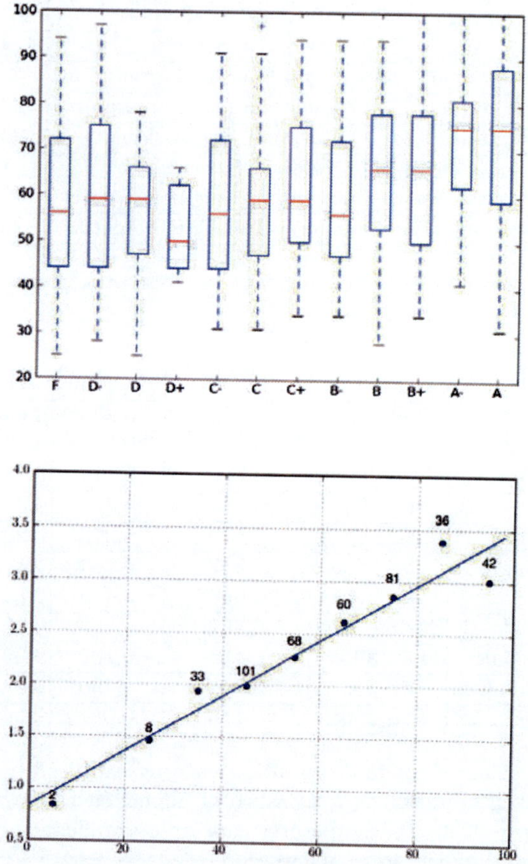


**Figure 3.6.** ALEKS Trigonometry Subscore, BusCalc, Fall 2008. Top: ALEKS Trigonometry Subscore (%) boxplots per Grade. Bottom: Trigonometry Subscore (%, width 10 buckets) vs. Grade Mean of bucketed scores (on 4-point scale), $r^2 = 0.94936$. The trigonometry and geometry subscore is clearly related to the final grade even though trigonometry is not used in the course.

port (using the aggregate score) correlates much more significantly than ACT scores (for instance, see Fig. 3.5). This is broadly true over all classes and semesters.

**3.5.6 ALEKS as a preparation assessment tool.** The core content of the placement exam mirrors the content in the PreCalc and so it is expected that ALEKS assessment scores will improve for students who successfully complete the Preparation for Calculus course. Because students are required to take an assessment to enter PreCalc and Calc, data are available for many students who progressed from PreCalc to Calc, taking assessments before and after the

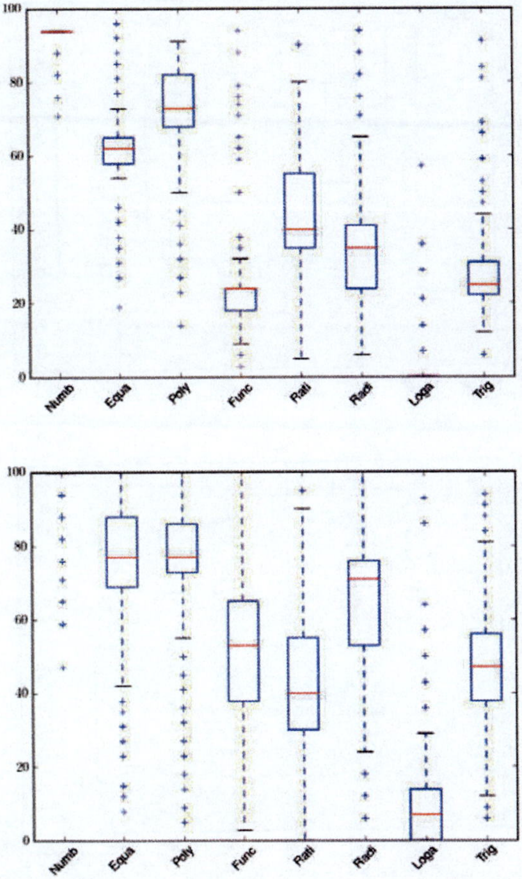


**Figure 3.7.** Effect of Policy change on Subscores in PreCalc. The vertical shift in every distribution between Fall 2007 (top) and Fall 2008 (bottom) is due to the higher placement requirement in 2007.

former course. The increase in ALEKS score for these students is shown in the Figure 3.9.

The 52.5% average relative improvement in assessment scores may be interpreted in at least two ways. Under the assumption that ALEKS accurately measures a student's state of knowledge, the improvement in assessment score indicates that the students who completed Preparation for Calculus learned a significant number of the concepts covered by the assessment. Conversely, assuming that the course effectively teaches students precalculus concepts, the improvement in assessment scores indicates that ALEKS is detecting the students' newly acquired knowledge.

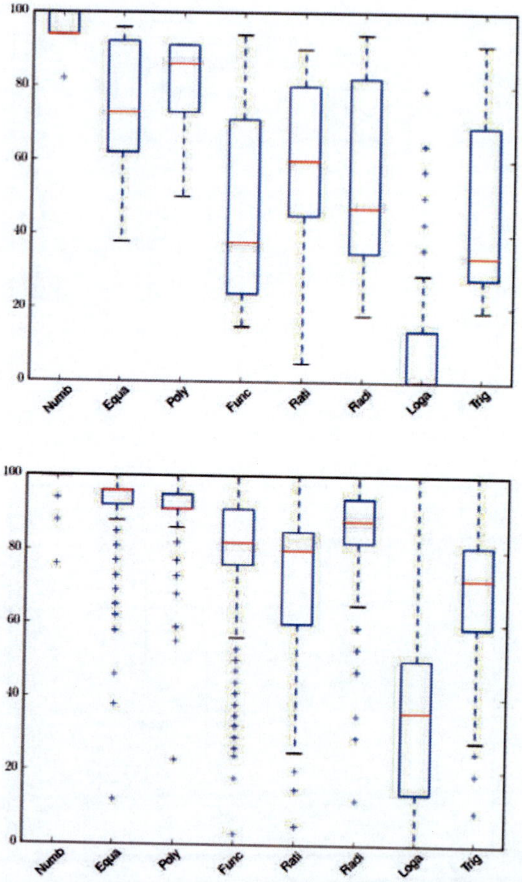


**Figure 3.8.** Effect of Policy change on Subscores in Calc. The vertical shift in every distribution between Fall 2007 (top) and Fall 2008 (bottom) is due to the higher placement requirement in 2007.

The improvement in student knowledge resulting in the successful completion of PreCalc is shown in Figure 3.10. Contrast with the incoming Calc students' abilities in Fall 2008 in Figure 3.11.

## 3.6 Discussion

The ALEKS-based mechanism used at the University of Illinois effectively reduces overplacement and is more effective than the previously used ACT-based

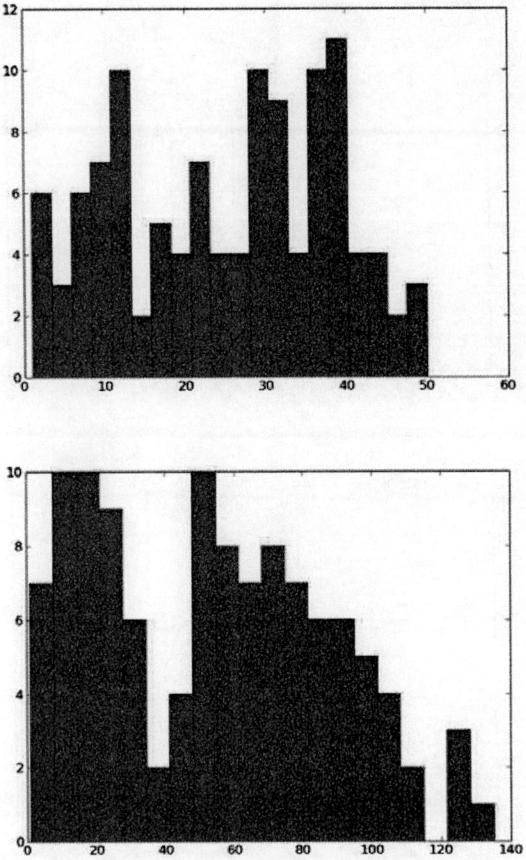


**Figure 3.9.** ALEKS Assessment difference Histograms for students passing through PreCalc to Calc. Top: Histogram of absolute differences in score. Bottom: Histogram of relative differences in score. Students passing through PreCalc into Calc in successive semesters had to retake the placement assessment. These students showed substantial increases in placement scores. Some students who took PreCalc could have already placed into Calc (if their score exceeded 70%), and so may not have improved much. Others went from ≈ 50% to at least 70%

mechanism. Significant enrollment distribution changes occurred as a result of the mechanism implementation. These changes are similar to the emulation implementation at Boise State University. The use of ALEKS as part of placement mechanisms is justified by the data, noting that preparation is one component of placement and is not expected to be a complete predictor.

ALEKS assessments provide more specific skill information than the ACT. Correlations of ALEKS subscores with student maturity and performance meet

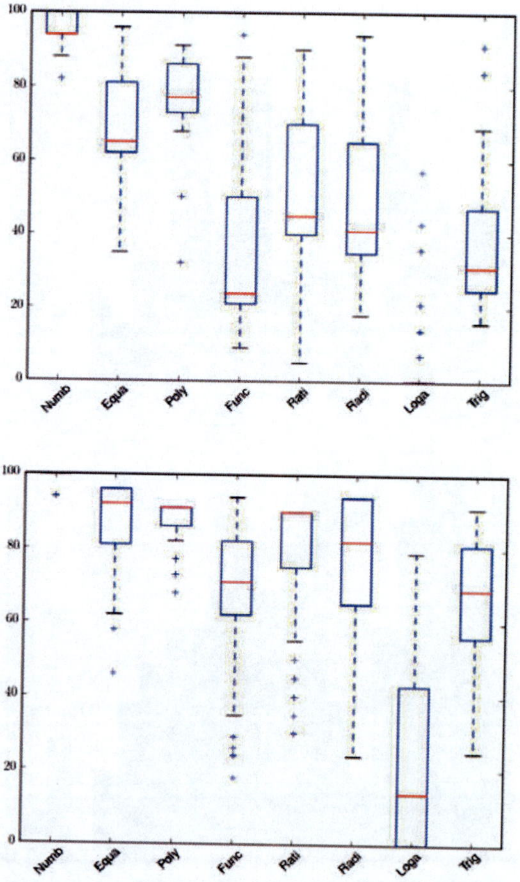


**Figure 3.10.** ALEKS Assessment Subscore differences from completion of PreCalc. Top: All assessments for the course for students before enrollment. Bottom: All students that subsequently enrolled in Calculus in the next semester.

expectations in many cases (all students in Calc in the Fall of 2008 had complete mastery of the most basic category), and reveal interesting characteristics of the student population in other (systematic weakness in exponentials and logarithms). ALEKS reveals skill bimodality in the population not captured by the previous placement mechanism which the authors believe is due to the high school math education policy of the state of Illinois. Specifically, mathematics is not a requirement for high school seniors in the state (of whom $\approx 90\%$ of the student body is derived), so many incoming students have not had direct exposure to mathematics in more than a year before enrollment.

The data show that preparation, as measured by ALEKS, correlates positively with course performance, and more strongly than the ACT in general.

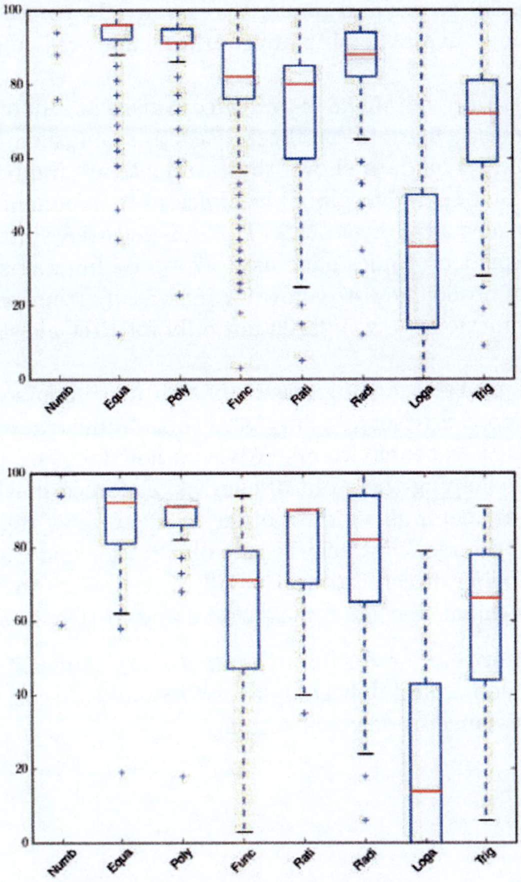


**Figure 3.11.** ALEKS Subscores for Calc, Fall 2008. Top: All Students. Bottom: All Students, excluding those from PreCalc.

Though a student may pass a course with a lower percentage of prerequisite concepts known, students receiving grades of A or B generally show greater preparedness. Longitudinal comparison of students taking PreCalc (Preparation for Calculus) shows that ALEKS assessments are an effective measure of knowledge increase. Calculus students with weaker skills can be brought to the skill level of their peers, as measured by ALEKS, by taking a preparatory course. Interestingly, the data provided by ALEKS provides a measure of course effectiveness when students' performance is aggregated and tracked longitudinally.

Policy changes in the second year of implementation improved the mechanism significantly and shows the need for consistent policy application that reduces student incentive to cheat. The results suggest the need for testing all students in the placement population because of incidences of high ACT scores but low ALEKS scores and poor performance. Data needs to dictate the details of the policy, such as setting appropriate cutscores, for a more effective mechanism.

The installation of the ALEKS-based mechanism at Illinois has had many tangential outcomes. The data has been used to strengthen and improve the PreCalc curriculum. The data shows that students are improving in the PreCalc course, but that there are global weaknesses in exponential and logarithmic functions, as measured by ALEKS. This has led to a shift in emphases as less time is now spent on topics such as polynomial functions (students enter with high levels of proficiency) to allow for increased attention on exponential and logarithmic functions, a weak area not only for PreCalc students, but also for Calc.

The campus culture regarding lower division math courses has shifted for the better as students, advisers, faculty, and the administration recognize that effective placement is a benefit to all. Advisers now focus on placing students into courses versus helping them out of courses. Enforced math placement has had a positive effect campus-wide, as other units are now reporting improvements in their core courses that heavily rely on PreCalc and Calc prerequisites. Attention is now being directed towards STEM initiatives to see what impact better course placement and outcomes are having in this area.

**Aknowledgments.** The authors thank Robert Muncaster, Jean-Claude Falmagne, Chris Doble, and Jeff Douglas for valuable discussions and Alison Champion for assistance in data collection.

4

# The Impact of a Mathematical Intelligent Tutoring System on Students' Performance on Standardized High-Stake Tests

Jeremiah Sullins[1]
Rob Meister[1]
Scotty D. Craig[1]
William M. Wilson[1]
Anna Bargagliotti[1]
Xiangen Hu[1]

## 4.1 Introduction

Recent research has suggested that standardized high-stakes tests, such as the SAT[2], have become increasingly important to policy makers, school districts, and society in general. Scoring well on these tests may determine the access to educational opportunities beyond high school. Unfortunately, recent reports have shown that Americans are falling behind their peers in other nations on comparable assessments (Gollub et al., 2002). Additionally, schools in the U.S. must adhere to the demands of the No Child Left Behind Act (NCLB). The policy states that federal district funding is dependent on student overall performance on standardized tests in mathematics, reading and other content areas. To alleviate the problem of U.S. students underachieving on standardized tests, educators must explore areas of pedagogy that have been empirically shown to be effective.

A potential alternative method is one-to-one human tutoring, which has often been put forth as a gold standard in instruction (Bloom, 1984; Cohen

[1]University of Memphis, TN.

[2]This acronym used to stand for 'Scholastic Aptitude Test.' The meaning then changed a few years ago to 'Scholastic Assessment Test.' Today, there does not seem to be any special meaning associated with the letters of the acronym.

J.-C. Falmagne et al. (eds.), *Knowledge Spaces: Applications in Education*,
DOI 10.1007/978-3-642-35329-1_4, 

et al., 1982). A meta-analysis conducted by Cohen et al. (1982) revealed that non-expert human tutors increased learning gains by .4 standard deviation units (this translates to approximately half a letter grade). Unfortunately, teachers do not have the time and districts do not have the resources to provide individualized tutoring to every student in a classroom. This is where an effective personalized teaching technology could play an essential role.

One area of research that has shown promise in the facilitation of math achievement is in the field of computer-based instruction and intelligent tutoring systems (ITS). Loosely defined, an ITS is a computer-based learning environment that provides individualized instruction to the learner. Some published research has shown the ALEKS[3] system to be effective in identifying and closing gaps in math knowledge and skills at the college level (Hagerty and Smith, 2005). Unfortunately, no research at the secondary and middle school levels using ALEKS currently exists. However, there are numerous researchers who have explored the possible influence of technology on achievement scores in varying domains (e.g., reading comprehension and mathematics).

Not everybody agrees that technology provides an advantage over traditional classroom instruction. For example, Angrist and Lavy (2002) compared computer skills training and computer aided instruction in relation to mathematical achievement among fourth graders. They define 'computer skills training' as teaching students how to use computers and 'computer aided instruction' as using computers to teach. Results revealed that computer aided instruction provided no benefit on academic achievement in math among fourth grade students, at least in the short term.

Additionally, Baker et al. (1994) evaluated the Apple Classrooms of Tomorrow and examined the impact of interactive technologies across five school sites in four different states (California, Tennessee, Minnesota, and Ohio). The Apple Classrooms of Tomorrow were designed to encourage instructional innovation, and to emphasize to teachers the potential of computers to support student initiative, long term projects, access to multiple resources, and cooperative learning. Although there were some positive outcomes as a result of the initiative, it was found that on standardized tests in various contents (e.g., mathematics, reading comprehension, and vocabulary) students performed no better than control groups or national norms.

Wenglinsky (1998) explored the effects of simulation and higher order thinking technologies on mathematics achievement on the National Assessment of Educational Progress. The sample size consisted of 6,227 fourth graders and 7,146 eight graders. While controlling for socioeconomic status, class size, and teacher characteristics, the results revealed that students who used such technologies performed worse on the National Assessment of Educational Progress than students who did not use those technologies.

[3] ALEKS is one example of an ITS. ALEKS is an acronym for *Assessment and LEarning in Knowledge Spaces*.

However, this view may not be the full story. More recent evidence has been reported in favor of technology. This research found that supplementing instruction with technology (in the ITS format) students' mathematical achievement can be increased significantly (Woolf et al., 2006; Melis and Siekmann, 2004). Woolf et al. (2006) have developed a web-based mathematical ITS called 'Wayang Outpost.' Wayang Outpost was designed to prepare students for the mathematics section of the SAT. Results from numerous studies have shown that Wayang Outpost has been beneficial for students in general with high improvements from pretest to post test (Woolf et al., 2006).

A meta-analysis conducted by Kulik (1994) examined over 500 studies exploring the effects of computer-based instruction. Results of this analysis revealed several positive benefits of using computer-based instruction. For example, students who used computer-based instruction scored in the 64th percentile on tests of achievement while students not using computer-based instruction scored in the 50th percentile. Additionally, students learned more in less time when using computer-based instruction.

Sivin-Kachala (1998) reviewed 219 studies to explore the effect of technology on educational achievement across various domains and ages. Students in environments that utilized technology showed increased achievement from preschool through higher education in all major subject areas.

Although there have been a large number of studies exploring the benefits of online learning and computer-based learning environments, there are only a few examples of studies focusing on K–12 students. This research aims to fill this gap in the literature by answering, for middle school students, the following question: is there a statistically significant positive correlation between the students' mastery of topics available in the ALEKS system and their achievement on equivalent items in the TCAP test[4]?

## ALEKS

ALEKS is a Web-based, artificially intelligent assessment and learning system that uses adaptive questioning to quickly and exactly determine what a student knows and does not know in a course. ALEKS then instructs the student on the topics that he or she is most ready to learn. As a student works through a course, ALEKS periodically reassesses the student to ensure that topics learned are also retained. ALEKS courses are very complete in their topic coverage and the system avoids multiple-choice questions. A student who shows a high level of mastery of an ALEKS course has the potential to do well in the actual course being taken.

[4]TCAP is the acronym for the *Tenessee Comprehensive Assessment Program.*

## 4.2 ALEKS in Memphis Area Schools

The schools in the Memphis Area began using ALEKS in 2004. The first group of users were college students taking a behavioral sciences statistics course at the University of Memphis. ALEKS was implemented in elementary and middle schools for the first time in 2005. Since 2005, eleven schools have used ALEKS during the academic year. In addition, students from over 170 schools used ALEKS in the framework of the Summer Online Learning Experience (SOLE) program. Overall, 5000 students have benefited from using ALEKS. The current report will analyze some preliminary findings of the data. We will only report student data from schools where an institutional review board (IRB) has approved our data collection protocol.

In this chapter, we describe two studies that we have conducted to assess whether providing individualized instruction to the learner in the form of a mathematical Intelligent Tutoring Systems (ITS) will increase student scores on a standardized high-stake test. The purpose of Study 1 is to determine whether mastery of topics in the online math tutorial system of ALEKS is correlated with the success in the corresponding overall math component of the Tennessee Comprehensive Assessment Program (TCAP). We have also explored the possible relationship between mastery of topics in ALEKS and the Standard Performance Indicators (SPIs) levels. The SPIs were developed by the Tennessee Department of Education to ensure that students in K–12 will learn the skills needed to succeed in the classroom and in the workplace with the ultimate goal being lifelong learning. The purpose of Study 2 is to partially replicate and expand on the results of Study 1 with a different sample (i.e., different grade level) in a different district. The two reports we present here focus on the correlation between the ALEKS assessment scores and the TCAP scaled scores.

## 4.3 Study 1

*Purpose of the study*

Studies have shown that ALEKS is efficient at providing individualized instruction to students at the college level (Hagerty and Smith, 2005). A question that warrants further investigation is whether this result also applies to younger students (i.e., middle school students). Specifically, the primary aim of our study concerns the association of ALEKS topics with the Tennessee state standards or SPIs. If significant positive correlations exist between a student's success in mastering ALEKS topics and a student's demonstration of mastery of the corresponding SPIs, then ALEKS may prove to be a potential predictor of future success on the Tennessee state standardized mathematics assessment (e.g., TCAP). In fact, ALEKS can be used to determine, regularly, well in advance of the administration of the TCAP test, each individual student's current strengths and weaknesses, and so the need for further targeted study. This would allow educators to redistribute their time and effort efficiently, and is likely to result in an improvement of the TCAP scores.

### 4.3.1 Method.

*Procedures*

ALEKS was available to the students for approximately three fourths of the school year (or about 27 weeks) from the beginning of the 2006–2007 school year up until the TCAP was administered. At the time of the standardized TCAP assessment, an ALEKS assessment was also administered online. Therefore, two types of mathematics performance measures were collected for each student: (1) the percent of ALEKS topics mastered; and (2) the mathematics TCAP scaled scores. The ALEKS percentage scores and the scaled scores in the TCAP were compared to evaluate the correlation between the two types of measures.

*Participants*

The study included students enrolled in the middle grades (grades six, seven, and eight) of the Memphis City and the Shelby County School systems. Students participated in their normal curriculum as determined by their respective school and school system. In addition, they were given access to the ALEKS system. One middle school from the Memphis City School system and one school from the Shelby County School system volunteered to use ALEKS as a significant part of their curriculum. The Memphis City middle school had a total of 267 students in 6th, 7th, and 8th grades. It had received a mathematics grade of 'D' in a "report card" from the Tennessee Department of Education. This grade means that its students were underperforming in mathematics. The Shelby County school had a total of 1073 students enrolled in its 3rd to 7th grades. Based on the "report card" from the Tennessee Department of Education, the mathematics category grade of that middle school was an 'A', meaning that its students were performing above grade level in mathematics.

The two schools participating in the ALEKS project agreed to have the ALEKS score of a student count for thirty percent of his or her overall course grade. Between these two schools, 218 students participated in the study in the spring of 2007.

*Hypotheses*

Our review of the literature was ambiguous. According to Angrist and Lavy (2002), Baker et al. (1994), and Wenglinsky (1998), we should expect no significant relationships between the students' proficiency in ALEKS and their achievement scores on the TCAP. On the other hand, based on Woolf et al. (2006), Melis and Siekmann (2004), and Sivin-Kachala (1998), a significant positive relationship between students ALEKS proficiency and their TCAP scores should be anticipated. Our working hypothesis is that the latter results will be confirmed.

**4.3.2 Results.** Our data analysis revealed a positive and statistically significant correlation ($r = .842$, $p < .01$, $n = 216$) between the TCAP scaled scores and the ALEKS assessment performance expressed as a percentage of the total number of concepts in the math complete curriculum. These results suggest that as students become more proficient in ALEKS they are more likely to score higher on the math portion of the TCAP. Figure 4.1 displays a scatter plot relating the two measures.

Our results indicate that similar correlations (between the overall math TCAP and ALEKS performance) are also positive and significant when the analysis is performed on the data of each school independently ($r = .716$, $p < .01$, $n = 146$ and $r = .697$, $p < .01$, $n = 72$).

Additionally, we computed correlations for each individual concept (strand) within ALEKS and SPI. The results of the correlation analysis indicate that a positive and statistically significant correlation ($r = .687$, $p < .01$, $n = 128$) exists between the TCAP and the ALEKS partial scores, for the 'Numbers and Operations' topic. Similar positive and significant correlations were also obtained for the other math topics: Algebraic Thinking, Graphs and Graphing, Data Analysis and Processing, Measurement, and Geometry. The results are displayed in Table 4.1.

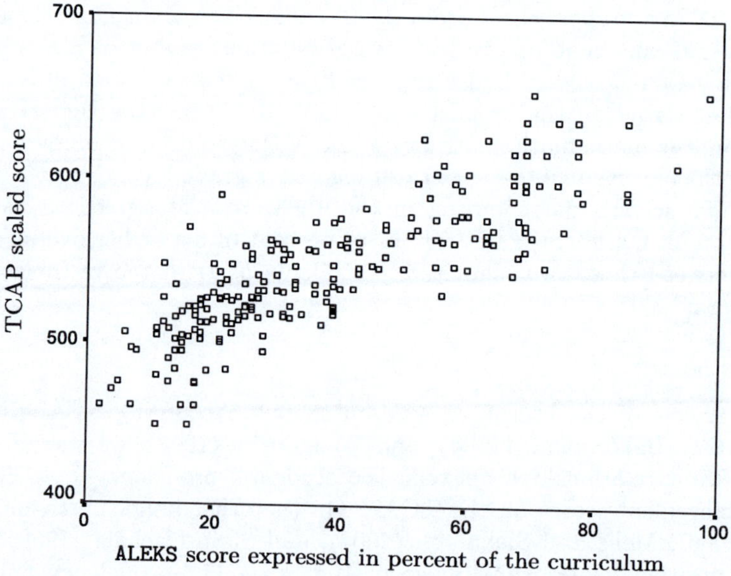


**Figure 4.1.** Scatter plot showing the relationship between TCAP scores and Assessment Performance. Each point represents a student.

| Topics | Pearson Correlation | | |
|---|---|---|---|
| Numbers and Operations | $r = .687$ | $p < .0001$ | $n = 128$ |
| Algebraic Thinking | $r = .522$ | $p < .0001$ | $n = 128$ |
| Graphs and Graphing | $r = .408$ | $p < .0001$ | $n = 128$ |
| Data Analysis and Processing | $r = .523$ | $p < .0001$ | $n = 128$ |
| Measurement | $r = .657$ | $p < .0001$ | $n = 128$ |
| Geometry | $r = .597$ | $p < .0001$ | $n = 128$ |

**Table 4.1.** Correlations Between ALEKS and Corresponding TCAP section. Note: There are only 128 records available for each of the TCAP sections.

**4.3.3 Remark.** The correlations found in the current study are determined by associating ALEKS's topics with the Tennessee state standards, which was established by ALEKS's developers, on the basis of face validity considerations.

## 4.4 Study 2

*Purpose of the study*

Study 2 was designed in order to partially replicate the results found in Study 1, using a sample of students from a different school district. The results of Study 1 suggest that performance in ALEKS is significantly correlated with performance on the standardized high-stakes test TCAP. Study 2 examines the relationship between students' mastery of the ALEKS math curriculum with their overall TCAP score. In this study, the relationship between ALEKS and TCAP scores is examined across all grades (fifth, sixth and seventh[5]) as well as each grade independently.

### 4.4.1 Method.

*Procedures*

Shortly after the start of the school year, students were given access to ALEKS and were allowed to interact with the system until the time of the administration of the TCAP, which took place between mid March and early April. As with the previous study, teachers used ALEKS as part of their regular mathematics instruction. Teachers allocated one day a week (usually Friday) as an "ALEKS day." Most of the students participating in the study had access to the internet at home. Therefore, such students could also use the system beyond the ALEKS day. As in Study 1, two outcome measures were collected for each student: (1) the percent of ALEKS math curriculum mastered; and (2) the TCAP mathematics scaled score.

[5]Due to the lack of availability of TCAP scores, the final sample did not contain any student from grade eight.

*Participants*

The sample of students used in the study was selected from a group of 1106 students (in grades five to seven) from an urban school in the Memphis area who had used ALEKS since 2005. From this group of students, this study selected only those who satisfied the following two conditions: (1) they were enrolled in grades five, six, and seven; and (2) their 2009 TCAP scores were available at the time of the study. Using these criteria, the resulting sample included 124 fifth graders, 98 sixth graders, and 99 seventh graders for a total sample of 321 students. Sample statistics for the grades are in Table 4.2.

| Grade | $N$ | Variable | Mean | Std Dev |
|---|---|---|---|---|
| 5 | 124 | TCAP | 535.33 | 34.1276 |
| | | ALEKS percentage[a] | 45% | 17.06% |
| 6 | 98 | TCAP | 556.28 | 36.5016 |
| | | ALEKS percentage[a] | 71% | 19% |
| 7 | 99 | TCAP | 567.69 | 32.6961 |
| | | ALEKS percentage[a] | 67% | 19.19% |

[a]Percentage Score: the part of the curriculum mastered by the student, expressed in percent.

**Table 4.2.** Sample Statistics

*Data Coding*

The data obtained from grades five through seven were first evaluated by combining individual classes within grade level. The criteria for combining individual classes within grade level were based on similarity of ALEKS topics covered by each individual class. To merge individual classes, a standardization of their respective TCAP scaled scores (ZTCAP), and ALEKS percentage scores (ALEKS %) was performed. Finally, to evaluate the strength of the relationship between the two variables ALEKS and ZTCAP a correlation was calculated for each individual grade level.

To this end, each of the grades' data were combined using their respective raw TCAP and ALEKS percentage scores. Next, the TCAP and ALEKS percentage scores were normalized into $z$ scores to analyze the strength of the relationship between the two variables.

**4.4.2 Results.** As suggested by the scatter plot of Figure 4.2, there is a statistically significant positive correlation between the standardized ALEKS percentage score and the standardized ZTCAP when all of the grades were combined ($r = .74$, $p < .0001$, $n = 321$).

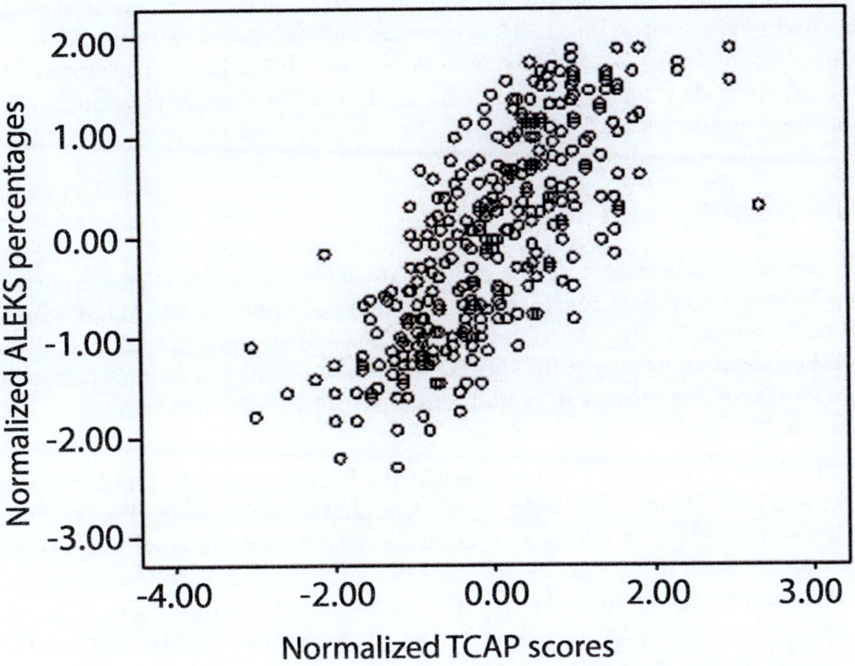


**Figure 4.2.** Scatter plot showing the relationship between TCAP scores and Assessment Performance in ALEKS. Each point represents a student.

The analysis at the individual grade levels revealed a statistically significant positive correlation between the ALEKS % and ZTCAP scores (Table 4.3). The results also revealed a statistically significant positive correlation between the standardized ALEKS % and the standardized ZTCAP. Note that, due to differences in topic mastery, the correlation between the variable ALEKS % and TCAP scores may be different.

| Grades | Pearson Correlation | | |
|---|---|---|---|
| 5 | $r = .63$ | $p < .0001$ | $n = 194$ |
| 6 | $r = .81$ | $p < .0001$ | $n = 98$ |
| 7 | $r = .72$ | $p < .0001$ | $n = 99$ |

**Table 4.3.** Normalized ALEKS percentages with TCAP Correlation.

**4.4.3 Remark.** In this study, we were unable to find out whether students who had exposure to the ALEKS system outside of the classroom did better on mastery of topics than those who only used ALEKS in the classroom. Accordingly, the significant results in the study may be due, at least in part, to students who used ALEKS at home.

## 4.5 Summary and Discussion

The two studies discussed in this chapter revealed significant positive correlations between students' mastery of mathematical topics in the ALEKS system and their scores on the TCAP in the corresponding domains. The results of an ALEKS assessment are quite specific, and may reveal a student's particular strengths and weaknesses in a mathematical topic. The importance of such results is twofold.

1. *Monitoring students' progress.* While the TCAP is administered once, between mid March and early April, the ALEKS system provides an online assessment given repeatedly through the school year. The predictive power of ALEKS with respect to the TCAP allows the teacher to follow closely the students' learning progress. So, the teacher may take immediate appropriate action if a student is shown to lag in some area of mathematics. This can be done in a variety of ways. The teacher can spend some time interacting with a student individually, or a student can be given homework targeting his or her weaknesses.
2. *Remedying students' weaknesses.* Another possibility to address a student's shortcomings is to use the ALEKS system itself. Indeed, the ALEKS system offers an efficient way to deal with a student's weaknesses in mathematics. At the end of an ALEKS assessment, the student is given a list of those concepts that he or she is ready to learn. Choosing to learn one of those concepts—by clicking the corresponding region of the computer screen—initiates a specific lesson focusing on this concept. Numerous data have demonstrated the efficiency of ALEKS as a teaching engine (see Chapter 5 in this connection).

## Acknowledgments

The research reported here was supported by the Institute of Education Sciences, U.S. Department of Education, through Grant R305A090528 to the University of Memphis. The opinions expressed are those of the authors and do not represent views of the Institute or the U.S. Department of Education. The authors thank the Memphis City Schools and the Shelby County Schools. Without their support, we would not have been able to obtain the data used in this report.

Corresponding Author: Xiangen Hu, University of Memphis, Memphis, TN 38152, xhu@memphis.edu

5

# A Potential Technological Solution for Reducing the Achievement Gap Between White And Black Students

Xiangen Hu[1]
Yonghong Jade Xu[1]
Charles Hall[2]
Kristin Walker[1]
Theresa Okwumabua[1]

## 5.1 Introduction

Disparities in achievement measures that exist between black and white students, commonly referred to as the 'black-white achievement gap', are widely discussed in educational literature (Lubienski, 2008). In fact, even before the advent of modern standardized tests, there was an extended analysis of performance gaps on tests between black and white students (e.g. Crowley, 1932; Bali and Alvarez, 2004; Ferguson, 2002; Harris and Herrington, 2006). Much has been written about performance disparities among blacks, whites, and other ethnic minority groups on the mathematical section of standardized tests, such as the Scholastic Aptitude Test (SAT). A common finding is that the scores of black students lag behind those of whites in the United States. For example, in 2005, the mean combined score on the mathematics and verbal sections of the SAT was 17% higher for white students than for black students (www.jbhe.com/features/49_college_admissions-test.html). Additionally, some research indicates that black students score at least one standard deviation below white students on standardized tests (Fryer and Levitt, 2004). These disparities remain despite the major gains made by both black and white students on exams such as the National Assessment of Educational Progress (NAEP), which assesses the educational achievement of elementary

[1]Dept. of Psychology, University of Memphis TN.
[2]Dept. of English, University of Memphis TN.

J.-C. Falmagne et al. (eds.), *Knowledge Spaces: Applications in Education*,
DOI 10.1007/978-3-642-35329-1_5, © Springer-Verlag Berlin Heidelberg 2013

and middle school students[3]. The most recent reports in this regard come from the National Education Association[4] and the Institute of Education Sciences[5].

Similar to the disparity in performance observed among blacks and whites on standardized exams, the black-white achievement gap is also evident in mathematics course grades (Lubienski, 2002). Differences exist between grades in mathematics courses at all levels (Lubienski, 2002). These disparities may start the kindergarten and persist across grade levels, and even widen over time (see www.jbhe.com/features/49_college_admissions-test.html.)

There is no clear explanation for the emergence of the black-white achievement disparity or its increases over time (Hedges and Nowell, 1999). Many factors have been proposed to explain these racial differences in student achievement test scores, with a significant amount of attention focused on students' socioeconomic status, family, peer, and school factors, as well as student learning characteristics, academic ability, and other environmental and structural factors (e.g. Brooks-Gunn et al., 1996; Jeynes, 2003; Phillips et al., 1998; Hanushek and Rivkin, 2006; Orr, 2003)[6].

In most of the educational literature that discusses the black-white achievement gap, socioeconomic status has been identified as one of the key contributors to the disparity in achievement observed among various racial groups (Hedges and Nowell, 1999). Students living in persistent poverty are more likely than other students to suffer from many conditions that impede their learning, including poor health, frequent changes in residence that require transfer to new schools repeatedly, lack of books and educational resources in their homes, parents or guardians with lower levels of education, and unstable family homes. However, Lubienski (2002) also examined socioeconomic status and its relation to the black-white achievement gap. The findings from this study indicated that students' socioeconomic status may not be the sole contributor to the achievement gap. In fact, white students from both low- and high-income families achieved at higher levels than their black counterparts (Lubienski, 2002). In a like manner, Fryer and Levitt (2004) discussed the socioeconomic status of the schools that students attend as an important factor in students' achievement levels.

Cultural attitudes and racism have also been shown to play a critical role in the achievement gap. In this regard, some researchers contend that many African American students simply stop trying to excel academically because they do not want to be thought of as "acting white" by their peers. Relatedly, some researchers suggest that black students are not motivated to excel because they perceive that others do not view them as capable or expect them to fail.

[3] http://nces.edu.gov/nationsreportcard, 2008.
[4] NEA, see http://www.nea.org/home/AchievementGaps.html.
[5] See IES, http://nces.ed.gov/nationsreportcard/studies/gaps/.
[6] See www.swcomputer.org/pdf/African_American_Overview.pdf.

Compounding these socio-cultural factors are several structural and institutional factors that are believed to contribute to the black-white achievement gap. Students from disadvantaged backgrounds often encounter school conditions that only exacerbate the problem. Limited availability of resources in the schools that serve mainly black students—larger class sizes, fewer qualified teachers, and limited access to technology in the classroom—are all factors that impede the academic development that is necessary for black students to perform at similar levels as white students (Lubienski, 2002; Fryer and Levitt, 2004; Judge et al., 2006) (see also footnote 5 on page 80).

## 5.2 Narrowing the Achievement Gap through Technology in the Classroom

What is needed to reduce the black-white achievement gap is not only complex but controversial. Many researchers and practitioners are using after-school tutoring sessions and remedial programs to help narrow the gap. Reorienting the students, so that equally qualified teachers, expectations, curriculum and resources are available within schools as well as tracking students by ability groups are also being used. As schools consider the students' ability levels, some researchers and practitioners are considering strategies to identify specific skill and knowledge deficits students might have, in order to respond in targeted ways. With the introduction of learning technologies in the classroom, many are exploring the utility of online intelligent tutoring systems as strategies for reducing the black-white achievement gap.

In actuality, computer technology for educational purposes ranges from the use of computers for in-class assignments and web-based tutorials to distance learning and virtual classrooms (Mitra et al., 2006). It is obvious that more and more classrooms are equipped with computers, that computer software packages are being adopted to enhance the learning experience, and that learning applications are more accessible through the Internet (Packard, 2007). Included in the ranks of these applications are online intelligent tutoring systems (ITS), like `ALEKS` (an intelligent tutor for mathematics), that can be used to enhance students' knowledge base and ultimately improve academic achievement (Packard, 2007).

As is shown frequently in the literature on the relationship between use of technology and academic achievement, students who use computers or engage in online learning experiences tend to perform better in school than students who do not have technology incorporated into their learning activities (Clements, 1999; Judge, 2005; Judge et al., 2006). This finding holds for children regardless of their ethnicity or racial and socioeconomic background. For example, Judge (2005) examined the impact of computer technology on academic achievement of African-American children and found that general use of computers was positively correlated with students' achievement scores. Likewise, the use of learning software (e.g., math, reading, and science) was positively correlated with achievement (Judge, 2005). Taken together, this research suggests that technology may be a viable option for increasing achievement levels of students in minority and majority groups. However, the empirical evidence that using technology can reduce the achievement gap between minority and majority groups is sparse. On-line tutoring systems, such as ALEKS, provide opportunities to individually address student deficits and, thereby, close the gap in student achievement.

## 5.3 Objective

This chapter presents some preliminary findings from an observational study conducted at a large urban university. The study explores the effectiveness of the online ITS ALEKS[7] for closing the racial score gaps in an undergraduate behavioral statistics course (rather than simply increasing absolute scores for all groups).

**Theoretical Framework.** Students majoring in psychology or social sciences find behavioral statistics one of the most challenging required undergraduate courses. The primary reason for the difficulty with the course is the prerequisite mathematical skills necessary to understand some of the basic concepts. This is especially true for students who have been out of school for a long time or who had a weaker mathematics background. Due to the wide range of those mathematical skills, lecture-based, traditional classroom instruction is very inefficient. Indeed, some students find it impossible to follow lectures, fall behind, and eventually fail.

While traditional lecture style instruction cannot help students who have fallen significantly behind, studies (e.g. Fletcher, 1990) have shown that learners do better with the help of appropriate technology. More specifically, Hagerty and Smith (2005) found that the Web-based ITS ALEKS helped college students learn college algebra. This research and other studies of the ALEKS system led us to explore the use of the ALEKS Behavioral Statistics package (http://www.behsci.aleks.com) as a possible solution to help those students who had fallen significantly behind in this specific course.

ALEKS Behavioral Statistics decomposes concepts in basic statistics, from frequency distributions to Analysis of Variance [ANOVA], into small units, called "items." There is a total of 19 mathematics readiness items and 109 statistics items. Each item is presented as a unique type of problem that students are required to solve. Items are then are grouped together into larger content areas that must be learned to demonstrate mastery in behavioral statistics. This collection of the 128 problem types forms a comprehensive course in behavioral statistics.

The power of ALEKS is rooted in its sound theoretical foundation, Knowledge Space Theory (Falmagne and Doignon, 2011; Falmagne et al., 2006),

[7] ALEKS is an acronym for *Assessment and LEarning in Knowledge Spaces.*

which enables the fine tuning of instruction for individual students. Every interaction between ALEKS and a student results in a reappraisal of the student's knowledge state, signaling the student's critical weaknesses or lacunae, formulated in terms of the curriculum, and allowing a precise gearing of teaching.

In addition to these features, there are other advantages to using a Web-based ITS such as ALEKS. For example, ALEKS is available anytime, anywhere with an internet connection. This flexibility is especially helpful for students who have difficulty following traditional lectures, due to a lack of basic mathematics knowledge. Such students may need additional time to catch up. Furthermore, ALEKS is designed such that students need to master prerequisite knowledge in order to learn subsequent, more advanced topics. It is through this design that ALEKS guides the learning progress of each student and prevents a student from becoming lost in concepts for which he or she is not ready. Thus, the primary reason to implement ALEKS in the undergraduate statistics course is that it provides a technological solution that allows students with lower levels of mathematics achievement to work more efficiently toward mastery of statistical concepts.

## 5.4 Methods

**5.4.1 Research design.** A nonequivalent control group design was used to compare the academic performance of students from online sections of a behavioral statistics course using ALEKS to a retrospective comparison group comprised of students enrolled in the same course taught in a traditional lecture format. The original data set included 2,329 students who enrolled in Behavioral Sciences Statistics at this urban institution between the Fall of 1994 through the Spring of 2008. Course grades were obtained from the university's Office of Institutional Research. Data from a smaller subset of students ($N = 1309$) were included in the current study. Students were excluded from the analysis due to one or more of the following reasons: (1) no ACT nor SAT mathematics score available; (2) received a grade "I" (incomplete) or early withdrawal; (3) neither black nor white; and (4) belonged to a class having fewer than 10 students. Hence, the study sample included 370 black students and 939 white students in 69 classes taught by 32 different professors. Among these 1,309 students, 112 students took the course online using ALEKS and 1197 students were enrolled in traditionally taught lecture sections (Table 5.1).

**5.4.2 Measures.** The analyses are based on a measure of grade performance. Several steps were taken in developing the grade performance measure. First, students' letter grades were retrieved from the university's database and those who withdrew early or had an incomplete were removed from the data. As mentioned earlier, only students enrolled in classes of size over 10 and having ACT or SAT scores are included in the analysis. Valid grades were then

| Students | Lecture | On line | Total |
|---|---|---|---|
| **Black** | 332 | 38 | 370 |
| **White** | 865 | 74 | 939 |
| **Total** | 1197 | 112 | 1309 |

**Table 5.1.** Numbers and types of students whose data are analyzed in the study.

transformed into numerical counterparts: A, B, C, D, and F into 4, 3, 2, 1, and 0, respectively. For students who took the course more than once, their highest grade was used. Because some instructors may consistently be more lenient or stricter in grading than others, valid students' grades from the same instructor were converted into standardized (i.e. normalized) $z$ scores in order to alleviate potential grading biases by individual instructors and to make the course grades from different instructors comparable.

**5.4.3 Procedures.** Each of the classes using the traditional lecture format was led by an instructor and met two days a week, with one hour devoted to a computer lab section (using SPSS). The online sections of the course used `ALEKS`, an online math tutorial system. For this study, `ALEKS` provided a self-paced distance-learning course. Students were required to spend approximately 8 hours per week on this 4 credit-hour class. Each lecture and online classes was led by a professor with a graduate student/teaching assistant. The online classes were offered to students, beginning during the Fall semester of 2004. Both course formats, online and lecture sections, were available to students between 2004 and 2008. Students self-selected to either of the classes.

**5.4.4 Analysis.** Three different analytical procedures were completed. The results are presented and compared below. First, a two-way ANOVA was performed with the $z$-score grade as the dependent variable and race (black vs. white) and course formats (online vs. traditional) as the independent variables. Second, an analysis of covariance (ANCOVA) was conducted by adding the ACT mathematics scores as the covariate. ACT mathematics scores are made available for all 1309 students. For those students who only have SAT scores, their SAT score is converted into equivalent ACT scores based on a conversion table from Collegeboard.com, which can be found on the web site

http://professionals.collegeboard.com/profdownload/pdf/rr9901_3913.pdf.

Third, an analysis of covariance (ANCOVA) was conducted by adding the cumulative GPA prior to enrolling in the course as the covariate variable. This is made into a separate model because, among the 1309 students, the University cumulative GPA is only available for 992 of the students.

## 5.5 Results

The data of the two-way ANOVA model, as presented in Table 5.2 and Figure 5.1, show that there is a significant interaction between Race and Class Format ($F(1,1305) = 6.04$, $p = 0.014$). Further tests show white students perform significantly better than black students in the lecture format (0.162 vs. 0.299, $F(1,1195) = 59.559$, $p = 0.000$) but exhibit comparable performance in the online (ALEKS ) format (0.085 vs. 0.100, $F(1,110) = 0.007$, $p = 0.934$, n.s.).

To examine whether the observed difference persists when racial differences in academic performance prior to entering college is controlled, an ANCOVA model with ACT Mathematics score is used as the covariate. The findings revealed that the ACT Mathematics scores of black students are significantly lower than that of the white students (18.16 vs. 21.43, $t(1307) = -13.554$, $p = 0$). The ACT Mathematics scores for students were not significantly different (20.53 vs. 20.20, $t(1307) = -0.805$, $p = 0.421$) in either the lecture and or online formats. The results of the ANCOVA model are presented in Table 5.3 and illustrated in Figure **??**.

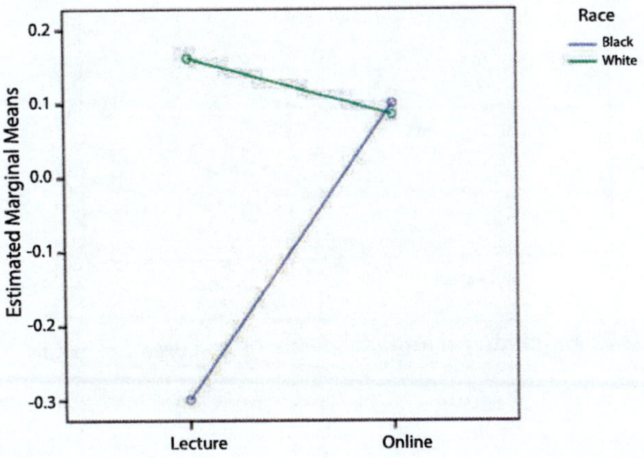


**Figure 5.1.** Estimated marginal means of $z$-scores for grade.

With ACT mathematics score as the covariate, the analysis show a significant interaction between Race and Format; we get ($F(1,1304) = 5.876$, $p = 0.006$). Furthermore, the ANCOVA reveals that the performance of black students is slightly better than white students in the Online condition (0.283 vs $-0.008$, $F(1,109) = 2.533$, $p = 0.114$, n.s). White students perform significantly better than black students in the Lecture condition (0.094 vs. 0.122, $F(1,1194) = 12.723$, $p = 0.000$). Another ANCOVA is then conducted in which the cumulative GPA prior to enrollment in the statistics course is used as the covariate. With a total of 1168 students (there are students missing accumulated GPA prior to the enrollment of the statistics class), we observed similar outcomes. Black students had significantly lower GPAs than did white students (2.75 vs 3.1, $t(990) = -9.9$, $p = 0.00$). The average GPA for lecture students (3.00) and online students (2.90) were not significantly different ($p = 0.102$, n.s). The results of the second ANCOVA model are summarized in Table 5.4 and Figure 5.2.

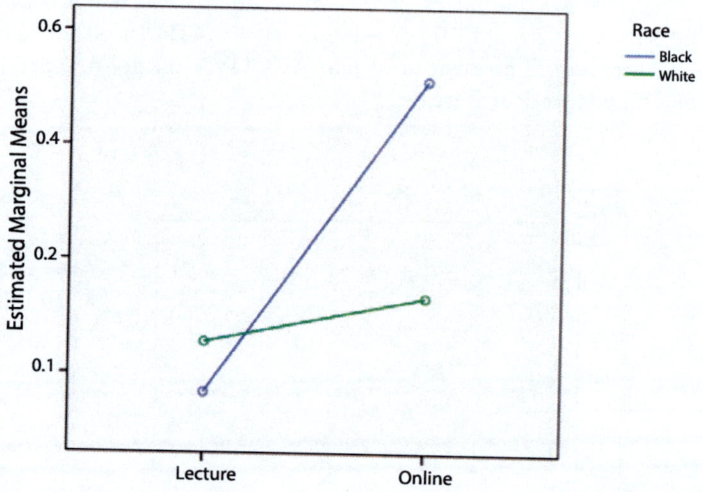


**Figure 5.2.** Estimated marginal means of $z$-scores for grade.

Test of Between-Subjects Effects

Dependent variable: *z*-score for grade

| Source | Type III Sum of Squares | df | Mean square | $F$ | Sig. |
|---|---|---|---|---|---|
| Corrected Model | 51.341[a] | 3 | 17.114 | 20.053 | .000 |
| Intercept | .053 | 1 | .053 | .062 | .803 |
| RACE | 4.521 | 1 | 4.521 | 5.297 | .022 |
| Format | 2.374 | 1 | 2.374 | 2.782 | .096 |
| RACE*Format | 5.154 | 1 | 5.154 | 6.040 | .014 |
| Error | 1113.692 | 1305 | .853 | | |
| Total | 1167.001 | 1309 | | | |
| Corrected Total | 1165.033 | 1308 | | | |

[a] $R$-squared = .44 (Adusted $R$-squared = .042)

RACE*Lecture Online

Dependent variable: *z*-score for grade

| RACE | Type of Course | Mean | Std Error | 99% Confidence interval | |
|---|---|---|---|---|---|
| | | | | Lower Bound | Upper Bound |
| Black | Lecture | −.299 | .051 | −.399 | −.200 |
| | Online | .100 | .150 | −.194 | .394 |
| White | Lecture | .162 | .031 | .100 | .223 |
| | Online | .085 | .107 | −.125 | .296 |

**Table 5.2.** Two-way ANOVA of the standardized course grade

Test of Between-Subjects Effects
Dependent variable: $z$-score for grade

| Source | Type III Sum of Squares | df | Mean square | $F$ | Sig. |
|---|---|---|---|---|---|
| Corrected Model | 168.176[a] | 4 | 42.044 | 54.998 | .000 |
| Intercept | 100.565 | 1 | 100.565 | 131.550 | .000 |
| RACE | .129 | 1 | .129 | .169 | .681 |
| ACT-MATH | 116.834 | 1 | 116.834 | 152.832 | .000 |
| Format | 2.848 | 1 | 2.848 | 3.725 | .054 |
| RACE∗Format | 5.876 | 1 | 5.876 | 7.687 | .008 |
| Error | 986.858 | 1304 | .764 | | |
| Total | 1167.001 | 1309 | | | |
| Corrected Total | 1165.033 | 1308 | | | |

[a] $R$-squared = .144 (Adusted $R$-squared = .142)

RACE∗Lecture Online
Dependent variable: $z$-score for grade

| RACE | Type of Course | Mean | Std Error | 99% Confidence interval | |
|---|---|---|---|---|---|
| | | | | Lower Bound | Upper Bound |
| Black | Lecture | −.124[a] | .050 | −.222 | −.026 |
| | Online | .307[a] | .143 | .027 | .587 |
| White | Lecture | .092[a] | .030 | .032 | .151 |
| | Online | .014[a] | .102 | −.185 | .214 |

[a] Covariates appearing in the model are evaluated at the following values: ACT-MATH = 20.5111.

**Table 5.3.** ANCOVA of the standardized course grade with ACT Mathematics Scores as covariante.

Test of Between-Subjects Effects

Dependent variable: $z$-score for grade

| Source | Type III Sum of Squares | df | Mean square | $F$ | Sig. |
|---|---|---|---|---|---|
| Corrected Model | 394.157[a] | 4 | 98.539 | 182.102 | .000 |
| Intercept | 308.952 | 1 | 308.952 | 570.948 | .000 |
| RACE | 1.832 | 1 | 1.832 | 3.386 | .068 |
| UCUMGPA | 347.201 | 1 | 347.201 | 641.633 | .000 |
| Format | 8.671 | 1 | 8.671 | 16.024 | .000 |
| RACE*Format | 4.913 | 1 | 4.913 | 9.079 | .003 |
| Error | 628.783 | 1162 | 541 | | |
| Total | 1026.176 | 1167 | | | |
| Corrected Total | 1022.940 | 1166 | | | |

[a] $R$-squared = .385 (Adusted $R$-squared = .383)

RACE*Lecture Online

Dependent variable: $z$-score for grade

| RACE | Type of Course | Mean | Std Error | 99% Confidence interval | |
|---|---|---|---|---|---|
| | | | | Lower Bound | Upper Bound |
| Black | Lecture | −.032[a] | .043 | −.117 | −.052 |
| | Online | .513[a] | .120 | .276 | .749 |
| White | Lecture | .056[a] | .027 | .002 | .110 |
| | Online | .134[a] | .086 | −.034 | .302 |

[a] Covariates appearing in the model are evaluated at the following values: UCUMGPA = 2.9903.

**Table 5.4.** ANCOVA of the standardized course grade with university accumulated GPA as covariate.

## 5.6 Summary

**5.6.1 Major findings.** The first noteworthy change from the ANOVA model to the ANCOVA model is that the explained variance of the course grade increases from 4.4% to 14.4% (ACT Math) and 38.5% (GPA), an indication that student performance prior to the course is a important factor to consider when studying racial gaps. The most interesting findings are the reversed order between white students and black students when considering the prior performances. These observations suggest that using ITS systems such as `ALEKS` help students who were behind relatively more efficiently. These analyses demonstrate the potential value of an online ITS as a tool for eliminating racial disparities in college students' academic performance. Differences in the final grades for passing black and white students enrolled in a behavioral statistics course were eliminated in this study. Our results indicate that the disparities in the performance of black and white students in the lecture-based behavioral statistics sections did not appear in online ITS-based sections that used `ALEKS`. This finding held even when the results were adjusted for prior GPA and ACT mathematics scores.

**5.6.2 Limitations of the current study.** While the findings from the study are encouraging, there are some limitations that preclude generalization of the findings. The study used a sample of 1300 students (out of 2300) taking behavior statistics in the last 15 years from an urban University. Generalizability is limited by the non-random sample, though the findings are still convincing that a technological solution is possible to solve the well documented black-white achievement gap in higher education. Another limitation is that the analyses and conclusions are based on observational data, while information on family background, student learning characteristics, and other factors that may play a role in racial academic gaps are unavailable for consideration. Nonetheless, the three analytical models yield consistent findings and they demonstrate the potential value of an online intelligent tutoring system (ITS) as a tool for eliminating racial disparities in college students' academic performance.

**5.6.3 Future Directions.** The findings of this study suggest that the racial gap in the final grades for black and white students enrolled in a behavioral statistics course may be reduced and even eliminated when a different learning environment, in the form of an online ITS, is used. The improvement in the performance of black students in online classes should not be interpreted as white students suffering disadvantages in non-traditional learning environment. Rather, it should lead to further examination of the roles that different learning styles and cultural backgrounds play in teaching and learning when continuous effort is made to close the academic performance gap between racial groups.

Even though the current study does not, by itself, provide data to support causal arguments, future research can explore the following hypotheses.

First, it is likely that black students, in general, may retreat from the competitive group learning as experienced in lecture format and feel more relaxed and efficient when working alone with an online ITS.

Second, it is also possible that the flexibility of an online class is a key advantage to black students in case they need longer time to complete learning tasks and wherein they experience less anxiety associated with class quizzes and exams.

Finally, numerous online learning systems/software are available and not all of them are effective. It is important to study and design flexible and effective learning enviroments to meet learners with different styles and preferences.

## Acknowledgements

The research reported here was supported by the Institute of Education Sciences, U.S. Department of Education, through Grant R305A090528 to the University of Memphis. The opinions expressed are those of the authors and do not represent views of the Institute or the U.S. Department of Education. The authors thank Drs. Jason Luellen and Lun Mo for their earlier work on the topic. They also thank Ms. Pamela Demmitt and Meihua Yang for their help during the preparation of the manuscript. Finally, the authors are grateful for the help received from the Office of Institutional Research at the University of Memphis. Without its support, we would not have been able to obtain the data used in this report.

# 6

# A Commercial Implementation of Knowledge Space Theory In College General Chemistry

Christopher J. Grayce[1]

## 6.1 Introduction

ALEKS Corporation was founded in 1996 to develop and market educational services based on Knowledge Space Theory (KST), and its earliest products were Internet-delivered assessment and tutoring services for algebra and related areas of mathematics. These remain the bulk of its business. However, in 2005 ALEKS began the release of a new line of assessment and tutoring services in college general chemistry, which we'll call "ALEKS Gen Chem." Experiences from the development and deployment of this product line will be described here.

### Why Chemistry?

From the point of view of the development of KST, building an assessment and learning service in chemistry can be expected to further illuminate the general challenges of constructing such services, as opposed to those that might be specific to math. While chemistry is certainly closely related to math, it has significant fundamental distinctions.

One of the more important may be that chemistry is a field of complex underlying principles—the quantum statistical mechanics of $10^{23}$ strongly interacting particles—that lead to simple and familiar consequences, which even beginning students are expected to be able to rationalize, such as that water boils at lower temperatures at higher altitudes, or that vinegar fizzes when added to baking soda. In many ways mathematics tends towards the reverse: simple underlying principles lead to complex consequences, the way recursive simple arithmetic leads to geometric fractals.

[1] ALEKS Corporation.

J.-C. Falmagne et al. (eds.), *Knowledge Spaces: Applications in Education*,
DOI 10.1007/978-3-642-35329-1_6, © Springer-Verlag Berlin Heidelberg 2013

In part to avoid overwhelming the beginning student, and to provide the student who only dips briefly into the field with something of practical utility, it is common to teach key concepts of chemistry via a sequence of increasingly accurate theoretical models, which we might compare to bringing a photograph slowly into focus (Figure 6.1). Each model, aside at most from the last,

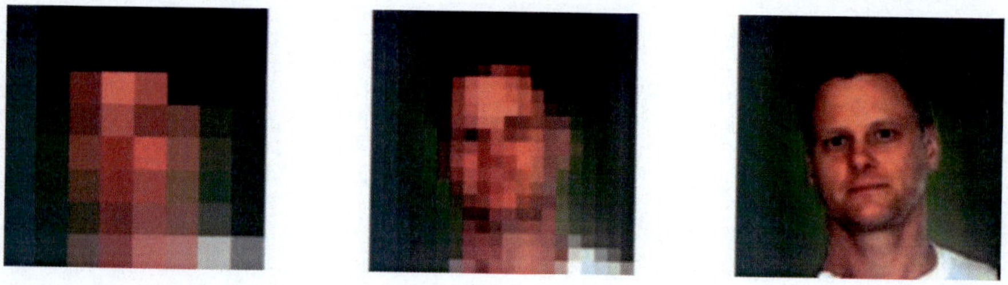

**Figure 6.1.** Three theoretical models of the author, which increase in accuracy from left to right.

will contain oversimplifications and overgeneralizations rarely based on sound mathematical principles of approximation, the purpose of which is to allow students to grasp the gist of the concept at increasing levels of sophistication without drowning in premature detail.

Unfortunately, little in the field of chemistry itself gives us guidance on the best possible such sequences. We may generally assume that some are better than others, i.e. lead to faster and better progress of the student, but the knowledge of which those might be belongs to the less well-developed fields of cognitive science generally, and chemical education specifically.

Worse, it can be hard to judge unambiguously the correlation between mastery (or partial mastery) of a given approximate model and progress towards genuine understanding of underlying principles.

To put this in the language of KST, while it is relatively straightforward to define knowledge states expressing *complete* ignorance or *complete* mastery of the major chemistry concepts, defining and ordering knowledge states expressing *partial* mastery is challenging, because those intermediate states are in practice defined by student progress along one or more chains of models simplified in various *ad hoc* ways, and we have as yet no unambiguous way to map degrees of mastery of these models to degrees of mastery of the underlying principles.

Since many fields of knowledge are also taught at least partly in the picture-into-focus way chemistry is taught, anything we can learn about the practical implementation of KST in that kind of teaching can be expected to be broadly useful.

**Why KST?**
On the other hand, from the point of view of improvements in chemical education, an advantage KST brings to assessment and learning is that in principle it can sidestep many issues about the optimal sequence of models, or of instructional sequence generally, by allowing instruction and learning to proceed along any logically consistent instructional path.

Indeed, we may hope to empirically measure such things as the typical paths by which students actually acquire mastery, or the dynamics by which they retain or lose that mastery over time, and how instructional methods and sequencing affect these things. These could, in principle, add a great deal of empirical rigor to debates about the optimal way to teach chemistry.

Also, chemistry instructors routinely debate the best sequences for teaching almost any important concept in chemistry, and significant variations in the preferred sequences exist between programs, textbooks and instructors. A successful commercial e-learning service needs to accomodate *any* reasonable variation, and implementing it within the framework of KST—which is specifically designed to accomodate the widest possible range of learning paths—confers a natural advantage.

## 6.2 ALEKS Gen Chem

The most recent version (4.0) of ALEKS Gen Chem was released on 22 July 2011. A complete educational service within this line of products consists of three components:

**Items.** An item represents a class of problems in chemistry or closely related mathematics, for example the class of problems that consist of balancing chemical equations. An example problem from the class, for example balancing a particular chemical equation, is called an *instance* of the item.

**A structure.** The structure defines a partial order on the items, which tells us their pre-requisite relationship, for example that students must be able to interpret chemical formulae before they can balance chemical equations.

**An implentation.** An implementation generally consists of a cycle of assessment and learning over the set of items, guided by the structure, with learning optionally confined to *intermediate objectives* (e.g. the material on this week's lecture).

Each component will be reviewed below.

**6.2.1 Items.** Version 4.0 of ALEKS Gen Chem defines knowledge states in general chemistry and closely related pre-requisite areas of mathematics in terms of mastery of 417 items.

These items may be classified in a number of ways. For example, Figure 6.2 shows a conceptual classification. As may be apparent to experienced chemistry instructors, the full general chemistry curriculum is not yet covered in

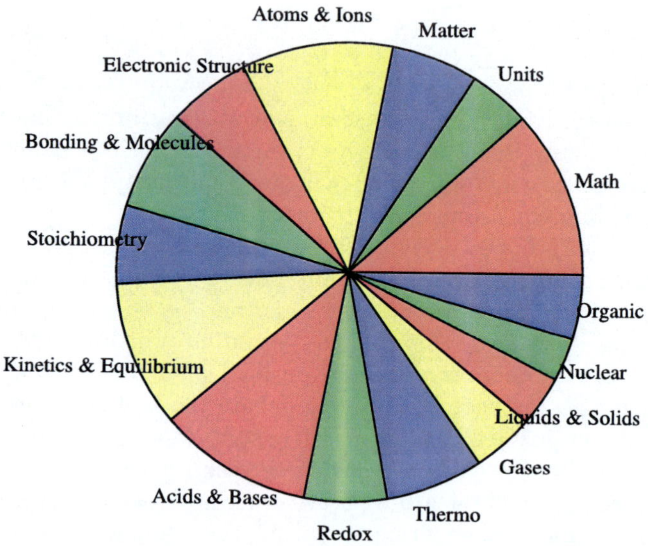


**Figure 6.2.** A conceptual partition of the 417 items in v4.0 of ALEKS Gen Chem.

ALEKS Gen Chem. Major areas still in development include coordination chemistry, organic reactivity, structure of the solid and liquid state, electronic band theory, and descriptive chemistry. Our most recent estimates are that the mature product line will contain 550-600 items. (This means that the system is prepared to pose around 10,000 unique problems to the student).

Items can also be classified by the types of solution students must generate:

**A quantity.** That is, the student must generate a number, measurement, chemical formula, algebraic expression, chemical equation, word, or phrase (71% of current items).

An important characteristic of items in this class, and the reason most ALEKS items are of this type, is that the answer space for each instance is very large. This has two benefits: it greatly reduces the chance of lucky guesses, or answers derived partly from "test taking" skills.

It also forces the student to begin his solution process with a "blank slate"—without any environmental hints at all about the conceptual class of the problem, or how it might begun to be solved. We believe this kind of unassisted practice memory retrieval is key in transforming short-term mastery into long-term.

Unfortunately, wide-open answer spaces present significant challenges to parsing and grading with a computer algorithm, which of course lacks the inferential acumen and flexibility of the human grader that allow the latter to understand what the student *meant* to answer, when that differs from what he actually answered.

For example, suppose the question posed is "What is the name of $FeSO_3$?" The correct answer is "iron (II) sulfite," and if the student enters that, we can grade him correct. But we must consider and program ALEKS to reasonably respond to a wide variety of near misses:

- The student might capitalize in unexpected ways. For example, he might enter "Iron (II) sulfite" because of the tradition of capitalizing the first words of a sentence, or "Iron (II) Sulfite" by (false) analogy with personal names, in which each word of the name is capitalized, or "Iron (II) SULFITE" because he pressed CAPS LOCK instead of SHIFT on his keyboard when entering the parentheses. We would take all these as correct, perhaps adding in the second case a "convention" feedback reminding the student that only the first word of chemical names is ever capitalized.
- The student may make any number of harmless simple typos, such as transposing two letters ("iron (II) sulifte") or dropping one letter ("irn (II) sulfite") or hitting a key one row up on the keyboard from the intended key ("iron (II) sulf8te"). Generally in these cases we produce a pop-up warning suggesting the student check his spelling before submitting the answer for grading.
- Unfortunately, some answers that very likely indicate a wrong answer based on faulty understanding of underlying principles ("iron (II) sulfide" or "iron (II) sulfate") are *also* only one letter off from the correct answer. So we must carefully eliminate these from the answers that trigger spelling warning pop-ups, so that students do not learn to use the presence or absence of pop-up warnings as a "terminal guidance system" that can be used to zero in on the correct answer from a position of mild error.
- Some harmless typographic idiosyncracies are very hard to spot, for example using lowercase L for uppercase I when writing Roman numerals, or using zero instead of capital O as the symbol for oxygen. (The zero and O keys are adjacent on normal keyboards, unfortunately.) In these cases providing a pop-up spelling warning can be frustrating, as the student can't *see* what's wrong with this answer. We have to simply anticipate these cases and take them as correct.
- Of course, we must also consider and allow student use of acceptable if less mainstream alternatives ("ferrous sulfite").
- We should also give useful feedback when we are sure of the error. For example, a student who enters "iron (III) sulfite" may be usefully told to think again about the charge on the iron cation and try to answer the problem again. But we *don't* want to give that same feedback if the student enters "iron (III) sulfate" or "iron (III) sulfoxide" because the charge on the Fe cation is *not* his major difficulty. As a rule, we only give specific feedback when there is only one error, and it is clearly identifiable, because we feel inappropriate feedback is worse than no feedback at all. (Students can always ask to see the complete and correct solution to any instance.)

**Multiple choice.** That is, the student must select one or a few of some number of possible words, phrases, values, or drawings with which he is presented (19% of current items).

This is the most versatile type of item, because answer parsing is easily done precisely by a computer algorithm, and the same answer-parsing algorithm will work for any kind of problem at all.

It is, however, resistant to the elimination of lucky guesses, because it is difficult to sufficiently disguise the correct solution among the offered answer choices without being deliberately misleading (e.g. adding "distractors" that tempt students towards making commonly-observed errors). One might well wonder whether such problems teach students to merely *recognize* instead of *generate* a correct answer.

Anecdotally, instructors often observe that a student who experiences serious difficulty with a problem complains of "not even knowing where to start," i.e. having a mental "slate" that is entirely blank. But if he has heretofore studied, or been assessed and advanced, largely through multiple-choice problems *the blank slate is an unfamiliar starting place,* and the student's difficulty with it is unsurprising. A student confronting a multiple-choice problem *never* starts with a blank mental slate. If nothing else, he knows the problem is soluble and the solution is hidden in one of the four or five choices in front of him. Furthermore, the language in which even the incorrect choices are couched will give him important clues about how to categorize the problem ("Is this a problem in stochiometry or equilibrium?"), which is arguably the essential starting point for solution.

Given that the real world of chemistry careers is not multiple choice, i.e. that there is profound practical utility in learning to successfully navigate towards a solution from a blank mental slate, `ALEKS` avoids multiple-choice items as much as possible.

**A drawing.** That is, the student must draw some visual representations, e.g. an energy level diagram, electron box diagram, Lewis structure, or graph (10% of current items).

These are powerful items, when debugged thoroughly, because the potential answer space is very large, like items in the *Quantity* category, and these items share the advantages of *Quantity* items noted above.

Furthermore, items that are answered by drawing instead of mathematical calculation can be designed to teach and test ideas and skills without the complication of co-requisite mathematical calculation skills, which can obscure the origin of student error; e.g. was it true conceptual error or mere mathematical slip?

For this reason this is the most rapidly growing category of item in `ALEKS` Gen Chem. Unfortunately, they are slow and expensive to develop, for a variety of technical and pedagogical reasons, e.g. they often requiring programming new sketching or construction tools, and complex answer analysis to ensure acceptance of all equivalent alternatives.

A classification of chemistry problems often discussed divides them into "conceptual" or "algorithmic" problems. A plausible example of each from ALEKS Gen Chem is given in Figures 6.3 and 6.4. However, aside from a

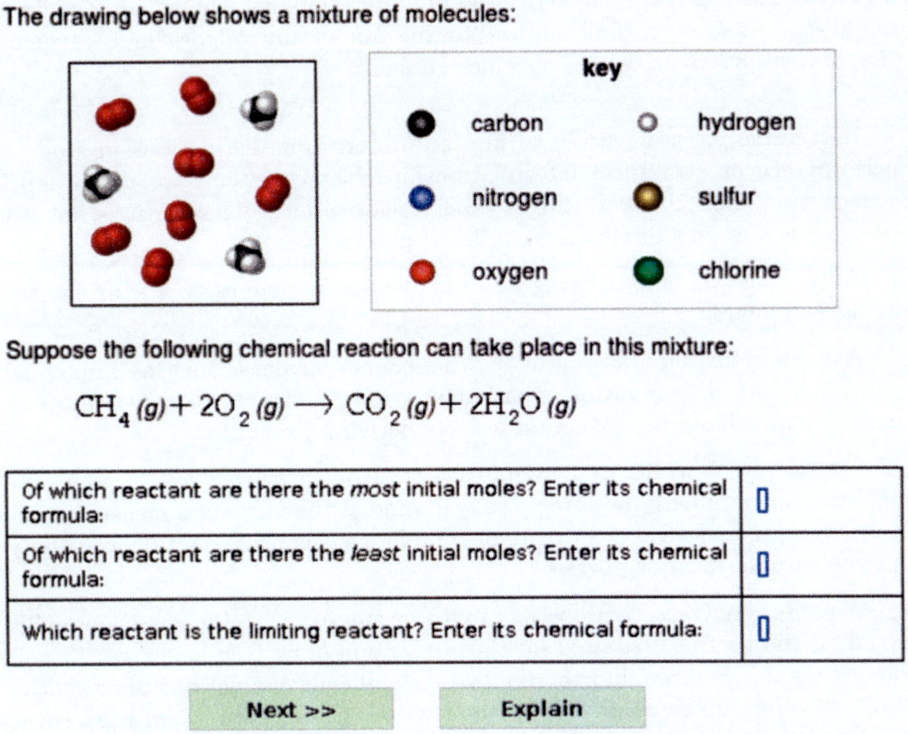


**Figure 6.3.** An example of a "conceptual" item from ALEKS Gen Chem.

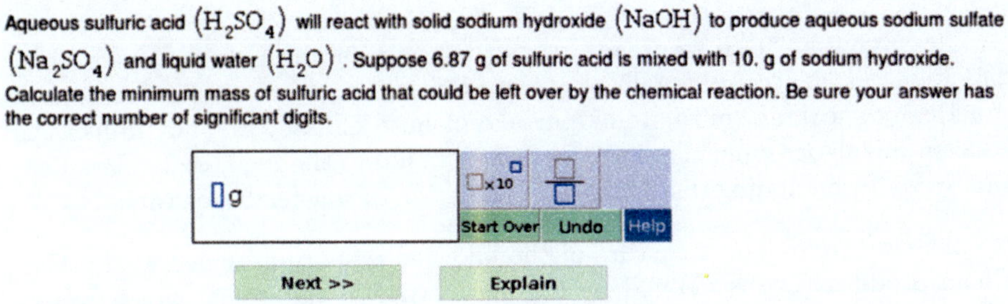


**Figure 6.4.** An example of an "algorithmic" item from ALEKS Gen Chem.

sprinkling of clear-cut cases such as the examples in Figures 6.3 and 6.4, we find that it is difficult to cleanly distinguish "conceptual" from "algorithmic" items. Even an item with substantial amounts of mathematical computation requires conceptual understanding to set up the computation, and even the most "conceptual" item requires some logically deductive ("algorithmic") process to generate one final answer that can be parsed by computer. ALEKS has no way to grade a narrative expressed in natural language by which a student might express "conceptual" understanding free of any "algorithmic" generation of a specific answer to a specific problem.

Furthermore, we have yet to find any interesting distinctions in student performance or structural interrelationships between items we might label "conceptual" and those we might label "algorithmic." That is, it is not yet clear to us how this distinction can be of practical use.

A final method of classifying items is to classify them as "read" or "write" items, as follows:

A **read** item requires a student to *understand* data, such as numerical data, a chemical formula or equation, a graph, a sketch, or a narration or description, usually by deduction from underlying principles.

Conversely, a **write** item requires a student to *predict* and present the implications of underlying principles, e.g. predict the value of a measurement, draw a graph, make a sketch, write or complete a chemical equation, or supply an appropriate word or phrase.

We call these "read" and "write" items by analogy with the analogous skills (reading and writing) required for the acquisition of written language. As will be discussed below, we find there may be significant distinctions between the mastery retention rates of "read" and "write" items, even when they cover nearly identical concepts and require similar types of answer entry.

If the analogy to actual reading and writing holds, this may not be surprising, as it is a familiar conclusion from neuroscience that actual reading and writing involve separate areas of the brain, and so presumably distinct mental processes.

Classifying an item as "read" or "write" is generally not any easier than classifying it as "conceptual" or "algorithmic," and for similar reasons—practically speaking, most items partake of both characters—but unlike the conceptual/algorithmic classification, we have found the read/write classification leads to intriguing distinctions in retention of mastery over time.

**6.2.2 Structure.** The structure of the knowledge space in which ALEKS Gen Chem products live is defined by a partial order of the items, which can in principle be diagrammed with a Hasse graph such as those shown in other chapters of this book.

Such a graph is not presented here because the large number of vertices (417) and very large number of edges (approximately $10^5$) rule out anything other than the vaguest of qualitative impressions, e.g. that it's big and complex. Automated analysis is required to interpret and improve it meaningfully.

The initial structure was constructed by expert intuition. That is, experienced instructors were asked which pairs of items seem likely to be connected by inference, forward or backward. However, records of actual student performance on assessments have now been monitored for five years, and the results used to discover new inferences and verify (or reject and remove) old inferences. The v4.0 structure is substantially different than the expert-derived v1.0 structure.

An interesting facet of ALEKS Gen Chem is that it actually employs two separate structures, for assessment and learning. The assessment structure is as "tight" as possible, meaning it contains as many inferential links as the data justify, to maximize the efficiency of the ALEKS assessment and the precision of the resulting predicted knowledge state.

The learning structure is a proper subset of the assessment structure (so that learning does not put the student in a state outside the knowledge space spanned by the assessment structure), but is quite a bit "looser," meaning it has significantly fewer inferential links. The missing links are generally those that connect items in substantially different conceptual areas (geometry and weak-acid equilibria, for example), and always represent the "empirical" links resulting from the typical order of instruction, or the demands of item mastery on general intelligence and sophistication.

That is, for example, nearly every student learns 7th grade algebra before he is taught to name alkanes in a college chemistry class. Hence the ability to name $CH_3CH_3$ may allow ALEKS to infer a student knows $7^{-2}$ is less than $7^2$. And we would certainly use this information on assessment.

However, there is no *logical* inferential connection between these two skills. In principle, students *could* learn to name alkanes in 6th grade, well before learning the properties of exponents, which would render the inferential link invalid. It's more or less just historical accident that this does not happen. It's this kind of link that may be missing from the learning structure.

The reason we remove empirical links that connect conceptually distinct regions of knowledge space is because chemistry instructors present material in many different orders, for reasons discussed in the Introduction, and we need to accomodate those various orders, so long as they are not actually illogical. By removing empirical links from the learning structure, we accomplish this without serious harm to student learning. At most, the lack of inference in the learning structure may mean ALEKS (conservatively) assigns a student an item in learning mode that he in fact already knows.

**6.2.3 Implementation.** Students go through ALEKS Gen Chem in a repeated cycle of assessment and learning:

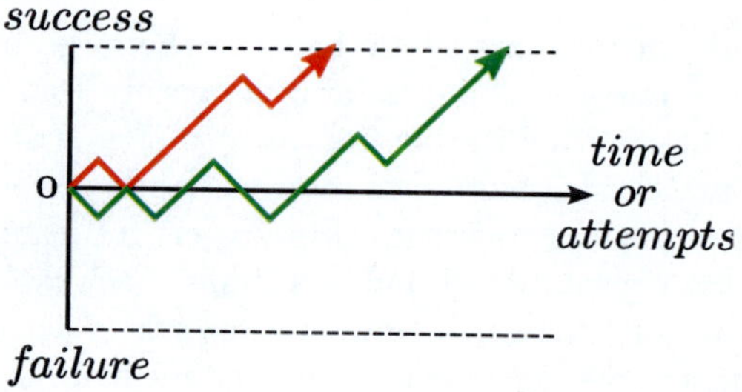


**Figure 6.5.** Two learning walks. The red walk is successful after 9 attempts, the green walk after 15.

1. An *initial assessment* on entry to any of the products assigns the student a knowledge state.
2. From the "outer fringe" of this knowledge state—items the pre-requisites of which are all in the student's knowledge state, but which are not themselves in the state—we populate a *learning navigation list* from which the student can begin learning. (In the `ALEKS` system this list is presented in the form of a clickable pie chart.)
3. The learning navigation list may be further constrained by an *intermediate objective*, established by the instructor, which consists of a list of items assigned to a particular due date in the near future. These might correspond, for example, to the items that are relevant to the next chapter of the textbook, or to the material to be lectured on this week.
4. Students then choose items from the learning navigation list and master them via a *learning walk*, a small cycle consisting of attempts to solve instances of an item, followed by grading (right or wrong) and any feedback `ALEKS` can reliably glean from the nature of a wrong answer.
   Students can also ask to see the detailed and complete solution to any instance of an item, and from these focussed solutions hyperlinks lead them to an inline text resource organized in the form of an encyclopedia. The learning walk is called a "walk" because for each correct solution of an instance, the student advances one step towards being credited with mastery of the item, and for each wrong solution, he advances one step towards being ejected back to the navigation list and being advised to pick a different item (Figure 6.5). Thus there is no fixed number of instances required to be assigned mastery, but the shortest possible walk is generally three correct answers in a row. Much longer walks are possible, and commonly observed.
5. When a student *masters* an item, it is added to his knowledge state, and the outer fringe recomputed, typically bringing several more items into

his navigation list. He picks another item and continues learning. (Failure of the learning walk does not affect the student's knowledge state.)

6. If the instructor has specified intermediate objectives with fixed deadlines, as is typical in general chemistry, then when the deadline passes *or* a student masters all the items in the current intermediate objective he is given a *progress assessment* that attempts to confirm his recently acquired knowledge. Students may gain or lose items from the knowledge state by virtue of this assessment, but the overwhelmingly typical result is a loss of 5-20% of recently-mastered items.
7. After a progress assessment that happens on a deadline, a student re-enters learning mode in the next instructor-assigned objective. If any items have been lost in the progress assessment that are pre-requisites for items in the current intermediate objective, they are "dragged forward" into the current objective. (Obviously this puts a significant premium on retention of recently-learned mastery.) If they are *not* required for the current or a subsequent intermediate objective, they remain out of the student's knowledge state and inaccessible to him unless he enters a free-learning period.
   After a progress assessment that happens *before* a deadline (because the student finished his assigned learning early), a student enters a free-learning period that lasts until the deadline. In a free-learning period all intermediate-objective constraints are removed from student learning paths. Students usually use these periods to re-learn items lost in progress assessments that were not pre-requisites for future objectives.
8. Instructor-structured incentives are of course a vital part of ALEKS courses. When ALEKS Gen Chem is used as a summer prep or placement and remediation tool, the incentive is usually gaining admission to a certain class in the fall. When ALEKS is used as an adjunct to a regular semester class, incentives are credit towards a class grade.
   We have found that best results come from assigning roughly equal weight to meeting intermediate objectives on time and to maximizing the size of the knowledge state at the end of the course. The former prevents procrastination, while the latter puts a useful premium on retaining mastery. (The regular progress assessments ensure that students cannot retain items in their knowledge state until the end of the course unless they repeatedly demonstrate mastery of the underlying skills and concepts.)

## 6.3 Results

In the fall of 2012 ALEKS Gen Chem was in use by about 15,000 students at about 90 universities. Almost all were enrolled in college general chemistry. In some cases ALEKS was used before the class began, e.g. during the last few weeks of the summer, as an advising, placement and remediation tool. Generally, however, ALEKS was used as an adjunct to a normal college class, i.e. as a supplement to regular lectures and recitations, assigned reading in a textbook, and possibly other forms of homework. A few interesting observations from both patterns of usage will be described.

### 6.3.1 Instructional needs are quite heterogeneous.

Students entering the main (Austin) campus of the University of Texas who wished to take general chemistry in the fall 2012 were required in July of 2012 to demonstrate a standard level of preparedness within the ALEKS system. This level was defined by a knowledge state containing 117 items, selected from those available in ALEKS Gen Chem by the Texas chemistry faculty, and corresponding to their detailed expectations for student preparedness on entry.

Students were in fact only required to achieve 85% mastery of these topics to enroll in general chemistry. There is no real purpose to the 85% figure aside from avoiding the psychological discomfort to students of knowing they have to score "100%" on what appears to be a placement exam.

After the initial assessment many students found their knowledge states already encompassed the 100 required items. They were done. Many other students, however, found they needed remedial refreshment or instruction in at least a few items. These students had the opportunity to immediately enter as many cycles of ALEKS learning and re-assessment as were required to enlarge their knowledge states with the missing items. Since doing so was a *sine qua non* for enrolling in general chemistry, almost all did so.

We generally find significant variations in the individual remedial effort required in this kind of ALEKS implementation. For example, at Texas many students easily exceeded the standard preparedness level in a single 90-minute initial-assessment session. But a substantial number required up to 60 hours of learning time, punctuated by at least three or four progress re-assessments. (These students essentially took or re-took high-school chemistry online, a real bargain at the $25 retail cost!)

Note the critical importance to the success and economy of this approach of the *individualized* learning paths easily assigned to each student in a KST-based assessment and learning service. Only this individuation allowed more than 3000 students to be brought to a reliably similar level of preparedness, despite significantly varying educational backgrounds, with a minimum of work assigned to each student.

A non-individualized remediation, e.g. a traditional summer review/prep class, would probably have required a great deal more work for the average student, without compensating gain, since the remediation assigned to *each* student in a non-individualized regime would necessarily be the union of the remediations actually needed by *every* student. Unless the students demonstrated a rather strong degree of uniformity in their needs, most students would find only a smallish portion of the assigned remediation useful.

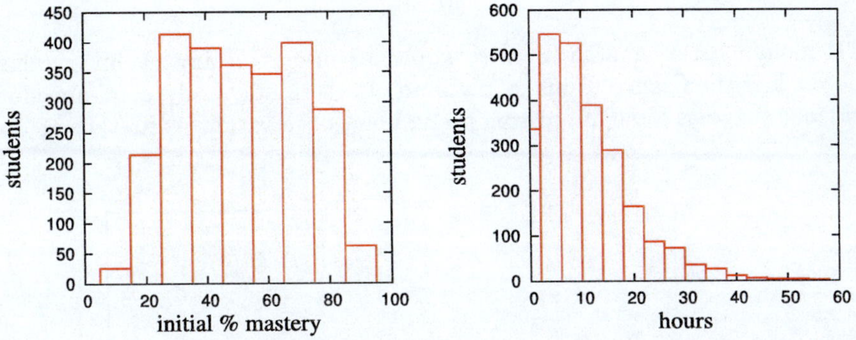


**Figure 6.6.** Students intending to take general chemistry at the University of Texas in fall 2012 exhibit an astonishingly broad range of initial preparedness. At left, the distribution of initial knowledge states, measured as a % of items in the Texas placement/remediation ALEKS course. At right, the distribution of time required by students to reach the level of preparedness required by Texas chemistry faculty to register for general chemistry.

In fact, our simplest characterization of the spectrum of remediation needs of the entering Texas class of 2016 (see Figure 6.6) demonstrated a heterogeneity even stronger than one might naively expect from the fact that each student has satisfactorily completed a chemistry class at a high-school adhering to state science education standards, and of sufficient stature to gain admission to a first-rank state university. These results might be interpreted to suggest that the level of heterogeneity in preparedness of incoming general chemistry students is *in general* far greater than has been generally believed (or hoped), and we might wonder whether this could be a factor in the 20-30% failure rates that are routine in general chemistry classes.

### 6.3.2 Individualized assessment and learning works.

In most cases ALEKS Gen Chem is used as an adjunct to an ordinary college chemistry class, i.e. for which there is also a textbook with assigned readings, lectures and recitations, exams, and sometimes other forms of homework.

An increasing number of programs are using *both* approaches, however: a summer prep to ensure students are brought to a relatively uniform level of preparedness, extending smoothly into a semester adjunct program. One of the major advantages of this combination approach is that a student's knowledge state is continually refined by ALEKS, and over a longer period assessment will be still more efficient, and learning better individually targeted.

As mentioned above, most semester adjunct programs use intermediate objectives to keep the pace and focus of student work in ALEKS lined up with the class lecture schedule.

Some most important early questions about the results of using ALEKS Gen Chem out in the real world are how and to what extent ALEKS judgments of student mastery match up with standard in-class exams. While we have not yet had the chance to make any kind of systematic study, semi-anecdotal evidence suggests some interesting preliminary observations. First, there *does*

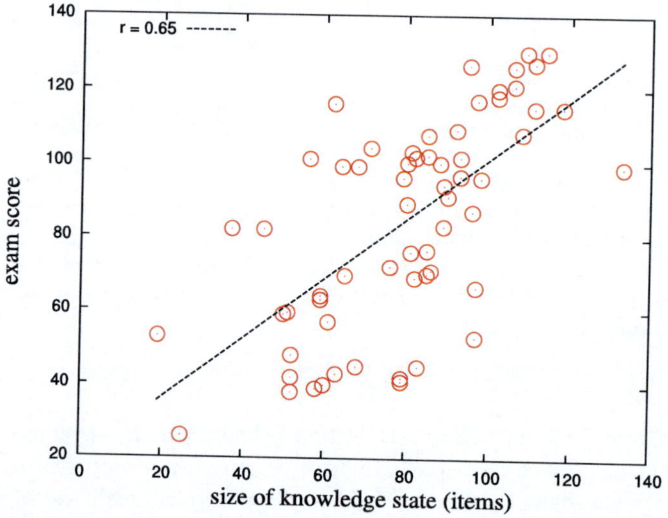


**Figure 6.7.** Performance on an in-class college general chemistry exam is correlated with mastery measured by `ALEKS` (*Exam data courtesy of Dr. Vally D'Souza, Dept. of Chemistry, University of Missouri St. Louis.*)

appear to be a reliable correlation between mastery measured by `ALEKS` and performance on in-class exams (Figure 6.7).

However, one might wonder whether this occurs for two trivial reasons that have nothing to do with any KST cleverness:

- It could be `ALEKS` mastery is just a proxy for student time on task, because `ALEKS` in effect logs student time working at home on chemistry, and instructors inspect those logs and reward (via credit) evidence of time spent, assignments completed, et cetera. We might imagine *any* system that forces the student to spend more time with his chemistry textbook and thinking about chemistry topics would improve test results. Perhaps `ALEKS` could be replaced with a camera that awarded points every time the student's head and the textbook were both in view at the same time!
- It could be that `ALEKS` is functioning essentially as just a test of general intelligence, motivation and/or prior experience, and that it simply picks out the most intelligent, well-prepared, or well-motivated students, and *that* is why `ALEKS` mastery corresponds to student performance on exams.

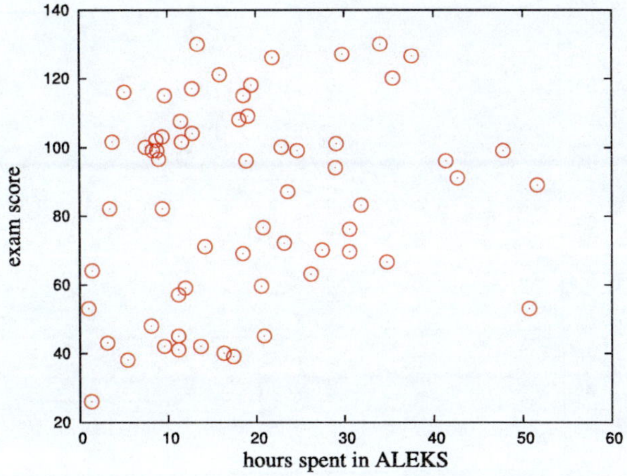


**Figure 6.8.** Performance on an in-class college general chemistry exam does *not* correlate with time spent in ALEKS (*Exam data courtesy of Dr. Vally D'Souza, Dept. of Chemistry, University of Missouri St. Louis.*)

However, we believe two more examinations of the data cast doubt on both these possibilities (Figures 6.8 and 6.9). If the first hypothesis were correct then there should also be a correlation between time in ALEKS (which is the "time on task" ALEKS is enforcing) and exam performance. There isn't.

On the other hand, if the second hypothesis were correct, then there should be also be a correlation between *initial* knowledge state (which, since it takes place before any ALEKS learning, can only reflect general intelligence, effort, or prior preparation) and exam performance. There is not.

While hardly proof, Figures 6.7, 6.8, and 6.9, taken together, suggest that ALEKS is actually successfully diagnosing and remediating important student weaknesses which other adjuncts, like the textbook or recitation, do not or cannot cure.

**6.3.3 Forgetting imposes a big overhead cost.** One of the valuable side-effects of a broad commercial usage of ALEK Gen Chem is the wealth of data on actual student performance that has accumulated. We recently analyzed some of this data with an eye to gaining insight into a persistent feature of student performance within the ALEKS system, which is that students routinely lose mastery of items when they are re-assessed some period of time after learning them (Figure 6.10). It is a well-known phenomenon to all of us that some information we can successfully retrieve from memory a *short* time after learning will be forgotten over a longer time.

It would seem clear that successful *long-term* mastery is the ultimate pedagogical goal. When we see it not happening, meaning students not demon-

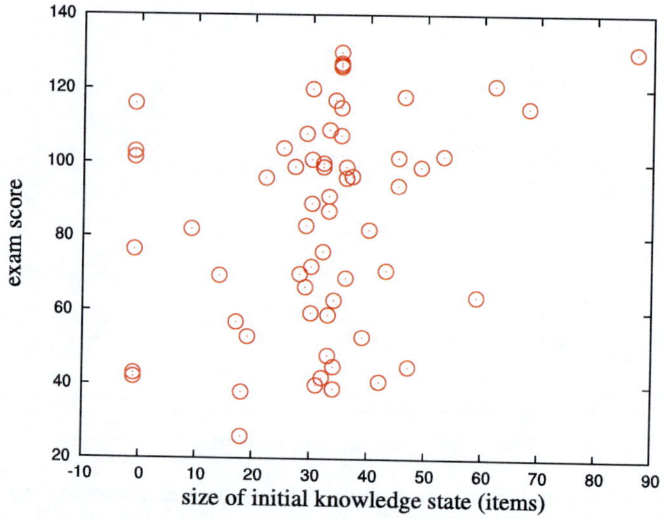


**Figure 6.9.** Performance on an in-class college general chemistry exam does *not* correlate with the *initial* level of student mastery measured by ALEKS (*Exam data courtesy of Dr. Vally D'Souza, Dept. of Chemistry, University of Missouri St. Louis.*)

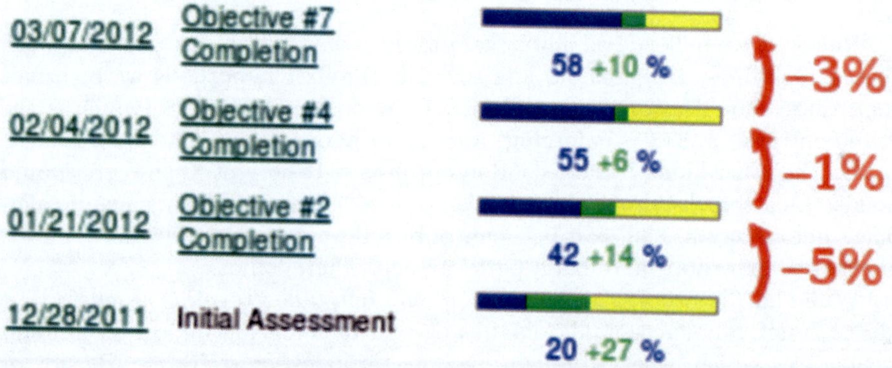


**Figure 6.10.** A portion of an ALEKS report on student performance on assessments for a particular student. The red arrows and percentages are added annotations, *not* part of the actual report, and show the slight slippage of confirmed mastery on each re-assessment. Assessment performance is given by the blue bars and numbers in each case, and stated as a percentage of the items in the course of which the student has demonstrated mastery.

strating mastery on a final exam, or in a subsequent course, we ask whether this is because the student never learned it in the first place, or because he learned it but then forgot. The empirical data gathered by ALEKS suggest the latter is more often the case (Figure 6.11). Figure 6.11 shows the probability

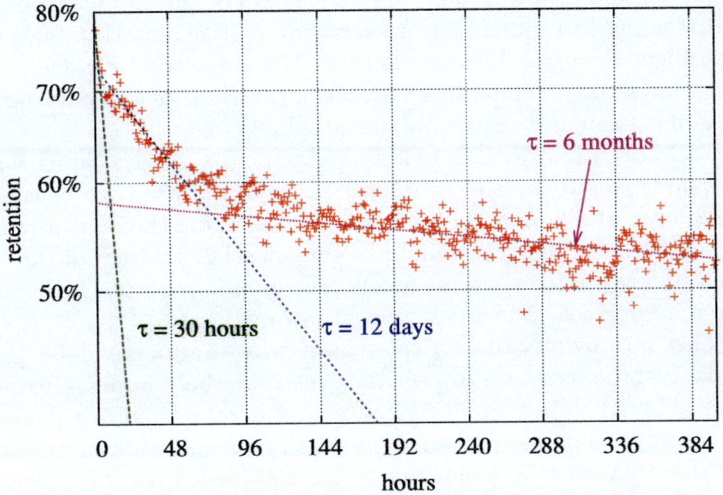


**Figure 6.11.** Student mastery retention as a function of time since initial mastery. The vertical axis is logarithmic. The three lines represent simple exponential decays with the indicated time constants.

of a student proving mastery of an item during assessment of which he has recently acquired mastery in learning, as a function of the time between the initial mastery in learning and when the assessment took place.

For example, if a student mastered an item on balancing chemical equations at 3 pm on February 1, and subsequently followed the average trend in Figure 6.11, then he would have about an 80% chance of balancing a chemical equation correctly on an assessment given at 9 pm the same day. On the other hand, if the assessment were given at 3 pm on February 3, forty-eight hours later, then the probability of his balancing a chemical equation correctly would have fallen to about 62%, and if the assessment were given on February 8, one week later, his probability of success declines to about 55%. It doesn't change much after that.

We compute the probability in Figure 6.11 by taking note, for each question in a progress assessment, of whether the student acquired mastery of the item of which the question is an instance during the most recent period of learning. If he did, we record the time since mastery of the item and whether the question is answered correctly. (Note that it is not likely the student will have studied the same *instances* in learning mode that he is given in assessment.)

Each point in Figure 6.11 then represents the probability that a question asked at a given time since mastery of the related item is answered correctly. The sample size at a given time varies between $3. \times 10^2$ and $2.5 \times 10^4$. Generally there are many more events at shorter times, but aside from this rough trend there is no simple variation in sample size with time. Additionally, since these data were gathered incidental to the commercial use of the product, and we have not yet attempted to control for or even characterize important parameters such as student demographics or where in the course the assessment was given, we have made no attempt to analyze the precision of the data points or the sources of error and variation in this graph.

On Figure 6.11 are also drawn three simple exponential decays, which when fit by least squares to data in the appropriate range produce time constants of 30 hours, 12 days, and 6 months. We have no compelling theoretical reason for thinking the decay of mastery should be exponential, aside from the fact that exponential decay is the simplest imaginable kinetic process. (If the data are drawn on a log-log scale, no simple power law emerges.)

There is also no obvious theoretical reason why there should be three exponential decay processes, aside from the fact that there appears to be a noticeable break in the data around 48 hours, and perhaps another at about 6 hours. We could speculate that what we are seeing is evidence of three distinct mental mechanisms by which mastery is retained, with three distinct characteristic decay times, which we might whimsically name "cramming," "short-term memory" and "long-term memory."

The study of the decay of memory and mastery has a long history (Ebbinghaus, 1885; Finkenbinder, 1913): Ebbinghaus first drew a curve like Figure 6.11 in 1885 and named it a "forgetting curve" (Ebbinghaus, 1885). Ebbinghaus claimed an exponential curve fit his data best. Later authors have made general arguments for a power-law curve (Averell and Heathcote, 2010), and the question of the mathematical shape of general forgetting curves, or whether a general forgetting curve can even be usefully defined is still a subject of debate (Brown et al., 2007; Roediger, 2008). We make no attempt to address these difficult general questions of theoretical psychology, and simply present Figure 6.11 as an interesting empirical measurement.

One fact that clearly emerges from our measured "forgetting curve" is the significant burden forgetting imposes on the normal educational process. If a typical student's retention declines from 80% to 55% over a week—this is often the difference between an A and an F on an exam!

Another way to put this is that it seems possible that a significant contribution to the difference between an A and an F on an exam is simply how much learning has been actually retained. *Most* students could perhaps be A students—if they were tested within hours of initial mastery, or if timely intervention could prevent the typical loss of mastery.

Incidentally, an ironic conclusion we could also draw is that the student practice of waiting until the last possible moment to study for an exam is actually a rational economization of time and mental resources, because that essentially eliminates the substantial "overhead" cost of forgetting: from Figure 6.11 a typical student can count on remembering over 80% of what he learns if the exam is taken within 24 hours, but only 60% if it's taken a week later. Or, from his outcome-oriented point of view, to achieve the same grade on the exam he'll have to study a lot more if he starts his studying earlier.

Of course this assumes the student's primary goal is passing the exam and not long-term mastery! Nor need this be a cynical observation, as students are not themselves qualified to define which short-term goals predictably lead to long-term mastery. They rely on teachers for that, and if the teacher has said (or implied by his actions) that long-term mastery can be guaranteed solely by achieving certain marks on mid-term and final exams, the students do not have the experience to conclude differently.

It's possible to regard mastery or knowledge lost over time as not having been *genuine* in some sense. That is, we might say that if a student forgets how to balance a chemical equation after 48 hours, he never really knew how to do it at all. To some extent the difference between this point of view and that taken above—that the learning was no less genuine for having been forgotten—is just a quibble over what is meant in detail by words like "mastery" and "forgetting."

But it's probably also important to point out that the initial mastery entering these data is *not* just answering one question correctly, something that might arguably involve a bit of luck. Initial mastery is the successful completion of the learning walk described above. The student has solved *several* different problems on the topic, over a small but meaningfully extended period of time (10–20 minutes is typical). It certainly looks like mastery at the time. It's hard to see how the student could demonstrate any more substantially reliable competence.

He could certainly demonstrate more lasting competence by repeating his performance later, but this begs the question, which is whether there is some evidence we could find *at the time of initial mastery* that would clearly distinguish "real" (i.e. lasting) mastery from what we might want to call the "illusion" of mastery because it doesn't last. (Indeed if we could think of any such evidence, we'd use it to extend the learning walk until we knew "real" mastery had been achieved.)

There are many ideas about what causes, and can prevent, forgetting. For example, one obvious hypothesis would be that it's harder to retain mastery of items that are inherently difficult—because they require building complex memories, are hard to relate to ordinary experience, require several math skills, et cetera.

If this were true, we would expect a negative correlation of retention with the effort required to acquire mastery in the first place, for example the length of the learning walk required to achieve initial mastery. We don't find this (Figure 6.12). (In these figures, and discussion following, "retention" at time $t$ is defined as the average over all items of the probability of getting a question on assessment correct $t$ hours after mastering the relevant item in learning mode.) We also find that *long-term* retention of mastery is not well correlated with *short-term* retention (Figure 6.12). This is another reason to doubt that long-term retention is related to the difficulty of initial mastery.

Incidentally, it also gives intriguing evidence that short- and long-term retention are dominated by very different mental processes, perhaps even by different brain structures. If this were true, then given the inherent individual variations in the human brain, the optimal study habits (meaning mix of short- and long-term memory reinforcement strategies) might be quite different for different students.

One influence on mastery retention is very clear: the nature of the item. In Figure 6.13, for example, we show the decay of mastery of two items with quite similar subject matter: the shorthand way in which chemists write the electronic structure of neutral atoms. The first item, labeled "read" in Figure 6.13, shows the student a short-hand electron configuration (e.g. $1s^2 2s^2 2p^5$) and then asks them to write the total number of electrons, the number of 1$s$ electrons, and the symbol for the element (in this case 7, 2 and F). The second item, labeled "write" in figure 6.13, gives the student a blank answer editor page and asks him to write the electron configuration of an atom, for example a neutral atom of F, in which case the answer is the electron configuration above. The possible electron configurations used in the two items are exactly the same.

These two items are clearly very closely related, in terms of the underlying concepts and even the detailed skills (counting electrons, interpreting orbital notation, knowing how many electrons the atom of a given element has). What can account for the significant difference in relative retention of mastery? Why do students forget how to do the "read" problem faster? The possibility that has come to our mind is suggested by the titles we give each item in

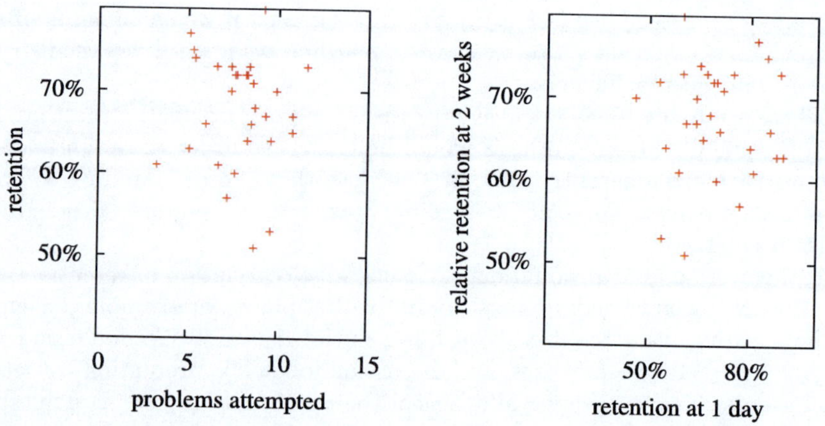


**Figure 6.12.** Long-term retention of mastery isn't strongly influenced by the inherent difficulty of the item, nor strongly correlated to *short-term* retention. At left, retention at 2 weeks versus the number of problems attempted before achieving initial mastery. At right, retention at 2 weeks versus retention at 1 day.

the plot, which correspond to the categorization of items mentioned in the Introduction: the "read" item requires students to interpret and explain a theoretical concept, while the "write" item asks them to actually *use* it to create an answer on an entirely blank page. It would seem that mastery of the latter skill, once achieved, is much better retained, lending some empirical weight to the proverbial hypothesis that one learns best by doing.

Since Figure 6.11 can be regarded in a broad sense as an averaging of curves like that in Figure 6.13 over all items in the product, it's also clear that the marked difference in individual item mastery decay rates may contribute significantly to the apparent noise in Figure 6.11. To properly characterize the *overall* "forgetting curve" we would have to be much more careful about understanding how the mix of *individual item* "forgetting curves" comprise it.

## Acknowledgements

The work presented here is a team effort, and the many essential contributions of Erin Chwialkowski, Malin Walker and Eric Gates are gratefully acknowledged.

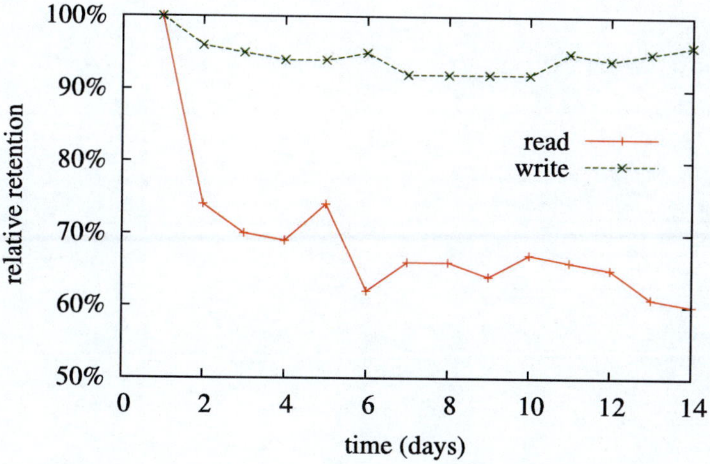


**Figure 6.13.** Retention of mastery is strongly related to the nature of the item. In this graph we show relative retention (retention at time $t$ divided by retention at $t = 1$ days) for two items on the notation for electronic structure. Mastery of the item in which students must write the notation on a blank input form ("write") is retained far better than mastery of an item in which students must interpret such a notation ("read").

# 7

# Using Knowledge Space Theory to Assess Student Understanding of Chemistry

Mare Taagepera[1]
Ramesh D. Arasasingham[2]

## 7.1 Introduction

The subject of chemistry is challenging to many secondary school and college students because it requires conceptualization and visualization skills as well as mathematical and problem solving skills. It further requires the ability to integrate different representations of the chemical phenomena at the macroscopic, molecular, symbolic, and graphical level. Many students have trouble making logical connections among the different representations and integrating them with underlying chemical concepts and principles (Kozma and Russell, 1997). These difficulties influence their success in general chemistry and their attitudes towards the science.

We have analyzed the ability of students to make logical connections between fundamental chemical principles and the various representations that characterize chemical phenomena for a number of chemistry concepts using the approach of knowledge space theory (KST). These studies include an examination of the concepts of density (Taagepera et al., 1997), bonding (Taagepera and Noori, 2000; Taagepera et al., 2002; Vaarik et al., 2008), stereochemistry (Taagepera et al., 2011), and stoichiometry (Arasasingham et al., 2004, 2005). The latter will be used as an example in this chapter to illustrate the overall approach. The concept of stoichiometry refers to the calculation of quantitative (measurable) relationships of the reactants and products in a balanced chemical reaction. It is a complex concept with aspects that particularly present difficulties for students at the beginning level. It requires an understanding of atomic structure and the ability to visualize at the molecular level. It also requires the use of symbols to indicate how two or more compounds combine to form new compounds and predict the relative ratios that would lead to a balanced chemical equation for the reaction. The symbolic

[1]Dept. of Chemistry, University of California, Irvine and Dept. of Chemistry, University of Tartu, Estonia.
[2]Dept. of Chemistry, University of California, Irvine.

J.-C. Falmagne et al. (eds.), *Knowledge Spaces: Applications in Education*,
DOI 10.1007/978-3-642-35329-1_7, © Springer-Verlag Berlin Heidelberg 2013

representation of the balanced chemical equation would make it possible to predict the amounts of new compounds that would be formed based on proportional reasoning. Handling stoichiometric problems requires mathematical and problem solving skills. It is tempting for students to undercut this procedural thinking and "plug in numbers" to a simple ratio by using a sequence of arithmetic steps. However, using such an algorithm will not always work. Unless students can construct the stoichiometric equations from basic principles, they will not be able to apply this knowledge to new situations, therefore limiting the usefulness of the acquired information. Instructors tend to assume that students have the underlying principles in place and that they are using more than memorized sequences of steps on their examinations. However, as many other authors have observed, this simple assumption does not often hold true (e.g. Nakleh et al., 1996). Students may know the basic principles, but do not seem to understand them well enough to make the logical connections. KST helps us analyze for this behavior. In fact, it has been stated that there would be more concern about students' mastery of chemistry if tests were given to probe for depth of understanding (Moore, 1999).

Our work described in this chapter examines how students' cognitive organization or thinking patterns changed during the learning process. Using the same test to track students' understanding of the concept makes it possible to assess changes in students' conceptual knowledge during the learning process. Here we are looking for meaningful patterns or a logical framework of understanding. Do students see the logical connections between fundamental chemical principles and the various representations that characterize chemical phenomena or are they a collection of unrelated facts that they memorize? These connections cannot be found by simply examining the percentage of correct responses on any test, particularly, if the test was not designed to follow conceptual development. Our study examined whether an instructional software program that emphasized the various relationships while presenting them concurrently could change learning outcomes. We are particularly interested in assessing its impact on students' understanding and on how they gain and retain skills important for understanding the concept of stoichiometry. We also examined whether using the computer-based instructional tools with molecular level visualizations could impact the construction of conceptual knowledge to assist students to understand chemical concepts and phenomena in more expert-like ways. Computer-based instructional software could make it easier for students to visualize molecular level phenomena by concurrently presenting multiple images or representations to visualize chemical phenomena. The materials in the software could provide logical links between various representations to aid students' understanding. Students could be given exercises that required them to convert one form of representation to another, to reflect on their underlying meaning, and to see how they supported the solution of quantitative problems.

## 7.2 Experimental Design

Our study explicitly examines whether an implementation of a Web-based instructional software program called Mastering Chemistry Web (MCWeb)[3] that emphasized the various relationships while presenting them concurrently could change learning outcomes. The study compared two sections of students in a yearlong general chemistry course for first year undergraduates at the University of California, Irvine (a total of 424 students). The students enrolled in the course were predominantly science and engineering majors and nearly all had one or more years of chemistry in high school. Both sections of students were taught by the same instructor and all aspects of the instruction were identical except for the type of homework. The *control group* (non-MCWeb students, 176 students) used the end-of-chapter problems in their textbook as homework. The *treatment group* (MCWeb students, 248 students) used the MCWeb software as homework. The MCWeb homework allowed students to practice problems that emphasized the development of molecular-level conceptualization and visualization, analytical reasoning, proportional reasoning, as well as to learn to recognize and relate different representations in chemistry.

Our application of KST depends on collecting student data from a set of questions that reflect different levels of understanding of the material. The test was developed by our research group using questions typically found in many textbooks and adding some when it was necessary to make logical connections (Figure 7.1). The questions involve some hierarchical ordering so as to progress in a logical fashion from fundamental concepts to more complex concepts.

The KST test was given before instruction and before completing any homework on the topic as a pretest and after instruction and after completing the homework as a posttest. It was scored in a binary fashion: each question was graded as either right or wrong. Questions requiring an explanation were marked right only if both the answer and reasoning were correct.

The expert-learning path in Figure 7.2 shows an example of a logical sequencing of the learning hierarchy. Experts notice meaningful patterns of information that are organized in ways that reflect deep understanding of the subject (Bransford et al., 1999). A number of approaches can be used to construct the expert learning path. These include going by item difficulty, by skills necessary, or by task analysis (Williamson and Abraham, 1995). We based our hierarchy on skills necessary to answer questions and solve problems. An expert would be expected to: (A) see representations (at the molecular, symbolic, and graphical level) that characterize chemical phenomena; (B) link, transform, and move fluidly among different chemical representations; and (C) transfer the knowledge and skill to solving problems.

[3]MCWeb is a software program developed by Patrick Wegner (California State University, Fullerton) as part of *Molecular Science*, one of five NSF systemic initiatives for the reform of chemical education. It is commercially available as *Catalyst*

1. Write a formula for the molecules represented in the following box:

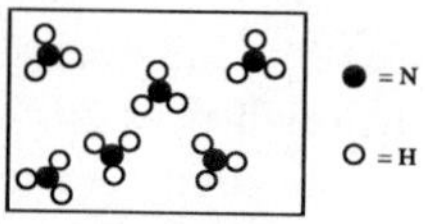


2. Write a balanced reaction equation for nitrogen ($N_2$) and hydrogen ($H_2$) reacting to give ammonia ($NH_3$).
3. What information is provided by the subscripts in the chemical symbols for the reactants ($N_2$ and $H_2$) and the product ($NH_3$) in the above reaction equation?
4. The equation for a reaction is: $2\,S + 3\,O_2 \rightarrow 2\,SO_3$. Consider a mixture of S (●) and $O_2$ (OO) in a closed container as illustrated below:

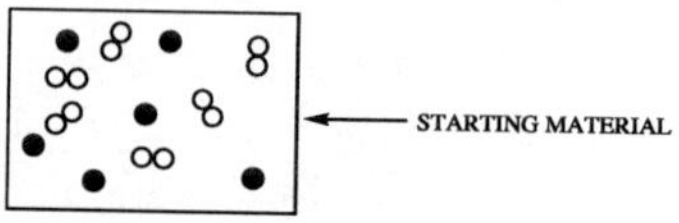


Assuming the reaction goes to completion, draw a representation of the product mixture below.

Explain how you arrived at this representation.

5. For the reaction: $2\,S + 3\,O_2 \rightarrow 2\,SO_3$, calculate the maximum number of moles of $SO_3$ that could be produced from 1.9 mol of oxygen and excess sulfur.
6. The reaction of element X (●) with element Y (O) is represented in the following diagram:

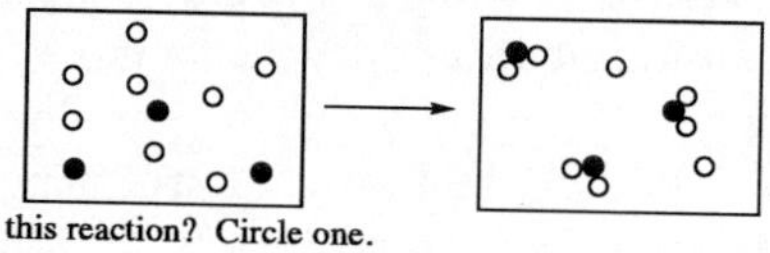

Which equation describes this reaction? Circle one.

(a) $3\,X + 8\,Y \rightarrow X_3Y_8$
(b) $3\,X + 6\,Y \rightarrow X_3Y_6$
(c) $X + 2\,Y \rightarrow XY_2$
(d) $3\,X + 8\,Y \rightarrow 3\,XY_2 + 2\,Y$
(e) $X + 4\,Y \rightarrow XY_2$

7. A reaction equation can be written to represent the formation of water from hydrogen gas and oxygen gas: $2\,H_2 + O_2 \rightarrow 2\,H_2O$. For a mixture of 2 mol $H_2$ and 2 mol $O_2$, what is the limiting reagent and how many moles of the excess reactant would remain unreacted after the reaction is completed?
8. For the chemical reaction: $2\,S + 3\,O_2 \rightarrow 2\,SO_3$, which of the following graphs best represents the formation of $SO_3$, if S is added indefinitely to a fixed amount of $O_2$? Circle one.

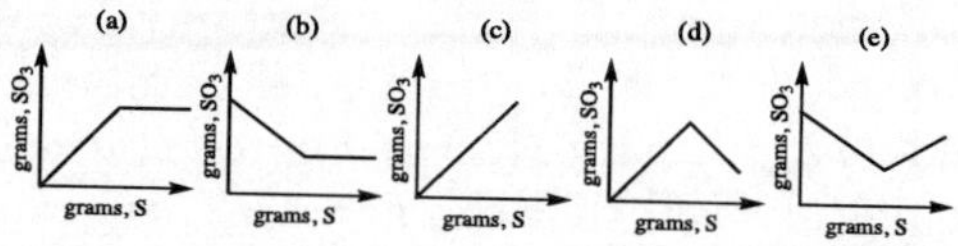


Explain your choice.

**Figure 7.1.** KST test on the fundamental aspects of stoichiometry. The hierarchical ordering is not sequential. (Figures 7.1 to 7.5 are adapted with permission from Arasasingham et al. (2004).)

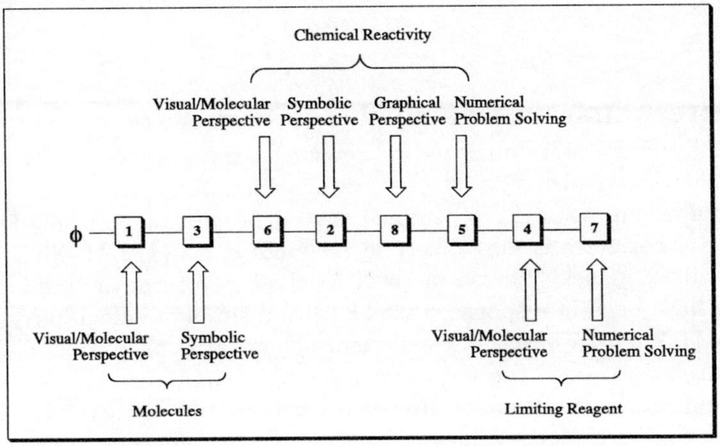


**Figure 7.2.** Rationale for constructing the expert learning path for the questions in Figure 7.1 ($\phi$ is the empty state or the state with no correct responses).

In the expert learning path, answering Questions 1 and 3 correctly requires knowledge of the visual and symbolic representations of atoms and molecules. Question 1 requires the student to translate a visual representation of the molecular perspective of a substance to its symbolic representation, while Question 3 requires the student to explain the information conveyed by the symbolic representations (or formulae) of substances.

The next level of complexity comes from Questions 6, 2, 8 and 5. These questions require knowledge at two levels. Knowledge of the visual and symbolic representations of atoms and molecules is required to determine the visual and symbolic representations of a chemical reaction, and to transfer that knowledge to numerical methods so as to extract quantitative information. In Question 6, the student must translate a visual representation of the molecular perspective of two substances being transformed into a different substance to its symbolic representation (chemical equation), and in Question 2 the student must write a symbolic representation of a chemical reaction (i.e., write a balanced chemical equation). Next comes Question 8, which is a graphical representation of the same phenomenon, and Question 5, which requires numerical problem solving in extracting quantitative information (stoichiometric ratios) from a balanced chemical equation. Finally, Questions 4 and 7 have the highest level of complexity in the hierarchy. In Question 4 students must devise a reaction in which one reactant is present in a limited supply at a molecular level and make predictions based on the balanced equation. Ques-

---

and is an automated online homework system. **Its design is not based on the fundamentals of KST.**

tion 7 requires numerical problem solving to extract quantitative information on the reactant that limits product formation and the reactant that is left over.

## 7.3 Determination of Knowledge Structure

Student responses were analyzed to determine the knowledge structure and critical learning paths. The set of questions answered correctly by a student is called a response state. For example, a student who answers only Questions 1, 2, and 4 correctly is most likely in response state $\{1, 2, 4\}$. Since there is a probability of lucky guesses as well as careless errors, the student might actually have been in response state $\{1, 2\}$ and simply made a lucky guess to fall into $\{1, 2, 4\}$. However, our experience shows that the probability of this happening is low since the tests are designed to minimize guessing, usually with requiring justification of answers. We tally the populations of all the occupied response states achieved by the students. Theoretically, for an 8-item test, $2^8$ or 256 response states, are possible. Typically 60 to 100 response states are observed. The more focused (structured) the learning, the fewer response states will represent the entire class.

From these student response states, the KST analysis identifies a subset of response states (called knowledge states) using the constraints illustrated in the analysis of the KST knowledge structure for the pretest (Figure 7.3). Each state has its probability value equal to its population divided by the total population, and likewise each learning path has its probability value, being the product of the state probabilities along that learning path. The highest probability learning paths are potential critical learning paths. Usually, just a few learning paths stand out from the rest, to become critical learning paths. The numerous learning paths with extremely small probabilities can be dropped, and the very few students in the dropped learning path states are reassigned to surviving states by doing the calculations again but without all 256 states. We have about 20–40 surviving states representing the student response set. The fit of the structure is evaluated by a chi-squared test after each cycle. Finally, the two or three most probable learning paths are identified as the critical learning paths consisting of response states that best define the class and now become representative knowledge states. Once the most probable critical learning path is determined, it is possible to contrast each individual student to the class in general, as well as to a learning path representing the sequencing of the questions in order of increasing complexity or the hypothetical expert learning path. The differing critical learning paths between content experts and student novices show how the thinking patterns of experts working on routine problems differ from novices for whom the tasks are unfamiliar (Bransford et al., 1999). The learning paths might have little, if any, similarity but provide insight for the instructors about their students' as well as their own logic structures. This information makes it possible for the instructors to reflect on their own logical framework of the material and to explicitly articulate their reasoning to students on how they see the logical connections. A simplified version of the optimization program that is used to construct the knowledge structure and the critical learning paths is available on the Web (see F. Potter: http://chem.ps.uci.edu/mtaagepe/KSTBasic.html).

## 7.4 Analysis and Discussion

Optimization of the pre- and posttest data gave well-defined knowledge structures. The overall knowledge structures are summarized in Figure 7.3. As shown in this figure, the knowledge structures for the combined pretest had 41 knowledge states, the posttest for the treatment group had 12 knowledge states, and the posttest for the control group had 13 knowledge states. The mathematical treatment of the formal concepts underlying the theory of knowledge spaces may be found in the books by Falmagne and Doignon (2011) or Albert and Lukas (1999). A special case of KST, called a 'learning space', is used in this paper. A knowledge space is a *learning space* if it satisfies two pedagogically sound principles discussed in Chapter 1 (see Subsection 1.2.4).

We recall these two principles here.

[L1*] LEARNING SMOOTHNESS. *Suppose that the state $K$ of the learner is included in some other, bigger state $L$. Then the learner in state $K$ can reach state $L$ by mastering the missing items one at a time, in some order, and so eventually reach state $L$.*

For example, in Figure 7.3, if the state of the learner is $\{1,2,5\}$, and the state $\{1,2,3,4,5,7,8\}$ exists in the learning space, then there must be a least one path in the structure, such as 3, 7, 4, 8, permitting the student in state $\{1,2,5\}$ to learn successively 3, then 7, then 4 and 8 and reach the most advanced state $\{1,2,3,4,5,7,8\}$. In other words, the states $\{1,2,3,5\}$, $\{1,2,3,5,7\}$ and $\{1,2,3,4,5,7\}$ exist in the structure.

The second principle starts the same way.

[L2*] LEARNING CONSISTENCY. *Suppose that the state $K$ of the learner is included in some bigger state $L$, and that the learner in state $K$ is ready to master some additional item $q$, then either $q$ is in $L$, or any student in state $L$ is also capable of mastering item $q$.*

In short, knowing more does not prevent one from learning something new. For example, if the state of the learner is $\{1,2\}$ and there exists another state $\{1,2,3\}$, then any student in state $\{1,2,5\}$ can master item 3. To avoid any confusion: our use of expressions such as '*a student in state* $\{1,2,5\}$ *can master item* 3' only means that the state $\{1,2,3,5\}$ exists in the structure. Whether such a student can in fact master item 3 is hypothetical. Our view is that the items should be framed in such a way that such a hypothesis is verified

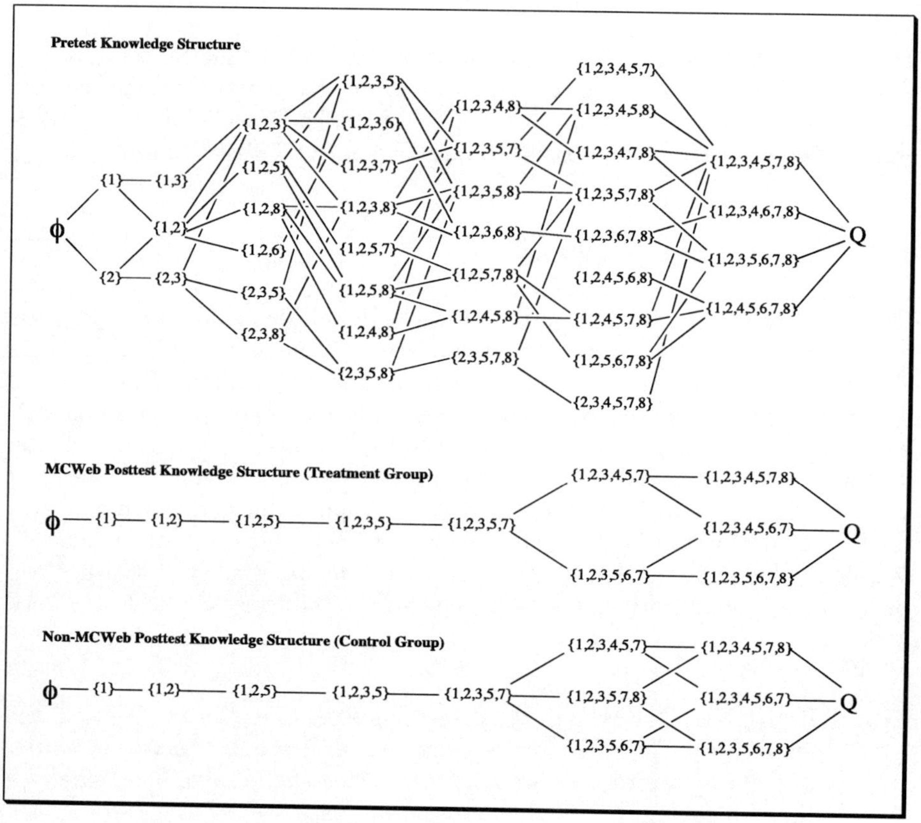


**Figure 7.3.** KST pretest and posttest knowledge structures.

in practice. These two principles put strong constraints on the collection of states. In particular, there should be no gaps.

Although the pretest data for the students in both the control and treatment groups (before instruction and before completing any homework) gave the same knowledge structure, the posttest data (after instruction and after completing the homework) did not (Figure 7.3). An examination of the fits for the knowledge structures revealed four interesting findings.

First, the students' posttest knowledge structures were not simply subsets of the pretest knowledge structure. Both posttest structures are considerably smaller than the pretest structures, but both posttest structures have the states $\{1, 2, 3, 5, 6, 7\}$ and $\{1, 2, 3, 4, 5, 6, 7\}$, which do not exist in the pretest structure.

Second, the posttests showed a more focused (or structured) learning when compared to their pretest knowledge structures for both groups. Moreover, the learning was more focused (or structured) for the treatment group over the control group as evidenced by the 12 knowledge states for the MCWeb group versus the 13 knowledge states for the non-MCWeb group.

Third, an examination of the relative distribution of the percent population in each of the knowledge states showed that while the distributions were very similar on the pretests for both groups at the beginning before completing any homework, this was not the case at the end after completing the homework. This data is summarized in the histograms shown in Figure 7.4.

The posttest knowledge structures revealed that even though content knowledge had increased for both groups and a greater population of students had progressed to knowledge states further along the knowledge structure, students in the treatment group had made far greater strides in moving further along the knowledge structure than the control group. The histograms compare the percent population of students on the pre- and posttests from each group in each of the knowledge states starting from the state with no questions correct (the $\phi$ state) to students in knowledge states with all questions correct (Q state). In the histograms, "0" represents the empty knowledge state (or the state with no correct responses, $\phi$), while "1" represents all the knowledge states with only one correct answer, "2" represents all the knowledge states with two correct answers, ...etc., to "8" representing the Q knowledge state (or the state with all questions correct). As seen from Figure 7.4, overall the treatment students made significant improvements over the control group. For example at the beginning (before completing any homework), 5% of the students in the treatment group as well as the control group were in the Q knowledge state. Similarly, 10% of the students in the treatment group as well as the control group were in knowledge states with seven questions correct. At the end (after completing the homework), 35% of the students in the treatment group and 25% of the control group were in the Q state, and 40% of the students in the treatment group and 30% of the control group were in knowledge states with seven questions correct.

Fourth, an examination of the pretest and posttest critical learning paths of the students in both groups showed that, overall, there were no major differences between the two groups on the pretests and there were no major differences between the two groups on the posttests. However, there were significant differences between the pretests and posttests for both groups as shown in Figure 7.4. The overall logical connections for the hypothetical expert in our study were from visualization, to symbolic representations, to numerical problem solving. The overall thinking patterns of the students, on the other hand, were mostly from symbolic representations, to numerical problem solving, to visualization. As shown in Figure 7.5 (see page 126), Question 1, which requires the lowest level of visualization, comes early, but the more cognitively demanding visualization questions (Questions 4 and 6) come later in the students' knowledge structure on both the pretests and posttests for both

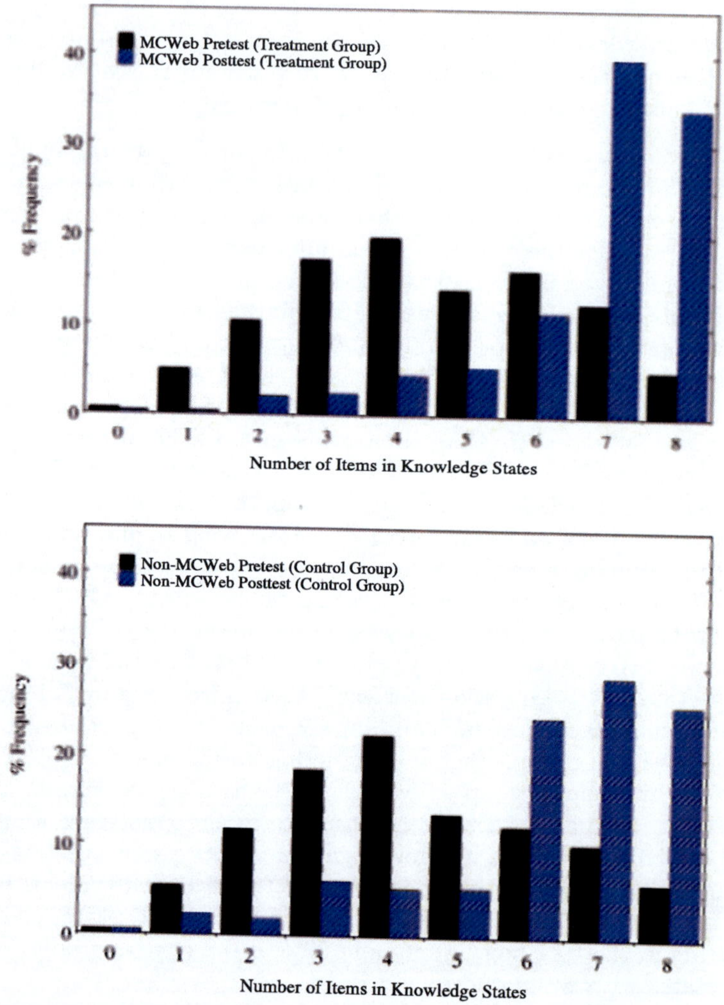


**Figure 7.4.** Comparison of the percentage of students in the knowledge states. In the histogram, 0 is the empty knowledge state or the state with no correct responses (φ); 1, knowledge states with only one correct answer; 2, knowledge states with two correct answers; 3, knowledge states with three correct answers...etc. to 8, all correct answers.

groups of students. Overall, the questions involving numerical problem solving to extract quantitative information (Questions 5 and 7) appeared earlier than expected from visualization or conceptual development (Questions 4 and 6).

Furthermore, the posttest critical learning paths revealed that even though the treatment students outperformed the control students on the KST instrument, their overall thinking patterns largely remained symbolic representations, to numerical problem solving, to visualization. Thus, there was little change in the students' overall thinking patterns over the quarter. In both groups, the algorithmic quantitative questions (Questions 5 and 7) appeared even earlier than the visualization or conceptual questions (Questions 4 and 6) on the posttest student learning paths. Novice students, as shown by numerous studies, are more likely to approach problems by searching for specific formulas or equations that could be used to manipulate the variables given in a problem, rather than reasoning conceptually in terms of core concepts or big ideas and by building mental models/representations of the problem. Our results indicate that the students in both groups were unable to make a transformation in their thinking patterns but that MCWeb was useful and successful in teaching visual and conceptual reasoning methods.

The posttest critical learning paths in Figure 7.5 showed that at the end of the quarter typical students (in both groups) progress through the questions in the following sequence: $1 \rightarrow 2 \rightarrow 5 \rightarrow 3 \rightarrow 7 \rightarrow 4 \rightarrow 8 \rightarrow 6$. The question that appeared first in the critical learning path was Question 1, which required students to translate a visual representation of the molecular perspective of gaseous ammonia to its symbolic representation. Next came Question 2, which required students to write a symbolic representation of a chemical reaction (i.e., write a balanced chemical equation for the reaction of N2 with H2 to provide NH3). Then came Question 5, which required numerical problem solving to extract quantitative information (stoichiometric ratios) from a balanced chemical equation. Thus, many students could balance a chemical equation (Question 2) and work a quantitative problem on stoichiometric ratios (Question 5), but could not interpret the significance of the subscripts in the formulae of that equation (Question 3).

The next level of difficulty in the learning path came from Questions 7 and 4, which involved stoichiometric problems in which one reactant was in limited supply. Question 7 was a quantitative problem-solving question that required a numerical solution and could be solved by a simple algorithm. Question 4 was a molecular level visualization or conceptual question that examined the chemistry behind the manipulations in Question 7.

The question that appeared next was Question 8 which asked students to pick the graph that best represented the formation of a product when one reactant was added indefinitely to a fixed amount of the other. This question appeared later than expected from the expert learning path in both pre- and posttests. Perhaps, the abstract and dynamic nature of this question made it particularly difficult for students. Finally, the question that appeared last in the knowledge structure was Question 6. In fact, Question 6 appeared last in all four critical learning paths in Figure 7.5 (pre- and posttest). This question presents students with a visual representation of a reactant mixture and a product mixture for the reaction $\mathbf{X} + 2\mathbf{Y} \rightarrow \mathbf{XY}_2$ using circles to represent atoms and asks them to identify the balanced chemical equation that best represents that reaction. A significant number of students gave

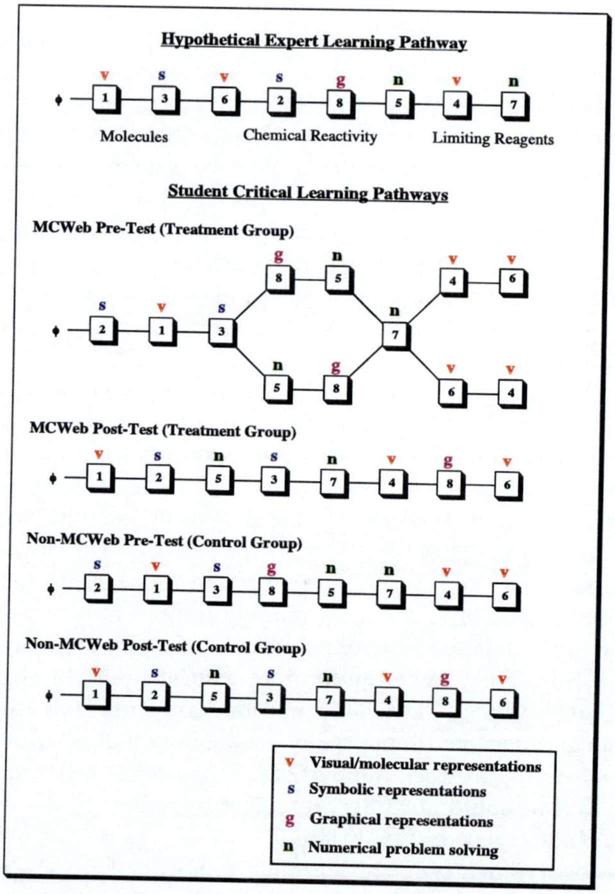


**Figure 7.5.** Comparison of student critical learning paths. The student (novice) critical learning paths represent the most probable sequence of response states in the knowledge structure and shows the order in which correct answers to questions were obtained. The expert learning path represents the hypothetical hierarchical sequencing of questions as determined by the research group; $\emptyset$ is the empty state (no correct responses) and Q is the state where all questions are correct.

$3X + 8Y \rightarrow 3XY_2 + 2Y$ as the answer. We attribute this to an alternate conception that perhaps arises from students' intuitive or fragmented knowledge. For these students the arrow in the balanced chemical equation was nothing more than an equal sign where the number of atoms on each side of the equation had to equal each other, rather than the chemical process expressed in that equation (Yarroch, 1985). Alternate conceptions have been described as conceptual ideas "that are not consistent with the consensus of the scientific community" (Wandersee et al., 1993). It is generally agreed that alternate conceptions are persistently held by students. While many strategies have been reported in the literature to be effective in overtly confronting students' alternate conceptions, the success rates of these have been far from perfect (Williamson and Abraham, 1995).

## 7.5 Conclusion

Our results reveal that the students' factual knowledge base increased after instruction, but their logic structure, as determined by KST analysis, did not develop. These results are similar to those in our studies on the concepts of bonding (Taagepera and Noori, 2000; Taagepera et al., 2002; Vaarik et al., 2008) and stereochemistry (Taagepera et al., 2011). Standard assessments typically measure the factual knowledge base as determined by the percentage of correct answers or, the percentage of students who got most of the questions correct after instruction. The depth of understanding or conceptual knowledge as reflected in the students' knowledge structure emerges when the connections in the factual knowledge are emphasized. Instructors are usually not aware of how limited their students' ability to make the necessary connections are - which in turn limits the effectiveness of instruction. Indeed, factual knowledge can often be stored just in short-term memory and is easily forgotten, if there is no structure for transferring the knowledge to long-term memory. It is the construction of this logic structure by emphasizing the connections between factual knowledge items that requires more attention.

The results strongly suggest the need for teaching approaches that pay more attention to helping students integrate their knowledge by emphasizing the relationships between the different representations (visual, symbolic and numerical) and presenting them concurrently during instruction. We need to explicitly articulate our reasoning on how we (as instructors) see the connections and the underlying meanings associated with the specific features of the representations and provide opportunities to practice through active-learning approaches.

These approaches could include integrating technologies that help students explore their misconceptions, emphasizing relationships, and presenting them concurrently. Williamson and Abraham (1995) found that students understood the particulate nature of matter better when using computer simulations than when using still diagrams such as pictures or transparencies. Our study revealed that using the MCWeb software in large-scale instruction provided an overall benefit to introductory chemistry students. The analysis of the KST pre- and posttests showed that both groups made significant improvements in their understanding of stoichiometry and limiting reagents, but that the treatment group showed more improvements than the control group. In both cases, the critical learning path analysis showed that the overall thinking patterns of the students went mostly from symbolic representations to numerical problem solving to visualization. This was the case for both the pretests and posttests. Thus, visual or conceptual reasoning at a molecular level came last in the students' knowledge structures on the more cognitively demanding questions even after completing homework that emphasized the relationships between the various representations. Many students found this type of reasoning to be difficult and our studies reveal that MCWeb was successful in teaching these methods to introductory chemistry students. However, a single quarter of instruction was probably not adequate to change their overall thinking patterns. Previous studies have shown that a change in students' thinking patterns can be effected by explicitly making the instructor's (expert) logic pattern much more transparent to the students than is commonly done by constantly repeating what seems obvious to the expert (Taagepera and Noori, 2000; Taagepera et al., 2002). There is also a need for textbooks to include more practice on molecular visualization and on emphasizing the relationships between the different representations (visual, symbolic and numerical).

The results indicate that KST is a useful tool for revealing various aspects of students' cognitive structure in chemistry and should be used more frequently, especially in large classes, where it is often not possible to probe for deep understanding by extensive discussions or tests requiring explanations of various phenomena. It can be utilized not only as an assessment tool but also as a pedagogical tool to address a number of student learning issues. As an assessment tool it can be used to track or monitor the development of students' conceptual understanding, to gauge how well students have integrated the concepts of the course, and to measure knowledge changes during the learning process. It can be used to assess the nature of the knowledge students bring to a course, to assess how well students integrate the concepts of the course after instruction, and to assess whether students have conceptualized the content as intended after the conclusion of the course. If students are not developing a logical framework of understanding, as intended, attention can be directed to remediating particular problem areas. As a pedagogical tool it can be used to examine conceptual change achieved through alternative instructional or study techniques such as computer software programs. It could also be considered by textbook authors to ensure that the necessary connections are made in presenting factual information.

**Acknowledgement.** We would like to thank Jean-Claude Falmagne for his unwavering willingness to guide us; Frank Potter, who did the KST calculations, for his enthusiasm and support during the project. We appreciate the technical support and helpful comments of Professors Patrick Wegner, Barbara Gonzalez and Matthew Wilken. We are indebted to the students in our research group who corrected and analyzed the data and shared their insights with us: Mohammad Anwar, Nicole Batard, Ai Bui, David Ford, Aaron Kearney, Jenivi Marucut, Susanne Spano and Jason Wu. We thank the U.S. Department of Education's Fund for the Improvement of Postsecondary Education (FIPSE Grant P116B001020) and the UC Irvine Division of Undergraduate Education for support.

# Part II

# RECENT THEORETICAL PROGRESS

# 8

# Learning Spaces: A Mathematical Compendium

Jean-Paul Doignon[1]
Jean-Claude Falmagne[2]
Eric Cosyn[3]

The core of an educational software based on learning space theory, such as the ALEKS system, is a combinatorial structure representing the cognitive organization of a particular curriculum, like beginning algebra or 4th grade arithmetic[4]. This structure consists in a family $\mathcal{K}$ of subsets of a basic set $Q$. The elements of $Q$ are the *types* of problems to be mastered by a student learning the curriculum. An example of a problem type in beginning algebra is:

[P] *Express the roots of the equation* $\alpha x^2 + \beta x + \gamma = 0$ *in terms of* $\alpha$, $\beta$ *and* $\gamma$.

When this type of problem is proposed to a student, either in an assessment or in the course of learning, ALEKS chooses an *instance* of [P], which may be for example

[I] *What are the roots of the equation* $4x^2 + 6x - 7 = 0$ ?

Typically, there are thousands of instances for a particular problem type. These instances are features of the implementation and do not play any role in the theory summarized in this chapter. The set $Q$ is called the 'domain' of the structure, and the elements of $Q$, the problem types, are referred to as 'items[5]' or 'questions.'

The family $\mathcal{K}$ is the 'learning space.' Its elements are called '(knowledge) states.' Every knowledge state is a set $K$ of items that a student in that state has mastered. To wit, a student in state $K$ cannot solve any instance of an item in $Q \setminus K$ (the instances are constructed so that lucky guesses are

[1]Dept. of Mathematics, University of Brussels.
[2]Dept. of Cognitive Sciences, University of California, Irvine, and ALEKS Corp.
[3]ALEKS Corp.

[4]Or geometry, precalculus, basic chemistry, etc. Parts of the text are excerpts of "Learning Spaces", a monograph by Falmagne and Doignon (2011).

[5]Note that 'item' is used here in a sense different from that used in standardized testing, where 'item' means what we call an 'instance.'

J.-C. Falmagne et al. (eds.), *Knowledge Spaces: Applications in Education*,
DOI 10.1007/978-3-642-35329-1_8, © Springer-Verlag Berlin Heidelberg 2013

impossible or very rare), and moreover the student can solve any instance of any item in $K$, barring careless errors. In principle, the family $\mathcal{K}$ contains all such feasible knowledge states in the population of students considered. We assume that both $Q$ and $\varnothing$ are in $\mathcal{K}$: the student may have mastered all the items in $Q$, or none of them; accordingly, $Q = \cup\mathcal{K}$.

By design, there are no educational gaps in the set $Q$ or in the learning space $\mathcal{K}$. This means, for example, that a student capable of solving all the items in the domain $Q$ of beginning algebra can be regarded as having mastered this curriculum as it is specified in the US schools. It also means that there is no gap in the learning sequence: whatever the student's knowledge state in beginning algebra at any moment, he or she can in principle learn the rest of the curriculum by gradually mastering the remaining items one by one.

Two kinds of arguments support the last statement. For one, the axioms [L1] and [L2] constraining the family $\mathcal{K}$ of knowledge states, which are specified in the next section, are consistent with the idea that, from a theoretical standpoint, gradual learning is feasible in all cases. The second argument is empirical: extensive data on student learning, based on millions of assessments, indicate that when a student is deemed by `ALEKS` ready to learn an item, then the estimated probability of successful mastery of that item is extremely high. Note that the "ready to learn" in the above sentence is given a mathematical meaning as the 'outer fringe' of a student's state (see Definition 8.3.1).

The number of items in a typical domain of school mathematics satisfying the educational standards of a U.S. state is around 650. The number of knowledge states in the learning space for such a domain is quite large, maybe on the order of $10^8$. Despite this large number of states, it is nevertheless possible to assess the knowledge state of a student, accurately, in the span of 25-35 questions.

The next few sections give a concise set-theoretical presentation of the concepts and results. No proofs of results are given here[6].

## 8.1 Axioms for Learning Spaces

**8.1.1 Definition.** A *partial knowledge structure* is a pair $(Q, \mathcal{K})$ in which $Q$ is a nonempty set, and $\mathcal{K}$ is a family of subsets of $Q$ containing at least $Q$. The set $Q$ is called the *domain* of the knowledge structure. The subsets in the family $\mathcal{K}$ are labeled *(knowledge) states*. The elements of $Q$ are called *items* or *questions*. A partial knowledge structure is a *knowledge structure* if $\mathcal{K}$ also

[6] For a comprehensive presentation of the theory, see Falmagne and Doignon (2011). An evolutive and searchable database of references on this topic is maintained by Cord Hockemeyer at the University of Graz, Austria: http://liinwww.ira.uka.de/bibliography/Ai/knowledge.spaces.html.

contains the empty set. The knowledge structure $(Q, \mathcal{K})$ is *finite* if $Q$ is a finite set. Note that since $Q = \cup\mathcal{K}$ we can without ambiguity refer to the family $\mathcal{K}$ itself as a knowledge structure.

The knowledge structure $(Q, \mathcal{K})$ is *discriminative* if for all items $q, p \in Q$ we have

$$(\forall K \in \mathcal{K})(q \in K \iff p \in K) \quad \Longrightarrow \quad q = p.$$

**Axioms.** A knowledge structure $(Q, \mathcal{K})$ is called a *learning space* if it satisfies the two following conditions.

[L1] Learning smoothness. If $K \subset L$ are two states, there is a finite chain of states

$$K_0 = K \subset K_1 \subset \cdots \subset K_n = L \tag{8.1}$$

with $K_i = K_{i-1} + \{q_i\}$ and $q_i \in Q$ for $1 \leq i \leq n$. We have thus $|L \setminus K| = n$.

[L2] Learning consistency. If $K \subset L$ are two states, with $q \notin K$ and $K + \{q\} \in \mathcal{K}$, then $L \cup \{q\} \in \mathcal{K}$.

Axiom [L1] implements the assumption that gradual, item by item, learning is always possible. Note that, by Axiom [L1], all learning spaces are finite. Axiom [L2] formalizes the idea that if some item $q$ is learnable by a student in some state $K$ which is included in some state $L$, then either $q$ is in $L$ or it is learnable by a student in state $L$.

An example of a learning space $\mathcal{H}$ on the domain $\{a, b, c, d\}$ is given by the equation

$$\mathcal{H} = \{\varnothing, \{a\}, \{c\}, \{a, b\}, \{a, c\}, \{c, d\}, \{a, b, c\}, \{a, b, d\}, \{a, c, d\}, \{a, b, c, d\}\}, \tag{8.2}$$

which is represented in Figure 8.1 by the Hasse diagram of its inclusion relation.

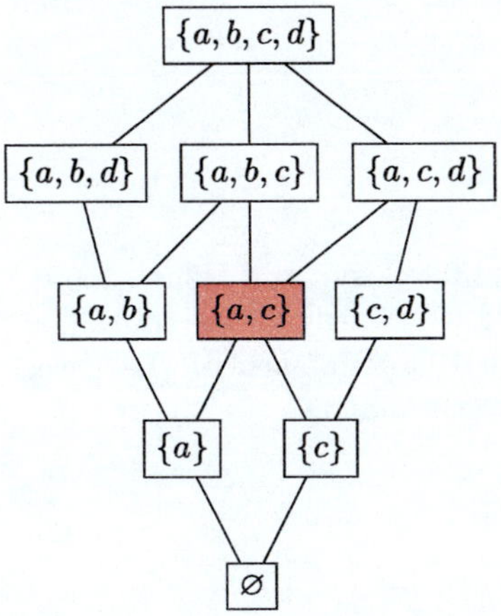


**Figure 8.1.** Hasse diagram of the inclusion relation of the learning space $\mathcal{H}$ defined by Eq. (8.2) on the domain $\{a, b, c, d\}$. Ignore the red shading for the moment (see Definition 8.3.1).

Note that the family $\mathcal{H}$ is closed under union. This is one of several important properties implied by Axioms [L1] and [L2]. We introduce them in the next definition.

**8.1.2 Definition.** Let $(Q, \mathcal{K})$ be a knowledge structure. When $\mathcal{K}$ is *closed under union*, that is, when $\cup\mathcal{A} \in \mathcal{K}$ whenever $\mathcal{A} \subseteq \mathcal{K}$, we say that $(Q, \mathcal{K})$ is a *knowledge space*, or equivalently, that $\mathcal{K}$ is a *knowledge space (on $Q$)*. Note that when a family is closed under union, we sometimes say for short that it is $\cup$-*closed*. The *dual* of a knowledge structure $\mathcal{K}$ on $Q$ is the knowledge structure $\overline{\mathcal{K}}$ containing all the complements of the states of $\mathcal{K}$, that is,

$$\overline{\mathcal{K}} = \{K \in 2^Q \mid Q \setminus K \in \mathcal{K}\}.$$

Thus, $\mathcal{K}$ and $\overline{\mathcal{K}}$ have the same domain.

We denote by $K \triangle L = (K \setminus L) \cup (L \setminus K)$ the *symmetric difference* between two sets $K$ and $L$, and by $d(K, L) = |(K \setminus L) \cup (L \setminus K)|$ the *symmetric difference distance* between those sets.

A family of sets $\mathcal{K}$ is *well-graded* if, for every two distinct sets $K$, $L$ in $\mathcal{K}$, there exists a finite sequence $K_0 = K, K_1, \ldots, K_n = L$ of sets in $\mathcal{K}$ such that $d(K_{i-1}, K_i) = 1$ for $1 \leq i \leq n$ and $n = d(K, L)$. We call the sequence of sets $(K_i)$ a *tight path* from $K$ to $L$. It is clear that a well-graded knowledge structure is discriminative. It is also necessarily finite since we can take $K = \varnothing$ and $L = Q$.

A family $\mathcal{K}$ of subsets of a finite set $Q = \cup\mathcal{K}$ is an *antimatroid*[7] if it is closed under union and moreover satisfies the following axiom.

[MA] If $K$ is a nonempty subset of the family $\mathcal{K}$, then there is some $q \in K$ such that $K \setminus \{q\} \in K$.

We may also say then that the pair $(Q, \mathcal{K})$ is an antimatroid. An antimatroid $(Q, \mathcal{K})$ is *finite* if $Q$ is finite. In such a case, $(Q, \mathcal{K})$ is a discriminative knowledge structure.

The next theorem specifies the relationship between these various concepts.

**8.1.3 Theorem.** *For every knowledge structure $(Q, \mathcal{K})$, the following three conditions are equivalent.*

(i) *$(Q, \mathcal{K})$ is a learning space.*
(ii) *$(Q, \mathcal{K})$ is an antimatroid.*
(iii) *$(Q, \mathcal{K})$ is a well-graded knowledge space.*

The equivalence of (i) and (iii) was established by Cosyn and Uzun (2009). The proof that (ii) is equivalent to (i) is straightforward and is contained in Falmagne and Doignon (2011). It is clear that, under each of the three conditions, the knowledge structure $(Q, \mathcal{K})$ is discriminative.

[7] Cf. Welsh (1995).

The large number of states in empirical learning spaces may create practical problems of manipulation and storage in a computer's memory. The fact that every learning space is closed under union established by Theorem 8.1.3 plays an important role in enabling an economical representation of any learning space in the form of its 'base.'

## 8.2 The Base and the Atoms

**8.2.1 Definition.** The *span* of a family of sets $\mathcal{G}$ is the family $\mathcal{G}'$ containing all sets that are the union of some subfamily of $\mathcal{G}$. In such a case, we say that $\mathcal{G}$ *spans* $\mathcal{G}'$. By definition, $\mathcal{G}'$ is then a $\cup$-closed family. A *base* of a $\cup$-closed family $\mathcal{K}$ is a minimal subfamily $\mathcal{B}$ of $\mathcal{K}$ spanning $\mathcal{K}$ (where 'minimal' is meant with respect to set inclusion: if $\mathcal{J}$ spans $\mathcal{K}$ for some $\mathcal{J} \subseteq \mathcal{B}$, then $\mathcal{J} = \mathcal{B}$). By a standard convention, the empty set is the union of the empty subfamily of $\mathcal{B}$. Thus, the empty set never belongs to a base. It is also clear that an element $K$ of some base $\mathcal{B}$ of $\mathcal{K}$ cannot be the union of other elements of $\mathcal{B}$.

**8.2.2 Theorem.** *Let $\mathcal{B}$ be a base for a knowledge space $(Q, \mathcal{K})$. Then $\mathcal{B} \subseteq \mathcal{F}$ for every subfamily $\mathcal{F}$ of states spanning $\mathcal{K}$. Consequently, a knowledge space admits at most one base. Every finite knowledge space has a base.*

Some knowledge spaces have no base, as for instance, the collection of all the open subsets of the real line.

The base of the learning space $\mathcal{H}$ of Eq. (8.2) and Figure 8.1 is the subcollection

$$\{\{a\}, \{c\}, \{a, b\}, \{c, d\}, \{a, b, d\}\}.$$

In the cases of learning spaces encountered in education, the cardinality of the base of a learning space $\mathcal{L}$ is typically much smaller than the cardinality of $\mathcal{L}$. The example of the family $2^A$ for any finite set $A$, in which the base is the collection $\{\{x\} \mid x \in A\}$ of all the singleton subsets of $A$, is suggestive in that regard.

Several efficient algorithms are available for the construction of the base of a knowledge space and for generating a knowledge space from its base (see in particular Dowling, 1993b; Falmagne and Doignon, 2011, Section 3.5, pages 49-50).

The states of the base have an important property.

**8.2.3 Definition.** Let $\mathcal{F}$ be a nonempty family of sets. For any $q \in \cup\mathcal{F}$, an *atom at* $q$ is a minimal set in $\mathcal{F}$ containing $q$, where 'minimal' refers to the inclusion relation. A set $X$ in $\mathcal{F}$ is an *atom* if it is an atom at $q$ for some $q \in \cup\mathcal{F}$.

**8.2.4 Theorem.** *Suppose that a knowledge space has a base. Then this base is formed by the collection of all the atoms.*

This property will play an essential role in the construction of an assessment algorithm for very large learning spaces. It will allow us to manufacture a state from any set of items by forming the union of some atoms of these items (see Step 9 in 8.8.1).

## 8.3 The Fringe Theorem

In the case of standardized tests the result of an assessment is a number regarded as measuring some aptitude. By contrast, the outcome of an assessment by a system such as ALEKS is a knowledge state, which may contain hundreds of items[8]. Displaying such an outcome by a possibly very long list of these items is awkward and not particularly useful. Fortunately, a considerably more concise representation of a knowledge state is available, which is meaningful to a student or a teacher. It relies on the twin concepts of the 'inner fringe' and the 'outer fringe' of a knowledge state.

**8.3.1 Definition.** The *inner fringe* of a state $K$ in a knowledge structure $(Q, \mathcal{K})$ is the subset of items

$$K^{\mathcal{I}} = \{q \in K \,|\, K \setminus \{q\} \in \mathcal{K}\}.$$

The *outer fringe* of a state $K$ is the subset

$$K^{\mathcal{O}} = \{q \in Q \setminus K \,|\, K \cup \{q\} \in \mathcal{K}\}.$$

For example, the inner fringe and the outer fringe of the state $\{a, c\}$ in the learning space $\mathcal{H}$, which is shaded red in Figure 8.1, are $\{a, c\}^{\mathcal{I}} = \{a, c\}$ and $\{a, c\}^{\mathcal{O}} = \{b, d\}$. The *fringe* of $K$ is the union of the inner fringe and the outer fringe. We write

$$K^{\mathcal{F}} = K^{\mathcal{I}} \cup K^{\mathcal{O}}.$$

Let $\mathcal{N}(K, n)$ be the set of all states whose distance from $K$ is at most $n$, thus:

$$\mathcal{N}(K, n) = \{L \in \mathcal{K} \,|\, d(K, L) \leq n\}. \tag{8.3}$$

We have then $K^{\mathcal{F}} = \big(\cup\mathcal{N}(K, 1)\big) \setminus \big(\cap\mathcal{N}(K, 1)\big)$. We refer to $\mathcal{N}(K, n)$ as the *n-neighborhood* of the state $K$. The importance of these concepts lies in the following result.

**8.3.2 Theorem.** *In a learning space, every state is defined by its two fringes; that is, there is only one state having these fringes.*

In fact, a stronger result holds: a finite knowledge structure is well-graded if and only if every state is defined by its two fringes (Falmagne and Doignon, 2011, Theorem 4.1.7).

---

[8] C.f. Chapter 1.

The fringes of the states play a major role in the ALEKS system. The fringes can be displayed at the end of an assessment to specify the knowledge state exactly. This is marked progress over the numerical score provided by a standardized test. Indeed, the importance of such a representation lies in the interpretation of the fringes. The inner fringe may be taken as containing the items representing the 'high points' of the student's competence in the topic. The outer fringe is even more important because its items may be regarded as those that the student is ready to learn. That feature plays an essential role in the 'learning module' of the ALEKS system. When an assessment is run as a prelude to learning and returns for a student a knowledge state $K$, the computer screen displays a window listing all the items in the outer fringe of $K$. The student may then choose one item in the list and begin to study it. A large set of learning data from the ALEKS system shows that the probability that a student successfully masters an item selected in the outer fringe of his or her state is very high. In beginning algebra, the estimated median probability, based on a very large sample of students, is .92.

## 8.4 Projections of a Knowledge Structure

The concept of projection for learning spaces is closely related to the concept bearing the same name for media introduced by Cavagnaro (2008) (cf. also Theorem 2.11.6 in Eppstein et al., 2007; Falmagne, 2008). This concept has been encountered in Chapter 1 where an example has been given (see page 19). A more precise discussion requires some construction.

Let $(Q, \mathcal{K})$ be a partial knowledge structure—thus, $\varnothing$ is not necessarily in $\mathcal{K}$—and let $Q'$ be any proper subset of $Q$. Define a relation $\sim$ on $\mathcal{K}$ by

$$\begin{aligned} K \sim L \quad &\iff \quad K \cap Q' = L \cap Q' \\ &\iff \quad K \mathbin{\triangle} L \subseteq Q \setminus Q'. \end{aligned}$$

Thus, $\sim$ is an equivalence relation on $\mathcal{K}$. The equivalence between the two right hand sides is easily checked.

We denote by $[K]$ the equivalence class of $\sim$ containing $K$, and by $\mathcal{K}_\sim = \{[K] \,|\, K \in \mathcal{K}\}$ the partition of $\mathcal{K}$ induced by $\sim$.

Let $(Q, \mathcal{K})$ be a knowledge structure and take any nonempty $Q' \subset Q$. We say that the family

$$\mathcal{K}_{|Q'} = \{W \subseteq Q \,|\, W = K \cap Q' \text{ for some } K \in \mathcal{K}\}$$

is the *projection* of $\mathcal{K}$ on $Q'$. We have thus $\mathcal{K}_{|Q'} \subseteq 2^{Q'}$. Note that the sets in $\mathcal{K}_{|Q'}$ may not be states of $\mathcal{K}$. For example the state $\{c, g, j\}$ of the projection pictured in Graph B of Figure 1.5 on page 19 is not a state of the original learning space (see Graph B on that figure). For every $K$ with $\varnothing \not\sim K \in \mathcal{K}$ and with $[K]$ as above, define

$$\mathcal{K}_{[K]} = \{M \mid M = \varnothing \text{ or } M = L \setminus \cap[K] \text{ for some } L \sim K\}.$$

So, if $\varnothing \in \mathcal{K}$, we have $\mathcal{K}_{[\varnothing]} = [\varnothing]$. The families $\mathcal{K}_{[K]}$ are called the *$Q'$-children* of $\mathcal{K}$, or simply the *children* of $\mathcal{K}$ when the set $Q'$ is obvious from the context. We refer to $\mathcal{K}$ as the *parent* structure. Notice that we may have $\mathcal{K}_{[K]} = \mathcal{K}_{[L]}$ even when $K \not\sim L$.

Here is the key result.

**8.4.1 Theorem.** *Let $\mathcal{K}$ be a learning space (resp. a well-graded $\cup$-closed family) on a domain $Q$ with $|Q| \geq 2$. The following two properties hold for every proper nonempty subset $Q'$ of $Q$:*

(i) *the projection $\mathcal{K}_{|Q'}$ of $\mathcal{K}$ on $Q'$ is a learning space (resp. a well-graded $\cup$-closed family);*
(ii) *in either case, the children of $\mathcal{K}$ are well-graded and $\cup$-closed families.*

Note that the singleton $\{\varnothing\}$ is vacuously a partial knowledge structure which is, also vacuously, well–graded and $\cup$-closed.

For a proof, see Falmagne (2008) or Falmagne and Doignon (2011, Theorem 2.4.8).

## 8.5 Building a Learning Space

At this time, the construction of a learning space is still a demanding enterprise extending over several months. It is based partly on the expertise of competent teachers of the topic. Their input provides a first draft of the learning space. Ideally, if the teachers were omniscient, we could ask them questions such as:

[Q] *Suppose that a student has failed to solve items $p_1, \ldots, p_n$. Do you believe this student would also fail to solve item $q$? You may assume that chance factors, such as lucky guesses and careless errors, do not interfere in the student's performance.*

Such a query is summarized by the nonempty set $\{p_1, p_2, \ldots, p_n\}$ of items, plus the single item $q$. Thus, all positive answers to the queries form a relation $\mathcal{P}$ from $2^Q \setminus \{\varnothing\}$ to $Q$. The expert is consistent with the (unknown) knowledge space $(Q, \mathcal{K})$ exactly when the following equivalence is satisfied for all $A \in 2^Q \setminus \{\varnothing\}$ and $q \in Q$:

$$A \mathcal{P} q \iff (\forall K \in \mathcal{K} : A \cap K = \varnothing \Rightarrow q \notin K). \tag{8.4}$$

The relation $\mathcal{P}$ could thus be used to remove from the family of potential states all the sets $K$ falsifying the implication in the r.h.s. of (8.4). In practice, however, it is only when $|A| = 1$ that a human expert can provide reliable responses to such queries. In this case, $\mathcal{P}$ is essentially a binary relation which, if (8.4) is satisfied, defines a quasi order on $Q$ (reflexive, transitive). The resulting learning space is then closed under both union and intersection[9]. It contains all the 'true' states, but also possibly many fictitious states which must be eliminated by further analysis by other means. This initial space is nevertheless acceptable and can be used with students since it contains all the 'true' states. A statistical analysis of student data allows then to refine the initial learning space and remove the fictitious states (for details, see Chapter 15 and 16 in Falmagne and Doignon, 2011).

The two theorems below play a key role.

**8.5.1 Theorem.** *Let $(Q, \mathcal{K})$ be a knowledge structure, and suppose that $\mathcal{P}$ is the relation from $2^Q$ to $Q$ defined by Equation (8.4). Then, necessarily:*

(i) *$\mathcal{P}$ extends the reverse membership relation, that is:*

$$\text{if } p \in A \subseteq Q, \text{ then } A\mathcal{P}p;$$

(ii) *for all $A$, $B \in 2^Q \setminus \{\varnothing\}$ and $p \in Q$:*

$$\text{if } A\mathcal{P}b \text{ and } B\mathcal{P}p \text{ for all } b \in B, \text{ then } A\mathcal{P}p.$$

**8.5.2 Definition.** An *entailment for* the nonempty domain $Q$ (which may be infinite) is a relation $\mathcal{P}$ from $2^Q \setminus \{\varnothing\}$ to $Q$ that satisfies Conditions (i) and (ii) in Theorem 8.5.1.

**8.5.3 Theorem.** *There is a one-to-one correspondence between the family of all knowledge spaces $\mathcal{K}$ on the same domain $Q$, and the family of all entailments $\mathcal{P}$ for $Q$. This correspondence is defined by the two equivalences*

$$A\,\mathcal{P}\,q \iff (\forall K \in \mathcal{K} : A \cap K = \varnothing \Rightarrow q \notin K), \tag{8.5}$$

$$K \in \mathcal{K} \iff \big(\forall (A, p) \in \mathcal{P} : A \cap K = \varnothing \Rightarrow p \notin K\big). \tag{8.6}$$

For the proof see Koppen and Doignon (1990) and Falmagne and Doignon (2011, Theorem 7.1.5).

A computer algorithm called QUERY has been designed to perform the actual construction of the space on the basis of an entailment (Koppen, 1993, 1994; Falmagne and Doignon, 2011, Chapters 15 and 16). However, as made clear by Theorem 8.5.3, the resulting structure is a knowledge space but not necessarily a learning space. The QUERY algorithm has been amended so as to allow the elimination of a set only when such an elimination does not invalidate Axiom [MA] of an antimatroid (cf. page 134 and Theorem 8.1.3). For details about such a construction, see Falmagne and Doignon (2011, Chapter 15 and 16).

A different possibility is to use the QUERY algorithm as such, which delivers a knowledge space, and then judiciously add some of the critically

[9] This results from a classical result from Birkhoff (1937).

missing states so as to obtain a learning space. Part of Chapter 14 is devoted to this technique.

## 8.6 Probabilistic Extension

The concept of a learning space is a deterministic one. As such, it cannot provide realistic predictions of students' responses to the problems of a test. There are two ways in which probabilities must enter into a realistic model. For one, the knowledge states will certainly occur with different frequencies in the population of reference. It is thus reasonable to postulate the existence of a probability distribution on the collection of states. For another, a student's knowledge state does not necessarily specify the observed responses. A student having mastered an item may be careless in responding, and make an error. Also, a student may guess the correct response to an item which is not in her state. This may happen, for example, when a multiple choice paradigm is used.

Accordingly, it makes sense to introduce conditional probabilities of responses, given the states. A number of simple probabilistic models are described in Chapter 13 of Falmagne and Doignon (2011). They illustrate how probabilistic concepts can be introduced within knowledge space theory. One of these models provides the context for the assessment algorithm outlined in the next section.

## 8.7 The Assessment Algorithm

An informal description of an assessment algorithm has been given in Section 1.3 on page 13. The general scheme sketched there is consistent with several formal interpretations. One of them is especially important and is currently used in many schools and colleges[10]. We give below its basic components and axioms[11].

**8.7.1 Concepts and notation.** Given a learning space $\mathcal{K}$ on a domain $Q$, any assessment is a realization of a Markovian stochastic process. Our notation is as follows. We use r.v. as an abbreviation for 'random variable.'

[10] As part of the ALEKS system.

[11] For details, see Falmagne and Doignon (2011, Chapter 13).

| | |
|---|---|
| $n$ | the step number, or *trial* number, $n = 1, 2, \dots$ ; |
| $\mathcal{K}_q$ | the subfamily of all the states containing $q$ ; |
| $\Lambda_+$ | the set of all positive probability distributions on $\mathcal{K}$ ; |
| $\mathbf{L}_n$ | a random probability distribution on $\mathcal{K}$; we have $\mathbf{L}_n = L_n \in \Lambda_+$ (so $L_n > 0$) if $L_n$ is the probability distribution on $\mathcal{K}$ at the beginning of trial $n$ ; |
| $\mathbf{L}_n(K)$ | a r.v. measuring the probability of state $K$ on trial $n$; |
| $\mathbf{Q}_n$ | a r.v. representing the question asked on trial $n$; we have $\mathbf{Q}_n = q \in Q$ if $q$ is the question asked on trial $n$ ; |
| $\mathbf{R}_n$ | a r.v. coding the response on trial $n$: $\mathbf{R}_n = \begin{cases} 1 & \text{if the response is correct} \\ 0 & \text{otherwise.} \end{cases}$ |
| $\mathbf{W}_n$ | the random history of the process from trial 1 to trial $n$ ; |
| $\iota_A$ | the indicator function of a set $A$ : $\iota_A(q) = \begin{cases} 1 \text{ if } q \in A \\ 0 \text{ if } q \notin A \end{cases}$ ; |
| $\zeta_{q,r}$ | with $1 < \zeta_{q,r}$ for $q \in Q, r = 0, 1$, a family of parameters specifying the updating operator (see Axiom [U] below). |

The process begins, on trial 1, by setting $\mathbf{L}_1 = L$, for some particular $L \in \Lambda_+$. So, the initial probability distribution is the same for all realizations. Any further trial $n > 1$ begins with a value $L_n \in \Lambda_+$ of the random distribution $\mathbf{L}_n$ updated as a function of the event on trial $n-1$. We write for any $\mathcal{F} \subseteq \mathcal{K}$,

$$L_n(\mathcal{F}) = \sum_{K \in \mathcal{F}} L_n(K). \tag{8.7}$$

Three general axioms specify the stochastic process $(\mathbf{L}_n, \mathbf{Q}_n, \mathbf{R}_n)$.

The version of these axioms given below is an important special case.

**8.7.2 Axioms.**

[U] **Updating Rule.** We have $\mathbb{P}(\mathbf{L}_1 = L) = 1$, and for any positive integer $n$, with $\mathbf{L}_n = L_n$, $\mathbf{Q}_n = q$, $\mathbf{R}_n = r$, and

$$\zeta_{q,r}^{K} = \begin{cases} 1 & \text{if } \iota_K(q) \neq r, \\ \zeta_{q,r} & \text{if } \iota_K(q) = r, \end{cases} \tag{8.8}$$

we have

$$L_{n+1}(K) = \frac{\zeta_{q,r}^{K} L_n(K)}{\sum_{K' \in \mathcal{K}} \zeta_{q,r}^{K'} L_n(K')}. \tag{8.9}$$

This updating rule is called *multiplicative with parameters* $\zeta_{q,r}$.

[Q] **Questioning Rule.** For all $q \in Q$ and all integers $n > 0$,

$$\mathbb{P}(\mathbf{Q}_n = q \,|\, \mathbf{L}_n, \mathbf{W}_{n-1}) = \frac{\iota_{S(L_n)}(q)}{|S(L_n)|} \tag{8.10}$$

where $S(L_n)$ is the subset of $Q$ containing all those items $q$ minimizing

$$|2L_n(\mathcal{K}_q) - 1|.$$

Under this questioning rule, which is called *half-split*, we must have $\mathbf{Q}_n \in S(L_n)$ with a probability equal to one. The questions in the set $S(L_n)$ are then chosen with equal probability.

[R] **Response Rule.** For all positive integers $n$,

$$\mathbb{P}(\mathbf{R}_n = \iota_{K_0}(q) \,|\, \mathbf{Q}_n = q, \mathbf{L}_n, \mathbf{W}_{n-1}) = 1$$

where $K_0$ is the *latent state* representing the set of all the items currently mastered by the student, that is, the state that must be uncovered by the Markovian procedure.

So, if the item selected by the process belongs to the latent state $K_0$, the probability of a correct response is equal to 1. In the more realistic versions of this axiom used in practice, one additional parameter is used which specifies the probability of a *careless error*. It may also be necessary to introduce a *lucky guess* parameter, for example if a multiple choice paradigm is used.

**8.7.3 Some Key Results.** These results follow from Axioms [U], [Q] and [R]. (For proofs, see Chapter 13 in Falmagne and Doignon, 2011).

1. The updating operator specified by Equation (8.9) in Axiom [U] is essentially a Bayesian operator. This can be shown by an appropriate transformation of the equation (see 13.4.5 on page 251 in Falmagne and Doignon, 2011).

2. This updating operator is *permutable*. This term is used in the functional equations literature (Aczél, 1966) to designate a function $F$ satisfying the equation
$$F(F(x, y), z) = F(F(x, z), y).$$
This property is essential because it means that the order of the questions has no import on the result of the assessment.
3. The stochastic process $(\mathbf{L}_n)$ is Markovian.
4. The stochastic process $(\mathbf{L}_n)$ converges to the latent state $K_0$ in the sense that
$$\mathbf{L}_n(K_0) \xrightarrow{\text{a.s.}} 1$$
(in which 'a.s.' means 'almost surely').

**Remark.** Another Markovian assessment procedure was also developed, which is based on a different principle (Falmagne and Doignon, 1988b and Chapter 14 in Falmagne and Doignon, 2011). The stochastic process is a finite Markov chain, the states of which are subsets of states of the learning space.

## 8.8 About Practical Implementations

The three axioms [U], [Q] and [R] are the foundation pieces of the assessment mechanism used by the ALEKS system. As mentioned in Chapter 1, however, a direct implementation of these axioms as an assessment software is not possible for two reasons.

One is that students commit careless errors. This means that Axiom [R] has to be modified by the introduction of a 'careless error parameter.' An obvious possibility is the axiom:

[R'] **Modified Response Rule.** For all positive integers $n$,

$$\mathbb{P}(\mathbf{R}_n = \iota_{K_0}(q) \,|\, \mathbf{Q}_n = q, \mathbf{L}_n, \mathbf{W}_{n-1}) = \begin{cases} 1 - \beta_q & \text{if } q \in K_0 \\ 0 & \text{if } q \notin K_0\,, \end{cases}$$

in which $\beta_q$ is the probability of making an error in responding to the item $q$ which is in the latent state $K_0$.

In some cases, 'lucky guess' parameters must also be used, for example to deal with the case of the multiple choice paradigm.

The second difficulty is that learning spaces formalizing actual curricula are always very large, with domains typically counting several hundred items. Such learning spaces may have many million states and cannot be searched in a straightforward manner. The adopted solution is to partition the domain $Q$ of the basic learning space $\mathcal{K}$—the *parent learning space*—into some number $N$ of subdomains $Q_1, \ldots, Q_N$. These subdomains are similar in that they contain approximately the same number of items and are representative of the domain, for example in such a way that each would be suitable as a placement test. Via Theorem 8.4.1, these $N$ subdomains determine $N$ projection learning spaces $\mathcal{K}^1, \ldots, \mathcal{K}^N$ of the parent learning space $\mathcal{K}$.

The general idea of the algorithm is to assess the student simultaneously on all the $N$ projection learning spaces, with mutual updating of the probability distributions, according to the scheme outlined below.

**8.8.1 The updating steps on trial $n$.** We write $\mathcal{K}^j_q$ for the subfamily of $\mathcal{K}^j$ containing $q$, with $1 \leq j \leq N$; so, $q \in Q_j$. For $K \in \mathcal{K}^j$, we denote by $L^j_n(K)$ the probability of the state $K$ in the learning space $\mathcal{K}^j$ on trial $n$.

1. On trial $n$, pick an item $q$ minimizing $|2L^j_n(\mathcal{K}^j_q) - 1|$, for all $1 \leq j \leq N$ and $q \in Q_j$. If more than one item achieves such a minimization, pick randomly between them[12].
2. Suppose that the chosen item $q$ belongs to $Q_j$. Record the student's response to item $q$. Update the probability distribution $L^j$ according to Axiom [U] and Equation (8.9), (with $L_n = L^j_n$ and $L_{n+1} = L^j_{n+1}$).
3. Add item $q$ to all the $N-1$ subdomains $Q_i \neq Q_j$, and write $Q^\star_i = Q_i \cup \{q\}$.
4. Build the $N-1$ projections $\mathcal{K}^{i\star}$ of $\mathcal{K}$ on the subdomains $Q^\star_i$. Note that $\mathcal{K}^{i\star}$ can in turn be projected on $Q_i$. For any state $K$ in $\mathcal{K}^{i\star}$, let $[K]$ denote its equivalence class with respect to the projection on $Q_i$, and let $K_{|Q_i}$ denote its projection on $Q_i$.
   That is, $M \in [K]$ if $M \cap Q_i = K \cap Q_i$, and
$$K_{|Q_i} = \begin{cases} K & \text{if } q \notin K, \\ K \setminus \{q\} & \text{if } q \in K\,. \end{cases}$$
5. For $1 \leq i \leq N$ and $i \neq j$, define the probability distributions $L^{i\star}_n$ on $\mathcal{K}^\star_i$ by the equation:
$$L^{i\star}_n(K) = \frac{L^i_n(K_{|Q_i})}{|[K]|}. \tag{8.11}$$
   So, the rule $L^{i\star}_n(K)$ splits equally the probability $L^i_n(K)$ among the states created by the addition of $q$ to $Q_i$.
6. Now, update all the $N-1$ probabilities $L^{i\star}_n$ according to Axiom [U] and Equation (8.9). For each learning space $\mathcal{K}^\star_i$, $1 \leq i \leq N$ and $i \neq j$, this results into a probability distribution $L^{i\star}_{n+1}$.
7. Remove item $q$ from all the domains $Q_i \cup \{q\}$, $1 \leq i \leq N$ and $i \neq j$, and normalize all the $L^{i\star}_{n+1}$ probability distributions. That means: compute the probability distribution $L^i_{n+1}$ on $\mathcal{K}^i$ from $L^{i\star}_{n+1}$ by the formula
$$L^i_{n+1}(K) = \sum_{M \text{ s.t. } M_{|Q_i}=K} L^{i\star}_{n+1}(M).$$

[12]This rule is consistent with Equation (8.10).

The above scheme can be altered in two major ways. One concerns Step 3, where more than one question can be added to the subdomains. For instance, on trial $n$, one can add the $n$ questions answered so far by the student to each subdomain, and then 'replay' the assessment on such extended subdomains. Another way concerns Step 5, where the rule $L_n^{i\star}(K)$ can split the probability $L_n^i(K)$ non-equally among the states created by the addition of $q$ to $Q_i$.

**8.8.2 The construction of the final state.** The procedure described below for the construction of the final state is only one of several possibilities. Suppose that the assessment ended at trial $n$. For each item $q$, we define its likelihood $\omega(q)$ as

$$\omega(q) = \sum_{K \in \mathcal{K}_q^j} L_n^{i\star}(K), \qquad (q \in Q_j).$$

Let $K_f$ denote the final knowledge state to be assigned to the student. This final state is built recursively. With $1 \leq i \leq |Q|$, let $(q_i)$ denote the sequence of items ordered by decreasing value of $\omega(q_i)$. The procedure starts by setting $K_f = \varnothing$ and proceeds recursively along $(q_i)$. On step $i$, the procedure examines each atom at $q_i$ and add it to $K_f$ if it passes the following simple criteria: for atom $A$ at $q_i$

$$\frac{1}{|A \setminus K_{f,i-1}|} \sum_{q \in A \setminus K_{f,i-1}} w(q) \geq \delta,$$

where $K_{f,i-1}$ is the state of the final knowledge state after item $q_{i-1}$ and $\delta, 0 < \delta < 1$, is a threshold parameter.

The procedure terminates when it reaches the end of the sequence $(q_i)$. Since $K_f$ is at all times either empty or the union of atoms, the procedure returns a knowledge state.

# 9

# Recent Developments in Performance-based Knowledge Space Theory

Ali Ünlü[1]
Martin Schrepp[2]
Jürgen Heller[3]
Cord Hockemeyer[4]
Gudrun Wesiak[4, 5]
Dietrich Albert[4, 5]

Knowledge Space Theory was founded by Doignon and Falmagne (1985). This paper initiated an extensive body of work[6], which is still in progress. In this chapter, we present recent examples regarding developments in the theory and its relationship to other approaches, methods and applications. As in other chapters of this volume (e.g., Ch. 10) it becomes obvious that there is a high potential for further developments of Knowledge Space Theory. Furthermore, for theoretical as well as practical reasons the relationship to other theoretical approaches such as Formal Concept Analysis, Latent Class Analysis, and Item Response Theory has to be taken into account.

## 9.1 Extensions of Knowledge Space Theory (KST)

Since its foundation by Doignon and Falmagne (1985), Knowledge Space Theory (KST) has been extended along several directions for at least two reasons. (a) For theoretical reasons several KST-extensions in cognition (see, e.g., Albert and Held, 1999; Albert and Lukas, 1999; Doignon, 1994b; Düntsch

[1]TUM School of Education, Technische Universität München, Germany
[2]SAP AG, Walldorf, Germany
[3]Department of Psychology, University of Tübingen, Germany
[4]Department of Psychology, University of Graz, Austria
[5]Knowledge Management Institute, Graz University of Technology, Austria
[6]See http://liinwww.ira.uka.de/bibliography/Ai/knowledge.spaces.html.

J.-C. Falmagne et al. (eds.), *Knowledge Spaces: Applications in Education*,
DOI 10.1007/978-3-642-35329-1_9, © Springer-Verlag Berlin Heidelberg 2013

and Gediga, 1995; Heller et al., 2006; Korossy, 1997; Steiner and Albert, 2008) and learning (see, e.g., Doignon and Falmagne, 1999; Falmagne et al., 2006; Falmagne, 2008) have been developed. The main aim was grounding of the knowledge structures. (b) For practical reasons some limitations have been overcome. Thus the cases of using more than two answer alternatives and more than one set of items have been analyzed and are presented in this section.

**9.1.1 Extension to domains with multiple answer alternatives.** A central assumption of the knowledge structure approach is that the answer of a subject to an item or a question is either correct or incorrect. But for many domains this assumption is too restrictive. For example, the quality of a solution is often rated by points or even by the time to find a correct solution.

To describe the quality of a solution in a formal way, let $Q$ be a finite set of items or problems. Furthermore, let us assume that the quality of a solution to $q \in Q$ can be rated by $l \in L$, where $(L, \leq_L)$ is a linearly ordered set containing at least two elements. The performance of a subject concerning $Q$ can be described by a mapping $K : Q \to L$, which we call a *knowledge state* on $(Q, L)$. A non-empty set of knowledge states on $(Q, L)$ is called a *knowledge structure* on $(Q, L)$[7].

Since surmise relations $\sqsubseteq$ and surmise functions $\delta$ describe the relative difficulty of problems their interpretations can be easily adjusted[8]. In the following, the expression $p \sqsubseteq q$ is interpreted as: *Every subject reaches a better (or equal) result in solving* $p$ *than in solving* $q$. The interpretation of $\delta(q)$ is similar: *Every subject reaches in all problems of at least one clause in* $\delta(q)$ *a better (or equal) result than in* $q$.

The following definition determines the dependency between knowledge structures on $(Q, L)$ and surmise relations $\sqsubseteq$ respectively surmise functions $\delta$.

**9.1.2 Definition.** Let $\sqsubseteq$ be a surmise relation on $Q$ and let $\delta$ be a surmise function on $Q$. We define:

$$\mathcal{K}(\sqsubseteq) = \{K : Q \to L \mid \forall p, q \in Q \; (p \sqsubseteq q \Rightarrow K(q) \leq_L K(p))\},$$
$$\mathcal{K}(\delta) = \{K : Q \to L \mid \forall q \in Q \; \exists R \in \delta(q) \; \forall p \in R \; (K(q) \leq_L K(p))\}.$$

We call $\mathcal{K}(\sqsubseteq)$ the *quasi ordinal knowledge space* corresponding to $\sqsubseteq$ and $\mathcal{K}(\delta)$ the *knowledge space* corresponding to $\delta$.

---

[7]Since every $K \subseteq Q$ can be identified with its characteristic function these are proper generalizations of the original concepts.

[8]A surmise function (also called a 'surmise system') is a mapping $\delta : Q \to 2^{2^Q}$, which fulfills the properties $K \in \delta(q) \Rightarrow q \in K$, $K \in \delta(q) \wedge p \in K \Rightarrow \exists K' \in \delta(p)(K' \subseteq K)$ and $K_1, K_2 \in \delta(q) \wedge K_1 \subseteq K_2 \Rightarrow K_1 = K_2$. A $K \in \delta(q)$ is called a clause of $q$. The interpretation of a surmise function $\delta$ is, that if a person is capable of solving a problem $q$ then this person is capable of solving all problems in at least one clause of $q$.

The following result shows how a surmise relation and a surmise function can be derived from a knowledge structure on $(Q, L)$.

**9.1.3 Lemma.** *Let $\mathcal{K}$ be a knowledge structure on $(Q, L)$. Then*

$$p \sqsubseteq q \Leftrightarrow \forall K \in \mathcal{K}\ (K(q) \leq_L K(p))$$

*defines a surmise relation $\sqsubseteq$ on $Q$ and*

$$\begin{aligned} R \in \delta(q) \Leftrightarrow \exists K \in \mathcal{K}\ (R = \{p \mid K(q) \leq_L K(p)\}) \\ \wedge\ \forall K' \in \mathcal{K}\ (\{p \mid K'(q) \leq_L K'(p)\} \not\subset R) \end{aligned}$$

*defines a surmise function on $Q$.*

For a proof of this Lemma see Schrepp (1997, Lemma 1). In the binary case ($L = \{0, 1\}$) knowledge spaces and quasi ordinal knowledge spaces are defined by closure properties. The following definition generalizes these closure properties.

**9.1.4 Definition.** A knowledge structure $\mathcal{K}$ on $(Q, L)$ is called *strictly closed* if for all $F : Q \to L$ the condition

$$\forall p, q \in Q\ \forall K \in \mathcal{K}\ (K(q) \leq_L K(p) \Rightarrow F(q) \leq_L F(p))$$

implies $F \in \mathcal{K}$. $\mathcal{K}$ is called *closed* if for all $F : Q \to L$ the condition

$$\forall q \in Q\ \exists K \in \mathcal{K}\ \forall p \in Q\ (K(q) \leq_L K(p) \Rightarrow F(q) \leq_L F(p))$$

implies $F \in \mathcal{K}$.

**9.1.5 Example.** Let $Q = \{q_1, q_2, q_3\}, L = \{0, 1, 2\}$ and $\leq_L = \leq$. Define:

$$\mathcal{K}_1 = \{(0,0,0), (2,1,0), (2,2,2)\},$$

$$\begin{aligned} \mathcal{K}_2 = \{&(0,0,0), (1,1,1), (2,2,2), (1,0,0), (2,0,0), (2,1,2), (2,1,1), (1,0,1), \\ &(2,0,2), (2,0,1), (1,2,2), (1,1,0), (0,1,1), (2,1,0), (2,2,0), (2,2,1), \\ &(0,2,2)\}, \end{aligned}$$

$$\begin{aligned} \mathcal{K}_3 = \{&(0,0,0), (1,1,1), (2,2,2), (1,0,0), (1,1,0), (1,0,1), (2,0,0), (2,1,0), \\ &(2,2,0), (2,0,1), (2,0,2), (2,1,1), (2,2,1), (2,1,2)\}. \end{aligned}$$

Let $K_1 = (0, 0, 0)$ and $K_2 = (1, 1, 1)$. Then for each $q \in Q$ we have $K_1(q) \leq_L K_1(p) \to K_2(q) \leq_L K_2(p)$ for all $p \in Q$, but $K_2 \notin \mathcal{K}_1$. Therefore, $\mathcal{K}_1$ is not closed. $\mathcal{K}_2$ is closed, but not strictly closed. Choose $p, q \in Q$ with $p \neq q$. From $(2, 1, 0), (1, 0, 1), (0, 1, 1) \in \mathcal{K}_2$ we can conclude $\neg\forall K \in \mathcal{K}_2\ (K(q) \leq_L K(p))$. For $K_3 = (0, 1, 0)$ this implies $K(q) \leq_L K(p) \to K_3(q) \leq_L K_3(p)$ for each $K \in \mathcal{K}_2$, but $K_3 \notin \mathcal{K}_2$. $\mathcal{K}_3$ is strictly closed.

It was shown that these closure properties are proper generalizations of the closure under union respectively closure under union and intersection in the binary case. In addition, there is a one-to-one correspondence between the strictly closed knowledge structures on $(Q, L)$ and the surmise relations on $Q$. There is also a one-to-one correspondence between the closed knowledge structures on $(Q, L)$ and the surmise functions on $Q$ (see Schrepp, 1997, Theorems 3 and 4).

**9.1.6 Example.** Let $Q, L, \leq_L, \mathcal{K}_2$ and $\mathcal{K}_3$ be defined as in the example above. Define a surmise relation $\sqsubseteq$ on $Q$ by $\sqsubseteq = \{(q_1, q_1), (q_2, q_2), (q_3, q_3), (q_1, q_2), (q_1, q_3)\}$. Then $\sqsubseteq$ is the surmise relation corresponding to $\mathcal{K}_3$. Define a surmise function $\delta$ on $Q$ by $\delta(q_1) = \{\{q_1\}\}$, $\delta(q_2) = \{\{q_1, q_2\}, \{q_2, q_3\}\}$, and $\delta(q_3) = \{\{q_1, q_3\}, \{q_2, q_3\}\}$. Then $\delta$ is the surmise function corresponding to $\mathcal{K}_2$.

We have described how the concept of a knowledge space can be generalized to problems with more than two answer alternatives. To be able to apply this generalization in practice we need to solve two important questions. First, it must be clarified how we can build such generalized knowledge spaces in given knowledge domains. In the case of more than two answer alternatives it is questionable if querying experts can be used for that purpose. However, it seems possible to use cognitive models of the problem solving processes in the domain to derive such knowledge structures, see for example Schrepp (1993, 1999c). Second, it should be clarified how effective procedures for the diagnosis of the knowledge state of a student can be formulated.

**9.1.7 Surmise relations between tests.**
For standard applications of KST, e.g. developing an adaptive test or selecting a learning path within a course, at least a surmise relation (or its corresponding knowledge structure $\mathcal{K}$) on the knowledge domain $Q$, represented e.g. by test items or learning objects, is sufficient. However in many applications the relation between subsets of items, forming a partition of $Q$, is also of interest. Two examples may briefly demonstrate this.[9]

In curriculum development (see, e.g., Albert and Hockemeyer, 1999), for instance in high school mathematics, the entire body of information (represented by $Q$) can be partitioned into such sub-bodies (represented by subsets of $Q$) as algebra, analysis, and geometry. An analysis of a student's knowledge at the level of subsets of items (tests) is useful because there are natural breaks in an academic subject, in particular in high school mathematics, around which curricula can be arranged. Work is also being done in the development of efficient tutoring systems that improve curriculum efficiency (see, e.g., Albert and Hockemeyer, 1999).

Secondly, as an example from child psychology, it might be necessary to access a child's general stage of cognitive development. The whole domain of cognition may be represented by all the items of the different tests regarding all

[9] Applications of surmise relations between tests are presented in Chapter 10

the specific cognitive abilities. Of course, in order not to over-burden the child each of the tests should be presented by an adaptive, personalized procedure. In case that a surmise or prerequisite structure exists on the set of tests, also sequencing the presentation of the tests should be personalized. If, for instance one cognitive functioning is a prerequisite of another one and the child fails in the first test, it is not necessary to apply the second test.

Based on this kind of reasoning we investigate some possible relationships among tests of items in a knowledge structure/domain, with a test being a subset of the respective domain. Here, as in the following, the term test is a generic one and may be replaced by, e.g., set, sub-domain of knowledge, or body of learning objects.

Surmise relations between tests (SRbT), based on the underlying surmise relation between items, are examined. In particular, relations defined as follows are discussed: For two tests $A$ and $B$,

$(B, A)$ *is in a relation* if and only if (iff) there is some $b \in B$ that is a prerequisite for some $a \in A$;
$(B, A)$ *is in a relation* iff each $a \in A$ has some prerequisite $b \in B$;
$(B, A)$ *is in a relation* iff each $b \in B$ is a prerequisite for some $a \in A$;
$(B, A)$ *is in a relation* iff each $a \in A$ has some prerequisite $b \in B$ and each $b \in B$ is a prerequisite for some $a \in A$.

These four types of relations are defined and discussed in more detail below (see Equations 9.1, 9.2 and 9.3).

Furthermore, given a partition made up of tests for a knowledge structure, a 'test knowledge structure' is obtained via intersections of the tests with the original states in the underlying knowledge structure (see details below, Page 155).

Thus, in common psychological assessment procedures we often deal with a set $\mathcal{T}$ of different tests that may be related. In the following, we call any element of a partition of $Q$ a test. On the background of Doignon and Falmagne's framework, Albert and his group (Albert, 1995; Brandt et al., 2003; Ünlü et al., 2004) extended the concept of a surmise relation between items $S \subseteq Q \times Q$ (i.e. *within* tests) to a surmise relation *between* tests $\dot{\mathcal{S}} \subseteq \mathcal{T} \times \mathcal{T}$. For a better understanding, we illustrate the approach for only two tests. However, the described concepts are applicable to any finite set of tests.

The interpretation of a surmise or prerequisite relationship between tests,[10] that is $(B, A) \in \dot{\mathcal{S}}$ or $B \,\dot{\mathcal{S}}\, A$, is that two tests $A, B \in \mathcal{T}$ are in a surmise relation from $A$ to $B$, if one can surmise from the correct solution of at least one item in test $A$ the correct solution of a non-empty subset of items in test $B$ (see Figure 9.1). In other words, solving item(s) in test $B$ (e.g. items $b_3$ and $b_4$ in

[10]Note that a surmise relation $\dot{\mathcal{S}}$ between tests is different from a surmise relation $S$ between items in two aspects. First, the former is a relation between subsets of items, whereas the latter is between individual items. Second, $\dot{\mathcal{S}}$ is not necessarily transitive (see Figure 9.2), as compared to $\mathcal{S}$ which is always transitive.

Figure 9.1) is a prerequisite for the solution of a given set of items in test $A$ (e.g. item $a_1$ in Figure 9.1). Formally, the relation $\dot{\mathcal{S}} \subseteq \mathcal{T} \times \mathcal{T}$ is defined by

$$B \,\dot{\mathcal{S}}\, A :\Leftrightarrow \exists a \in A, B_a \neq \emptyset \quad \text{with } B_a = B \cap \bigcap \mathcal{K}_a, \tag{9.1}$$

with $A, B \in \mathcal{T}$ and $\mathcal{K}_a$ denoting the set of all knowledge states $K \in \mathcal{K}$ containing item $a$ (Brandt et al., 2003, Definition 5).

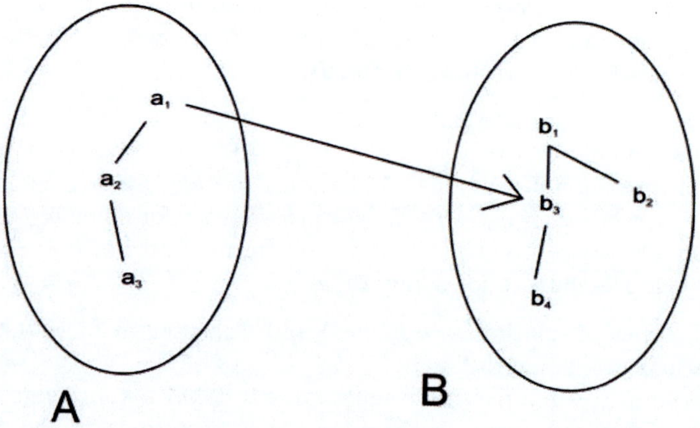


**Figure 9.1.** Two tests $A$ and $B$ in a surmise relation from $A$ to $B$ ($B \,\dot{\mathcal{S}}\, A$).

For a set of tests $\mathcal{T} = \{A, B, C, \dots\}$, a surmise relation between tests has the property of reflexivity but not necessarily of transitivity (see Figure 9.2), that is, in general it is not a quasi order (Brandt et al., 2003, Propositions 6 and 7). However, there are special subsets, for which transitivity holds, namely left-, right-, and total-covering surmise relations between tests.

Two tests $A, B \in \mathcal{T}$ are in a *left-covering surmise relation* from test $A$ to test $B$ ($B \,\dot{\mathcal{S}}_l\, A$, see Figure 9.3), if for each item $a \in A$ there exists a nonempty subset of prerequisites in test $B$. This means that a person who doesn't solve any item in $B$ will not be able to solve any item in $A$, either. There is no need to test this person on test $A$. The relation is defined (Brandt et al., 2003, Definition 8) by

$$B \,\dot{\mathcal{S}}_l\, A :\Leftrightarrow \forall a \in A, B_a \neq \emptyset \qquad \text{for all } A, B \in \mathcal{T} \tag{9.2}$$

Two tests $A, B \in \mathcal{T}$ are in a *right-covering surmise relation* from test $A$ to test $B$, if for each item $b \in B$, there exists at least one item $a \in A$ for which $b$ is a prerequisite ($B \,\dot{\mathcal{S}}_r\, A$). Failing to solve any item in test $B$ implies a failure

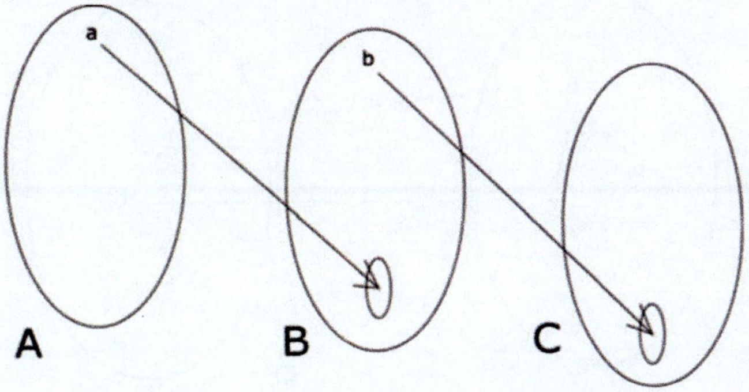


**Figure 9.2.** $C \, \dot{\mathcal{S}} \, B$ and $B \, \dot{\mathcal{S}} \, A$, but $C \, \not\dot{\mathcal{S}} A$

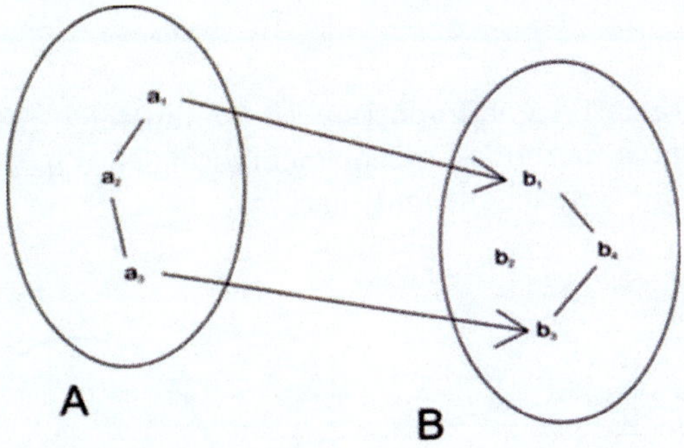


**Figure 9.3.** Left-covering surmise relation from test $A$ to test $B$ ($B \, \dot{\mathcal{S}}_l \, A$).

on a subset of items in test $A$ (see Figure 9.4). In other words, a person who solves all items in test $A$ is also able to solve all items in test $B$. Thus, there is no need to test this person on test $B$. The relation is defined (see Brandt et al., 2003, Definition 12) by

$$B \, \dot{\mathcal{S}}_r \, A :\Leftrightarrow \bigcup_{a \in A} B_a = B \qquad \text{for all } A, B \in \mathcal{T} \tag{9.3}$$

Finally, we speak of a *total-covering surmise relation* from test $A$ to test $B$ ($B \, \dot{\mathcal{S}}_t \, A$), if the ordered pairs of tests are in a left- as well as right-covering relation. See Figure 9.5 for an example.

Simple proofs show that the previously defined prerequisite relations (based on the same underlying surmise relation or corresponding knowledge structure on $Q$) are in subset relationships $\dot{\mathcal{S}}_t \subseteq \dot{\mathcal{S}}_l, \dot{\mathcal{S}}_r$ and $\dot{\mathcal{S}}_l, \dot{\mathcal{S}}_r \subseteq \dot{\mathcal{S}}$ (Brandt et al., 2003, Lemmata 9 und 13). Furthermore, left-, right-, and total-covering surmise relations are quasi orders on the set $\mathcal{T}$ of tests (Brandt et al., 2003, Corollary 10 to Lemma 15), i.e. they are reflexive and transitive. Especially

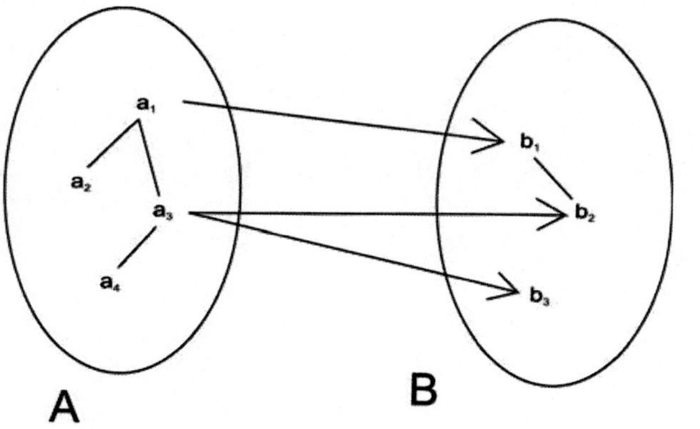


**Figure 9.4.** Right-covering surmise relation from test $A$ to test $B$ ($B \dot{\mathbb{S}}_r A$).

the property of transitivity is important to administer e.g. tests and courses adaptively. The same methods for sequencing items are useful for sequencing tests and courses for assessment and teaching.

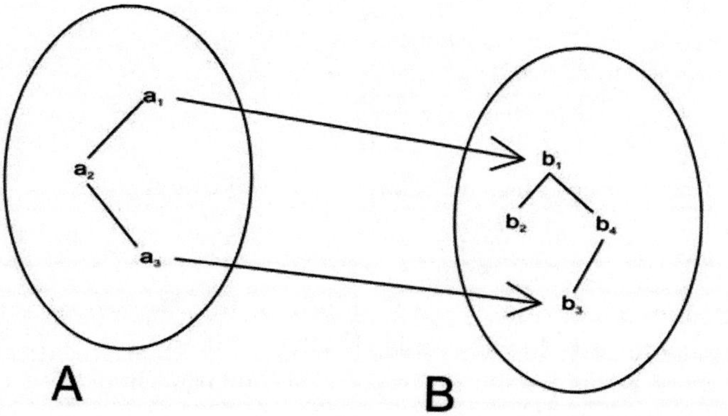


**Figure 9.5.** Total-covering surmise relation from test $A$ to test $B$ ($B \dot{\mathbb{S}}_t A$).

For adaptive assessment and teaching in general, for empirical validation of surmise relations between tests, and for computational reasons we differentiate between various subsets of the surmise relation on the entire set $Q$ of items. $S_{QxQ}$ denotes the surmise relation on $Q$ and is referred to as surmise relation *between items (SRbI)*. The disjoint subsets of $S_{QxQ}$ for two tests $A$ and $B$ (with $A \cup B = Q$) are denoted $S_{AxA}$, $S_{BxB}$, $S_{AxB}$, and $S_{BxA}$. The subset $S_{AxA}$ united with $S_{BxB}$ is called surmise relation *within tests (SRwT)*. The union of the subsets $S_{AxB}$, $S_{BxA}$, and the set $\Delta = \{(q, q) | q \in Q\}$ is called a surmise relation *across tests (SRxT)*. Each subset is defined as follows:

$$
\begin{aligned}
SRwT &= S_{AxA} \cup S_{BxB} \\
SRxT &= S_{AxB} \cup S_{BxA} \cup \Delta \\
SRbI &= SRwT \cup SRxT
\end{aligned}
\tag{9.4}
$$

where

$$
\begin{aligned}
S_{AxA} &= \{(a_i, a_j) \mid a_i, a_j \in A \wedge a_i S a_j\} \\
S_{BxB} &= \{(b_i, b_j) \mid b_i, b_j \in B \wedge b_i S b_j\} \\
S_{AxB} &= \{(a, b) \mid a \in A, b \in B \wedge aSb\} \\
S_{BxA} &= \{(b, a) \mid a \in A, b \in B \wedge bSa\}
\end{aligned}
$$

If either the set $S_{AxB}$ or the set $S_{BxA}$ are not empty, then there exists a surmise relation between the tests $A$ and $B$ ($\dot{\mathcal{S}} \subseteq \mathcal{T} \times \mathcal{T}$; cf. Equation 9.1).

Extending the concepts of Doignon and Falmagne's approach, a *test knowledge state* $\dot{K}$ is defined as the $|\mathcal{T}|$-tuple of items per test a person in state $K$ is capable of mastering: $A \cap K, B \cap K, \ldots$ The collection of all test knowledge states, that are in agreement with the $SRbI$, is called the *test knowledge structure*, which is defined as the pair $(\mathcal{T}, \dot{\mathcal{K}})$ and abbreviated by $\dot{\mathcal{K}}$.

If a test knowledge structure $\dot{\mathcal{K}}$ is closed under union it is called *test knowledge space*, if it is also closed under intersection we speak of a *quasi ordinal test knowledge space.*[11]

A mathematical proof shows an interesting relationship between the test knowledge structure $\dot{\mathcal{K}}$ and the underlying knowledge structure $\mathcal{K}$: $\dot{\mathcal{K}}$ is a (quasi ordinal) test knowledge space if and only if $\mathcal{K}$ is a (quasi ordinal) knowledge space (Brandt et al. 2003, Lemma 19; Ünlü et al. 2007). Furthermore a base $\dot{\mathcal{B}}$ of the test knowledge structure $(\mathcal{T}, \dot{\mathcal{K}})$ can be derived from a base $\mathcal{B}$ of the knowledge structure $(Q, \mathcal{K})$. By means of the base $\dot{\mathcal{B}}$ we can infer the test knowledge space $\dot{\mathcal{K}}$, the corresponding knowledge space $\mathcal{K}$ and the surmise relation between items; moreover, as an important conclusion we can infer the surmise relation between tests and its properties, namely antisymmetry, transitivity, left-, and right coveringness, by means of the base (Brandt et al., 2003, Propositions 23 to 25).

The surmise relation between tests can be derived from its corresponding test knowledge space, however, the reverse inference is not valid for a set of tests. As for the surmise relations we differentiate between the knowledge spaces between items ($KSbI$), within tests ($KSwT$), and across tests ($KSxT$). The subsets are derived by applying the Birkhoff-Theorem to the corresponding surmise relation (cf. Equation 9.4), that is the $SRbI$, the $SRwT$, and the $SRxT$ respectively.

Because left- and right-covering surmise relations are transitive, they allow an efficient assessment of a person's test knowledge state. The knowledge

[11] The union of two $|\mathcal{T}|$-tuple test knowledge states $\dot{K}$ and $\dot{K}'$, is defined as $\dot{K} \dot{\cup} \dot{K}' = (T \bigcap (K \cup K'))_{T \in \mathcal{T}}$, the intersection as $\dot{K} \dot{\cap} \dot{K}' = (T \bigcap (K \cap K'))_{T \in \mathcal{T}}$.

state of a person in one test can give information about the expected performance on a set of related tests. For existing tests, it might be useful to repartition the items into a set of tests, that are in a transitive relation. The prerequisites between the tests can then be used as personalized curricula and developmental paths. In psychological diagnostics, such paths help to avoid too much strenuousness, which is particularly important when working with children, for instance. The test knowledge states can also be used to classify individuals according to their already acquired knowledge, which leads to classes with similarly capable members and therefore to an adequate training for each member. Further potential applications are in the fields of educational psychology (e.g., curriculum development; Albert and Hockemeyer, 1999) or computer science (e.g., structuring hypertext documents; Albert and Hockemeyer, 1997).

Summarizing, the concept 'surmise relation between tests' allows us to specify prerequisite relations not only between single items but between subsets of items, as for example, tests of cognitive or developmental functioning where the possession of one ability is prerequisite for the acquisition of some other ability. Regarding the area of diagnostics, the number of employed tests can be reduced and we can develop adaptive testing systems covering a wide range of items.

Following an approach of Kim and Roush (1984), we can also use Boolean matrices for the representation of knowledge structures. A knowledge structure $\mathcal{K} = \{K_1, \ldots, K_n\}$ on the set $Q = \{x_1, \ldots, x_m\}$ can be represented by an $n \times m$ Boolean matrix $X$, whose entries are defined by $X_{ij} = 1$ if knowledge state $K_i$ contains the item $x_j \in Q$, and $X_{ij} = 0$, otherwise, for $i = 1, \ldots, n$ and $j = 1, \ldots, m$. Partitioning the columns of $X$ into tests establishes the relation between the knowledge structure and the corresponding test knowledge structure. For more details see Brandt et al. (2003, Lemmata 26 to 33). Using the matrix representation by Kim and Roush is stimulating further research: the range of applications of surmise relations between tests is enlarged, computations may be realizable by more efficient procedures, and proofs may become more elegant.

In addition, efficient algorithms have to be derived from the SRbT-approach for partitioning a set of items into tests and courses regarding mathematical criteria like antisymmetry, transitivity, and left-, and right-coveringness as well as content-oriented criteria. For this purpose, analyzing the relationships between surmise relations between tests and test knowledge spaces more deeply may be helpful. Taking also semantic technology into account, further interdependencies and parallelism of tests and courses can be investigated based on the SRbT-approach. The aims are to reduce the large amount of tests labeled differently, and to restructure tests and courses so that the property of transitivity can be used for adaptive, personalized applications. The SRbT-model and the derived algorithms are a basis for future software tools that analyze tests and courses, and that partition a set of items into tests and courses.

Of course, it seems natural to generalize the concept of surmise relations between tests to surmise systems/functions between tests, which allow, e.g., different ways or abilities for performing a test. Also, in extension of the approaches introduced in Chapter 11, the establishment of surmise relationships between sets of skills and competencies and their mappings to test performances is aimed for.

Apart from that establishing principles for handling data—especially noisy data—is necessary. In general, empirically obtained data are noisy, e.g. because of careless errors and lucky guesses or because of missing data. Thus, a probabilistic extension for SRbT has to be developed.

The applicability of the SRbT concepts is not restricted to psychological tests (see, e.g., Chapter 10 in this volume). In addition to the relationships between courses already mentioned interpretations and applications may be in (re-)structuring different types of objects, e.g., hypertexts, the organization of companies, and sequencing works.

## 9.2 Relationships between KST and other Theoretical Approaches

Compared with the long tradition of developing psychological tests and grounding them by psychological test theories, Knowledge Space Theory (KST) has a short history. With growing up of KST considering its relationships with other approaches has become a research topic of its own. The selection of to be compared approaches is guided by different aspects. For instance, Formal Concept Analysis (FCA) has the same mathematical background (lattice theory) and is also aiming at ordering two types/sets of elements simultaneously. The same is true also of a special case of KST, the Guttman scale. The Guttman scale can be viewed as both the origin of KST and Item Response Theory (IRT). Both are generalizations albeit of different types. Beyond these examples, there is the Bayesian Network approach which has become very popular recently and which is sometimes seen as a strong competitor for KST although it should rather be seen as complementary (see, e.g., Desmarais and Gagnon, 2006; Villano, 1992). Although the following investigations of the relationships between KST and its alternatives or competitors might be motivated by different reasons, all of them have in common deep mathematical analyses. Some of their results may stimulate further developing KST, others may offer new or alternative methods which can be used for KST applications or they may offer arguments for complementing KST in applied settings. In any case, the analyses and comparisons seem to be fruitful.

### 9.2.1 Knowledge space theory and formal concept analysis.

Formal concept analysis (Ganter and Wille, 1996; Wille, 1982) is a set-theoretic approach which shows that any characterization of a set of objects by distinctive features induces a hierarchy of concepts. In fact, the induced conceptual structure is that of a complete lattice (see also Davey and Priestley, 1990). After sketching the basics of formal concept analysis it is shown that the approach is closely related to knowledge space theory.

FORMAL CONTEXTS AND CONCEPT LATTICES

We start by defining the notion of a formal context.

**9.2.2 Definition.** A relational structure $(G, M, I)$ with $G$ and $M$ sets and $I \subseteq G \times M$ a binary relation, is called a *(formal) context*. The elements $g \in G$ are interpreted as *objects* and the elements $m \in M$ are interpreted as *attributes*[12]. We say the object $g$ has the attribute $m$, if $g\ I\ m$ holds.

The following example is chosen to demonstrate the universality of the approach.

**9.2.3 Example.** Let $G = \{R_1, R_2, R_3, R_4, R_5\}$ be the set of the binary relations

$$\begin{aligned} R_1 &= \emptyset \\ R_2 &= \{(a,a),(b,b)\} \\ R_3 &= \{(a,b),(b,a)\} \\ R_4 &= \{(a,a),(b,b),(a,b)\} \\ R_5 &= S \times S \end{aligned}$$

on the set $S = \{a, b\}$, and consider $M = \{r, c, t, a, s\}$ as set of attributes representing the properties of reflexivity, connectedness, transitivity, antisymmetry and symmetry, respectively. Then we get a context as shown in Table 9.1 where the rows correspond to the objects $g \in G$ and the columns to the attributes $m \in M$.

**Table 9.1.** Formal context of binary relations $R_1$ to $R_5$.

| | reflexive | connected | transitive | antisymmetric | symmetric |
|---|---|---|---|---|---|
| $R_1$ | | | × | × | × |
| $R_2$ | × | | × | × | × |
| $R_3$ | | | | | × |
| $R_4$ | × | × | × | × | |
| $R_5$ | × | × | × | | × |

[12]The letters $G$ and $M$ refer to the initial letters of the German “Gegenstände” for objects, and “Merkmale” for attributes, respectively.

The relation $I \subseteq G \times M$ induces relations between the subsets of $G$ and $M$. Let $gI$ denote the set of all attributes which are assigned to the object $g \in G$, thus $gI = \{m \in M \mid gIm\}$. Similarly, the set of all objects that have a particular attribute $m \in M$ is denoted by $Im$, i.e. $Im = \{g \in G \mid gIm\}$. Now, for all $A \subseteq G$ and $B \subseteq M$ consider the assignments

$$A \mapsto A' = \bigcap_{g \in A} gI, \quad \text{and} \quad B \mapsto B' = \bigcap_{m \in B} Im.$$

By this definition, $A'$ is the set of attributes which the objects $g \in A$ have in common, while $B'$ is the set of objects to which all the attributes $m \in B$ are assigned to. Referring to Example 9.2.3 for the subset $A = \{R_1, R_4\}$ of the set of binary relations we derive $A' = \{t, a\}$.

Both relations $R_1$ and $R_4$ are transitive and antisymmetric, and share none of the other properties. Conversely, for $B = \{t, a\}$ we obtain $B' = \{R_1, R_2, R_4\}$. Notice that here $B'' = (B')' = \{t, a\} = B$ holds.

The subsets $A \subseteq G$ and $B \subseteq M$, which are related to each other by the equations

$$A' = B \quad \text{and} \quad B' = A$$

are of particular interest.

**9.2.4 Definition.** Let $(G, M, I)$ be a context. An ordered pair $(A, B)$ with $A \subseteq G$, $B \subseteq M$ and $A' = B$, $B' = A$ is called a *(formal) concept* of the given context. $A$ is said to be the *extent* of the concept and $B$ is said to be the *intent* of the concept.

The extent $A$ of a concept $(A, B)$ is the set of objects, which share the attributes in $B$. The intent $B$ of $(A, B)$ is the set of attributes assigned to all objects in $A$. As we have already seen, the pair $(\{R_1, R_2, R_4\}, \{t, a\})$ is a concept of the context of Example 9.2.3. This is also true of the pairs $(\{R_2, R_4, R_5\}, \{r, t\})$ and $(\{R_4\}, \{r, c, t, a\})$, whereas the pair $(\{R_1, R_4\}, \{t, a\})$ is not a concept of the context of Example 9.2.3, since $\{t, a\}' = \{R_1, R_2, R_4\}$ is a proper superset of $\{R_1, R_4\}$.

Now, if $\mathfrak{B}_{GMI}$[13] denotes the set of all concepts of a context $(G, M, I)$, then an ordering on $\mathfrak{B}_{GMI}$ arises in a natural way.

**9.2.5 Definition.** Suppose $(A_1, B_1), (A_2, B_2) \in \mathfrak{B}_{GMI}$. A binary relation $\sqsubseteq$ on $\mathfrak{B}_{GMI}$ is defined by

$$(A_1, B_1) \sqsubseteq (A_2, B_2) \quad \text{iff} \quad A_1 \subseteq A_2\,.$$

Whenever $(A_1, B_1) \sqsubseteq (A_2, B_2)$ holds then $(A_1, B_1)$ is said to be a *subconcept* of $(A_2, B_2)$, and conversely, $(A_2, B_2)$ is said to be a *superconcept* of $(A_1, B_1)$.

[13] The letter $\mathfrak{B}$ refers to the German "Begriff" for concept.

Notice that the condition $(A_1, B_1) \sqsubseteq (A_2, B_2)$ iff $A_1 \subseteq A_2$ is equivalent to the condition $(A_1, B_1) \sqsubseteq (A_2, B_2)$ iff $B_2 \subseteq B_1$. Due to the equivalence of $A_1 \subseteq A_2$ and $B_2 \subseteq B_1$ it suffices to consider either the extents, or the intents of the concepts $(A_1, B_1)$ and $(A_2, B_2)$. From Definition 9.2.5 it is evident that $(\mathfrak{B}_{GMI}, \sqsubseteq)$ is a partial order. For $(\mathfrak{B}_{GMI}, \sqsubseteq)$ being a complete lattice, the existence of least upper bounds (or, suprema) and greatest lower bounds (or, infima) for all subsets of $\mathfrak{B}_{GMI}$ has to be shown. This is done by defining

$$\inf\{(A_j, B_j) \mid (A_j, B_j) \in \mathfrak{B}_{GMI}, j \in J\} = (\bigcap_{j \in J} A_j, (\bigcup_{j \in J} B_j)''),$$
$$\sup\{(A_j, B_j) \mid (A_j, B_j) \in \mathfrak{B}_{GMI}, j \in J\} = ((\bigcup_{j \in J} A_j)'', \bigcap_{j \in J} B_j),$$

with $J$ an appropriate index set. According to these definitions the extent of the infimum of a subset of $\mathfrak{B}_{GMI}$ is simply the intersection of the extents of the respective concepts, and in the same way the intent of the supremum of a set of concepts results from intersecting the respective intents. This means that both the collections of all extents as well as all intents of $\mathfrak{B}_{GMI}$ are closed under intersection: The intersection of extents (intents) of $\mathfrak{B}_{GMI}$ is again an extent (intent) of $\mathfrak{B}_{GMI}$. This is not true of the set-theoretic union of extents and intents of $\mathfrak{B}_{GMI}$. The intent of the infimum of a set of concepts, for example, is the least intent of $\mathfrak{B}_{GMI}$ that contains the union of the respective intents. Notice that $(\mathfrak{B}_{GMI}, \sqsubseteq)$ being a complete lattice does not put constraints on the relation $I$ of the context $(G, M, I)$.

The complete lattice $(\mathfrak{B}_{GMI}, \sqsubseteq)$ can be illustrated by a Hasse diagram. Assigning the names of objects and attributes to the concepts in the following way assures that the context $(G, M, I)$ can be reconstructed from the corresponding *concept lattice*: The name of an object $g \in G$ is assigned to the concept $(\{g\}'', \{g\}')$, which is the concept with least extent including $g$. The name of a attribute $m \in M$ is assigned to the concept $(\{m\}', \{m\}'')$, which is the concept with least intent including $m$. The extent of a concept is the set of objects assigned to that particular concept or to a concept which is a subconcept of the given one. The intent of a concept is the set of all attributes assigned to that particular concept or to a concept, which is a superconcept of the given one. For reconstructing the context from the corresponding concept lattice, define $g\, I\, m$ if and only if $g$ and $m$ are assigned to the same concept, or the concept $m$ is a superconcept of the concept to which $g$ is assigned.

**9.2.6 Example.** The diagram in Figure 9.6 illustrates the concept lattice derived from the context of Table 9.1.

### LINKING FORMAL CONCEPT ANALYSIS TO KNOWLEDGE SPACE THEORY

There is an intimate relation between formal concept analysis and knowledge space theory (Rusch and Wille, 1996). Both the collection of extents as well

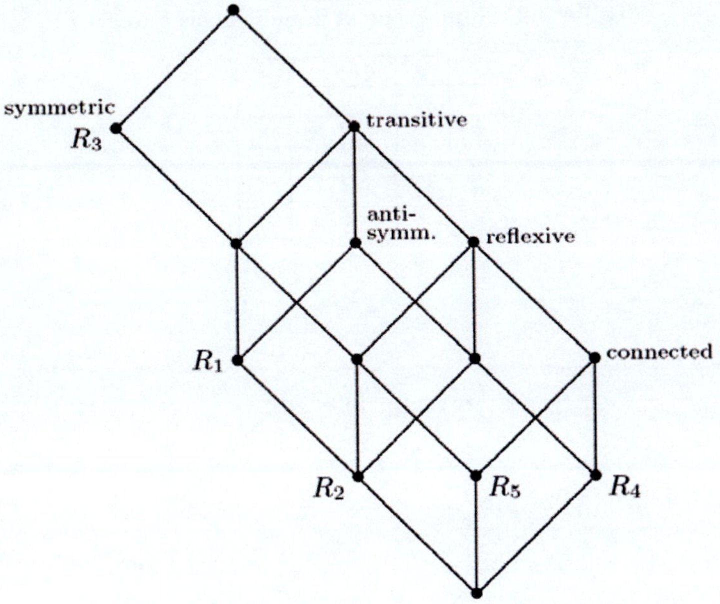


**Figure 9.6.** Hasse diagram of the concept lattice derived from the context of Table 9.1.

as the collection of intents of the concepts in $\mathfrak{B}_{GMI}$ are closed under intersection, a property which is at the core of the link between concept lattices and knowledge spaces. Collections of subsets closed under intersection are also known as closure systems, and were shown to be equivalent to closure operators (see, e.g., Caspard and Monjardet, 2003; Davey and Priestley, 1990). Associating to each subset $A \subseteq G$ ($B \subseteq M$) the set $A''$ ($B''$) establishes a closure operator, and the resulting collection of all subsets $A''$ ($B''$) then form a closure system on $G$ ($M$). Conversely, if we associate to each subset $A \subseteq G$ ($B \subseteq M$) the least extent (intent) of a closure system on $G$ ($M$) that contains $A$ ($B$) then we obtain a closure operator on $G$ ($M$).

Consider a formal context consisting of a set $A$ of individuals and a knowledge domain $Q$. For all $a \in A$ and $q \in Q$ set $a\, I\, q$ if individual $a$ is *not* capable of solving item $q$. Following Rusch and Wille (1996) such a context $(A, Q, I)$ is called a *knowledge context.*

**9.2.7 Example.** Let $A = \{a, b, c, d, e\}$ be a set of individuals and $Q = \{p, q, r, s\}$ a knowledge domain, and let the solution behavior be characterized by the relation $I$ in the formal context defined in Table 9.2.

Determining the concepts $\mathfrak{B}_{AQI}$ induced by the context $(A, Q, I)$ of Example 9.2.7 provides the concept lattice which is depicted in Figure 9.7.

**Table 9.2.** Formal context of individuals $a$ to $e$

| | $p$ | $q$ | $r$ | $s$ |
|---|---|---|---|---|
| $a$ | | | × | × |
| $b$ | × | | × | × |
| $c$ | | × | | |
| $d$ | × | × | × | × |
| $e$ | × | × | × | |

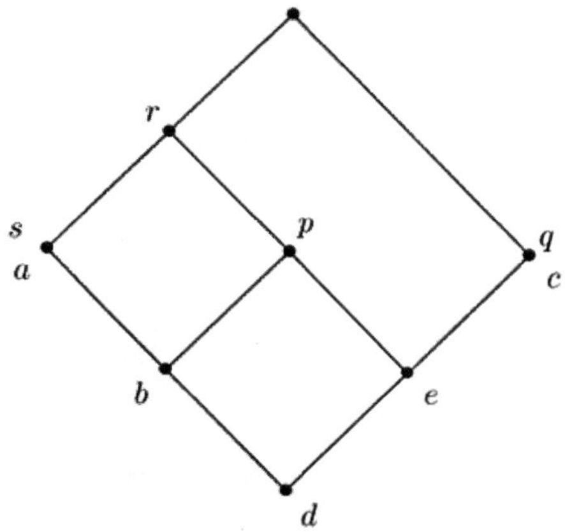


**Figure 9.7.** Hasse diagram of the concept lattice derived from the knowledge context of Example 9.2.7.

The collection of intents can be read off directly from the diagram of the concept lattice $(\mathfrak{B}_{AQI}, \sqsubseteq)$, providing

$$\mathfrak{I}_{AQI} = \{\emptyset, \{r\}, \{q\}, \{p, r\}, \{r, s\}, \{p, r, q\}, \{p, r, s\}, Q\}.$$

The reader may easily verify that the collection of intents $\mathfrak{I}_{AQI}$ is closed under intersection. Notice that there is $Q \in \mathfrak{I}_{AQI}$ because the intent $Q$ results by definition from the intersection of the empty collection of intents. However, the empty set $\emptyset$ is not necessarily an intent. Consider a case where there exists an item that is not solved by all the individuals in $A$. Then this item is contained in each of the intents, all of which are therefore nonempty.

If we now take the set-theoretic complements of all the intents with respect to the knowledge domain $Q$ then we obtain a family of subsets of $Q$ which is closed under union, and contains the empty set $\emptyset$. Whenever the resulting family of subsets also contains the domain $Q$ then it forms a knowledge space. Since this is true in the given example, we obtain the knowledge space

$$\mathcal{K}_{AQI} = \{\emptyset, \{q\}, \{s\}, \{p, q\}, \{q, s\}, \{p, r, s\}, \{p, q, s\}, Q\}.$$

A collection $\mathcal{F}$ of subsets of $Q$ is said to *span* the collection $\mathcal{F}'$, whenever $\mathcal{F}'$ is formed by taking unions of some members of $\mathcal{F}$ (see Definition 8.2.1).. Then it is easily seen that $\mathcal{K}_{AQI}$ is spanned by the collection $\{Q \setminus aI \mid a \in A\}$ of subsets of solved items for all individuals.

Conversely, we may associate a knowledge context to any knowledge space $\mathcal{K}$ by defining $(\mathcal{K}, Q, \not\ni)$. The intents of the concepts induced by this knowledge context are the complements of the states in $\mathcal{K}$ with respect to $Q$.

Further details on the relations between formal concept analysis and knowledge space theory can be found in Rusch and Wille (1996). They also consider, for example, the correspondence between certain sets of attribute implications and entail relations as introduced by Koppen and Doignon (1990). A valid attribute implication of the knowledge context $(A, Q, I)$ is a pair $(S, T)$ of subsets of $Q$ (usually denoted by $S \to T$), such that each person who fails on all items in $S$ does also fail on all items in $T$. This means that we have $S' \subseteq T'$, or, equivalently, $S \subseteq T''$ (Rusch and Wille, 1996). This parallels the interpretation of an entail relation in the process of building a knowledge space by querying an expert (Gediga and Düntsch, 2002; Koppen and Doignon, 1990; Koppen, 1993; Kambouri et al., 1994; Stefanutti and Koppen, 2003). Caspard and Monjardet (2003, 2005) also consider notions like these from an even more general perspective. They show that implicational systems of this sort are in fact another way of looking at closure systems (or, equivalently, closure operators) that not only appear in formal concept analysis and, via taking set-theoretic complements, in knowledge space theory, but in various other fields in pure or applied mathematics and computer science. By recasting the sometimes independently developed concepts within a unified formal framework they take a first step in exploring carefully the links between all of these approaches.

### 9.2.8 Knowledge space theory and latent class analysis.

CLASSICAL UNRESTRICTED LATENT CLASS MODEL

*Latent class analysis* (LCA; Andersen, 1982; Goodman, 1978; Hagenaars and McCutcheon, 2002; Lazarsfeld and Henry, 1968; McCutcheon, 1987; Vermunt and Magidson, 2004) is a statistical approach for examining unobserved categorical variables.[14]

The population of reference is assumed to be partitioned into $T$ mutually exclusive and exhaustive subpopulations (classes) $C_1, \ldots, C_T$, with unknown proportions $p(C_t) > 0$. The latent classes can be viewed as the realizations of a random variable $X$. Let $I_1, I_2, \ldots, I_L$ be the dichotomous manifest variables (indicators), with realizations $i_l \in \{0, 1\}$ $(1 \leq l \leq L)$, respectively. In vector notation, $I = (I_l)_{1 \leq l \leq L}$ and $i = (i_l)_{1 \leq l \leq L}$. For each of the $T$ latent classes,

[14]Numerous examples can be found in the cited references.

there is a set of conditional probabilities for every indicator; $0 \leq r(I_l = i_l | X = C_t) \leq 1$ for any $1 \leq l \leq L$, $i_l \in \{0, 1\}$, and $1 \leq t \leq T$.

Within any of the $T$ latent classes, the observed indicator scores are assumed to be independent. This is the assumption of *local independence* which is fundamental in (classical) LCA.

The *classical unrestricted* ($T$*-class*) *latent class model* (*LCM*) is a *multinomial probability model*:

$$\rho(i) = \sum_t \left\{ p(C_t) r(I = i | X = C_t) \right\}$$
$$= \sum_t \left\{ p(C_t) \prod_l r(I_l = i_l | X = C_t) \right\},$$

where $\rho(i)$ is the occurrence probability of the response pattern $i = (i_l)$, and the data (the observed absolute counts of the response patterns) are assumed to follow a *multinomial distribution*.[15]

Imposing (e.g., equality or fixed value) restrictions on the parameters of the LCM (some examples below), gives *restricted* LCMs. In particular, the LCM is a *finite mixture model* (Everitt and Hand, 1981; McLachlan and Basford, 1988; McLachlan and Peel, 2000; Titterington et al., 1985), and LCA is often referred to as *finite mixture modeling*.

### KST AS UNRESTRICTED AND RESTRICTED LCA

This section integrates KST probabilistic models into the latent class modeling framework.

Let the indicators $I_1, I_2, \ldots, I_L$ be dichotomous items; that is, let the domain (of questions or problems) be $Q = \{I_l : 1 \leq l \leq L\}$. Scoring 1 or 0 to an item is interpreted as solving or failing to solve that item, respectively. Let $\mathcal{K}$ be a knowledge structure on $Q$. Let the latent classes be represented by the knowledge states; the random variable $X$ now assumes values in the knowledge structure $\mathcal{K}$. Note the two equivalent formulations for a response pattern, $R \in 2^Q$ (set-notation) and $I = i$ (vector-notation).

The *basic probabilistic model* (Doignon and Falmagne, 1999, p. 144)

$$\rho(i) = \sum_{K \in \mathcal{K}} \left\{ p(K) r(I = i | X = K) \right\}$$

obviously is an unrestricted latent class (finite mixture) model. It is a general model relying on the basic idea underlying any type of latent class model. The probability of observing a response pattern $i$, $\rho(i)$, is a finite weighted average of the $|\mathcal{K}|$ class-specific conditional probabilities $r(I = i | X = K)$.

[15] Note that the equality in the second line follows from the assumption of local independence.

Under the assumption of local independence, the basic probabilistic model specializes to the classical unrestricted LCM

$$\rho(i) = \sum_{K \in \mathcal{K}} \left\{ p(K) \prod_{l} r(I_l = i_l | X = K) \right\}.$$

This model assumes that the item responses of an examinee are independent given the knowledge state of the examinee, and that the response probabilities $r(I_l = i_l | X = K)$ are attached to the items (item-specific) and knowledge states (state-specific). This model is implicitly referred to in Doignon and Falmagne (1999, p. 146).

Imposing restrictions on the parameters of the classical unrestricted LCM gives the following *restricted latent class (scaling) models* (in the order of increasing generality). For a description of these models, see, for instance, Heinen (1996).

1. A single response error rate across all items and knowledge states. That is, for a constant (error rate) $\tau \in [0, 1)$,

$$\begin{aligned} r(I_l = 0 | X = K) &= \tau \qquad && (I_l \in Q, K \in \mathcal{K}, I_l \in K), \\ r(I_l = 1 | X = K) &= \tau \qquad && (I_l \in Q, K \in \mathcal{K}, I_l \notin K). \end{aligned}$$

   This restricted LCM is a 'de-linearized' Proctor (1970) model.
2. Two response error rates across all items and knowledge states. That is, for constants (error rates) $\beta \in [0, 1)$ and $\eta \in [0, 1)$,

$$\begin{aligned} r(I_l = 0 | X = K) &= \beta \qquad && (I_l \in Q, K \in \mathcal{K}, I_l \in K), \\ r(I_l = 1 | X = K) &= \eta \qquad && (I_l \in Q, K \in \mathcal{K}, I_l \notin K). \end{aligned}$$

   This restricted LCM is a 'de-linearized' Dayton and Macready (1976) *intrusion-omission model* (*intrusion rate* $\eta$ and *omission rate* $\beta$).
3. For each item, two response error rates across all knowledge states (item-specific). That is, for constants (error rates) $\beta_l \in [0, 1)$ and $\eta_l \in [0, 1)$ $(1 \leq l \leq L)$,

$$\begin{aligned} r(I_l = 0 | X = K) &= \beta_l \qquad && (I_l \in Q, K \in \mathcal{K}, I_l \in K), \\ r(I_l = 1 | X = K) &= \eta_l \qquad && (I_l \in Q, K \in \mathcal{K}, I_l \notin K). \end{aligned}$$

   This restricted LCM is called the *basic local independence model* (BLIM) in KST (Doignon and Falmagne, 1999, pp. 144–145). The interpretation of the BLIM as a restricted LCM is discussed in Schrepp (2005) and Ünlü (2006). It is a sort of 'de-linearized' Lazarsfeld and Henry (1968) *latent distance model*.

INFERENTIAL STATISTICS

As a statistical method, LCA comes with corresponding inference methodology. Parameter estimation and model testing are typically performed using *maximum likelihood*. Estimates of the model parameters are obtained maximizing the likelihood function. The goodness-of-fit of an estimated LCM can be tested based on the log-likelihood ratio statistic (deviance). Since it is not valid to compare LCMs with different numbers of latent classes by general likelihood ratio tests, model selection information criteria such as AIC or BIC are often used. The software package *Latent GOLD* (Vermunt and Magidson, 2000), for instance, implements these procedures.

This statistical inference methodology in particular applies to the KST probabilistic models, as they are special latent class models. Albeit without a reference to LCA, this has been demonstrated by Doignon and Falmagne (1999), and the statistical techniques presented there are consistent with latent class modeling methodology. In other words, KST and LCA are related also at the level of inferential statistics used in both theories. For more information about inferential statistics for this class of models, see, for instance, Ünlü (2006, Section 7).

#### Applications of LCA in KST

The following publications about knowledge structures explicitly point out or build upon the outlined connection between KST and LCA.

Schrepp (2005) proposes an exploratory method for constructing knowledge structures from data. Candidate knowledge structures are constructed, they are considered as restricted latent class models and fitted to the data, and the BIC criterion is used to choose among them. The method is applied to empirical data. For a note on this paper, adding additional information about the relationship between KST and LCA, see also Ünlü (2011). Similar to Schrepp (2005), Stefanutti and Robusto (2009) discusses a special case of the BLIM where the response error rates are constrained, in order to recover a probabilistic knowledge structure.

Ünlü (2006) investigates latent class modeling with random effects for the estimation of response error rates for dichotomous test items in KST. In particular, this approach is extended to give a generalization of the BLIM (and of a number of other LCMs). This allows for local dependence among the items given the knowledge state of an examinee and/or for the incorporation of covariates. The procedures used for parameter estimation and model testing are discussed. The approach is illustrated with simulation data.

#### Outlook: Latent variable modeling in KST

An interesting and important direction for future research is investigating models based on stochastic processes. Dynamic KST probabilistic models such as the *stochastic learning paths systems* (Doignon and Falmagne, 1999; Falmagne, 1993; Falmagne and Doignon, 2011) have not yet been compared with statistical approaches to latent variable modeling. For instance, *latent Markov* (*hidden Markov* or *latent transition*) *models* (Hagenaars, 1990; MacDonald and Zucchini, 1997; Vermunt, 1997) are an interesting approach to be used in further comparisons of KST and statistical analyses of latent variables.

**9.2.9 Knowledge space theory and item response theory.** The Guttman scale (Guttman, 1944)[16] can be viewed as a common origin of both, *item response theory* (IRT; e.g., Boomsma et al., 2001; Fischer and Molenaar, 1995; van der Linden and Hambleton, 1997) and *knowledge space theory* (KST; e.g., Doignon and Falmagne, 1985, 1999). They generalize the Guttman scale in probabilistic, statistical and deterministic, combinatorial directions, respectively.

In KST, persons are represented by collections of items (of a representative and fully comprehensive domain) they are capable of mastering. Persons can be incomparable, with respect to set-inclusion. Items are assumed to be ordered, for instance, with respect to a hierarchy of mastery dependencies. Items can be incomparable, with respect to that hierarchy. In IRT, on the other hand, persons and items are, for instance, represented by single real numbers, ability and difficulty parameters, respectively. Persons and items are linearly ordered, with respect to the natural ordering of the real numbers. Conceptually speaking, KST may be viewed as a more 'qualitative, behavioral' approach (mainly based on combinatorics and stochastic processes), unlike IRT, as a 'quantitative, statistical' approach (mainly based on calculus and statistics). Further technical and philosophical differences between the two theories are discussed in Falmagne et al. (2008, see also Chapter 1 of this volume) and Ünlü (2007).

A nonparametric, as opposed to a parametric, approach is pursued. Nonparametric IRT includes a broad range of parametric IRT models. Nevertheless, parametric IRT-type modeling strategies in KST are important directions for future research. For a logistic approach, see Stefanutti (2006); for a generalized normal ogive approach, see Ünlü (2006).

NONPARAMETRIC IRT: AXIOMS AND PROPERTIES

This section reviews the axioms underlying the Mokken (1971) nonparametric IRT models of monotone homogeneity (MHM) and double monotonicity (DMM) (see also Mokken, 1997; Sijtsma, 1998; Sijtsma and Molenaar, 2002).[17] The properties of monotone likelihood ratio (MLR) and stochastic ordering (SO) justifying the use of Mokken's models as measurement models for persons are also reviewed.

**9.2.10 Axioms.** Let $X_l$ with realization $x_l \in \{0, 1\}$ be the *item score variable* for item $I_l$ $(1 \leq l \leq m)$, and let $X_+ = \sum_{l=1}^{m} X_l$ with realization $x_+ \in \{0, 1, \ldots, m\}$ denote the *total score variable*. A function $f : \{0, 1, \ldots, m\} \to \mathbb{R}$ is *nondecreasing* iff

[16] See also Ducamp and Falmagne (1969).

[17] Throughout, only dichotomous items are considered.

$$\forall x, y \in \{0, 1, \dots, m\}, x \leq y : f(x) \leq f(y).$$

Let the *latent trait* be denoted by $\theta$, $\theta \in \Theta \subseteq \mathbb{R}$; this is referred to as the axiom of *unidimensionality*. A function $f : \Theta \to \mathbb{R}$ is *nondecreasing* iff it satisfies an obvious analog of the above condition. Let the conditional positive response probability $P(X_l = 1|\theta)$ as a function of $\theta \in \Theta$ be the *item response function* (IRF) of the item $I_l$. The axiom of *local independence* states that

$$P(X_1 = x_1, \dots, X_m = x_m|\theta) = \prod_{l=1}^{m} P(X_l = x_l|\theta)$$

for any $x_l \in \{0, 1\}$ and $\theta \in \Theta$. The axiom of *monotonicity* holds iff any IRF $P(X_l = 1|.)$ is nondecreasing. The axiom of *invariant item ordering* states that the IRFs $P(X_l = 1|.)$ can be ordered such that

$$\forall \theta \in \Theta : P(X_{l_1} = 1|\theta) \leq \dots \leq P(X_{l_m} = 1|\theta)$$

where $1 \leq l_i \leq m$ $(1 \leq i \leq m)$.

Mokken's *MHM* is based on the axioms of unidimensionality, local independence, and monotonicity. His *DMM* further adds the axiom of invariant item ordering.

**9.2.11 Properties.** MLR for the total score variable and latent trait plays an important role in IRT. It implies SO properties that can be interpreted in an IRT context (e.g., Hemker et al., 2001; Ünlü, 2007; van der Ark, 2001, 2005).

The total score variable $X_+$ has *MLR* in $\theta$ iff, for any $0 \leq x_{+,1} \leq x_{+,2} \leq m$,

$$\frac{P(X_+ = x_{+,2}|\theta)}{P(X_+ = x_{+,1}|\theta)}$$

is a nondecreasing function of (unidimensional) $\theta \in \Theta$. Similarly, the latent trait $\theta$ has *MLR* in $X_+$ iff, for any $\theta_1 \leq \theta_2$,

$$\frac{P(\theta_2|X_+ = x_+)}{P(\theta_1|X_+ = x_+)}$$

is a nondecreasing function of $x_+ \in \{0, 1, \dots, m\}$ (where $P(\theta|X_+ = x_+)$ is the conditional probability for ability $\theta$ given total score $X_+ = x_+$).

The fundamental result (Ghurye and Wallace, 1959; Grayson, 1988; Huynh, 1994; Ünlü, 2008) states that under the axioms of unidimensionality, local independence, and monotonicity, the total score variable has MLR in the (unidimensional) latent trait.

The property of MLR implies that $X_+$ is stochastically ordered by $\theta$. The *stochastic ordering of the manifest variable* $X_+$ by $\theta$ (SOM) means that, for any $0 \leq x_+ \leq m$,

$$P(X_+ \geq x_+|\theta)$$

is a nondecreasing function of (unidimensional) $\theta \in \Theta$. The MLR property also implies that $\theta$ is stochastically ordered by $X_+$. The *stochastic ordering of the latent trait* $\theta$ by $X_+$ (SOL) means that, for any $\theta_0 \in \Theta$,

$$P(\theta \geq \theta_0|X_+ = x_+)$$

is a nondecreasing function of $0 \leq x_+ \leq m$. The property of SOL is very important for practical measurement, because it justifies the use of the total score variable to estimate the ordering of subjects on the latent trait. This is the key result that justifies the use of the MHM and DMM as measurement models for persons.

#### Application of nonparametric IRT in KST

Ünlü (2007) generalizes the unidimensional nonparametric IRT axioms and properties to quasi ordered person and indicator spaces, and applies the extended IRT concepts in KST.

**9.2.12 Axioms.** Let $Q = \{I_l : 1 \leq l \leq m\}$. Let $\mathcal{K}$ be a knowledge structure on $Q$, partially ordered with respect to set-inclusion $\subseteq$. The IRT concepts can be formulated for $(\mathcal{K}, \subseteq)$. For instance, a function $f : \mathcal{K} \to \mathbb{R}$ is *isotonic* iff

$$\forall K_1, K_2 \in \mathcal{K}, K_1 \subseteq K_2 : f(K_1) \leq f(K_2).$$

The *item response function* (IRF) of an item $I_l \in Q$ is

$$P(X_l = 1|.) : \mathcal{K} \to [0, 1], K \mapsto P(X_l = 1|K).$$

The axioms of *local independence* and *isotonicity* are formulated accordingly.

Let $\mathcal{S}$ be a surmise relation on $Q$. Let $\mathcal{K}_\mathcal{S}$ be the quasi ordinal knowledge space derived from it according to Birkhoff's theorem (e.g., Doignon and Falmagne, 1999, Theorem 1.49). The axiom of *invariant item ordering* states that the IRFs $P(X_l = 1|.)$ can be ordered such that

$$\forall K \in \mathcal{K}_\mathcal{S} : P(X_{l_2} = 1|K) \leq P(X_{l_1} = 1|K) \tag{9.5}$$

for any $(I_{l_1}, I_{l_2}) \in \mathcal{S}$ $(1 \leq l_1, l_2 \leq m)$.

**9.2.13 Properties.** The properties of *MLR*, *SOM*, and *SOL* can be formulated for $(\mathcal{K}, \subseteq)$. For instance, the *stochastic ordering of the latent 'trait'*[18] $K \in \mathcal{K}$ by $X_+$ (SOL) means that, for any $K_0 \in \mathcal{K}$,

$$P(K \supseteq K_0|X_+ = x_+)$$

is a nondecreasing function of $0 \leq x_+ \leq m$.

[18] Albeit the knowledge states may be rather viewed as latent classes, we want to keep the terminology common in nonparametric IRT and speak of a latent 'trait' $K \in \mathcal{K}$.

As presented in Ünlü (2008), the fundamental result on MLR of the total score variable in unidimensional IRT is extended to quasi ordered latent trait spaces, including, as special cases, partially ordered knowledge structures. In particular, for $(\mathcal{K}, \subseteq)$, under the axioms of local independence and isotonicity, the total score variable has MLR in the (discrete-dimensional) latent trait $K \in \mathcal{K}$.

The generalized MLR property implies the generalized SOM property, but may fail to imply the generalized SOL property. The reason for this is the order-theoretic completeness property. Conditions can be specified under which the MLR property implies the SOL property, in the framework of the Mokken-type nonparametric KST formulation.

In future endeavors, the nonparametric concepts discussed in this section have to be applied in practice. An artificial numerical example is discussed below (Example 9.2.16).

PARAMETRIC VERSUS NONPARAMETRIC KST

The nonparametric KST axioms and properties are compared with the assumptions underlying the parametric *basic local independence model* (BLIM; Doignon and Falmagne, 1999, pp. 144–145). The BLIM satisfies the axiom of local independence by definition. Since the BLIM assumes item-specific, state-independent *careless error* and *lucky guess probabilities*, respectively, $\beta_l$ and $\eta_l$, at any item $I_l \in Q$, the IRF of an item $I_l \in Q$ is (as a function of $K \in \mathcal{K}$)

$$P(X_l = 1|K) = \begin{cases} 1 - \beta_l & : \text{ if } I_l \in K, \\ \eta_l & : \text{ if } I_l \notin K. \end{cases}$$

A characterization of the axiom of isotonicity under the BLIM is as follows.

**9.2.14 Theorem.** *Let $\mathcal{K}$ be a knowledge structure on $Q$. In general, a set $Q$ of BLIM IRFs does not satisfy the axiom of isotonicity. A set $Q$ of BLIM IRFs satisfies the axiom of isotonicity if, and only if, $\eta_l \leq 1 - \beta_l$ for any $I_l \in Q$ $(1 \leq l \leq m)$.*

PROOF. See Ünlü (2007, Theorem 6). □

The axiom of invariant item ordering (see Axiom 9.2.12 on Page 169) can be characterized as follows.

**9.2.15 Theorem.** *Let $\mathcal{S}$ be a surmise relation on $Q$, and let $\mathcal{K}_\mathcal{S}$ be the corresponding quasi ordinal knowledge space. In general, a set $Q$ of BLIM IRFs does not satisfy the axiom of invariant item ordering. A set $Q$ of BLIM IRFs exhibits an invariant item ordering if, and only if, for any $(I_{l_1}, I_{l_2}) \in \mathcal{S}$ $(1 \leq l_1, l_2 \leq m)$,*

$$\beta_{l_1} \leq \beta_{l_2},$$
$$\eta_{l_1} \geq \eta_{l_2},$$

*and if* $(I_{l_2}, I_{l_1}) \notin \mathcal{S}$, *in addition,*

$$1 - \beta_{l_1} \geq \eta_{l_2}.$$

PROOF. See Ünlü (2007, Theorem 7). □

**9.2.16 Example.** Let $Q = \{I_1, I_2, I_3\}$, and $\mathcal{K} = \{\emptyset, \{I_1\}, \{I_1, I_2\}, \{I_1, I_3\}, Q\}$. One can specify values for the BLIM parameters such that (a) the BLIM IRFs are isotonic, assuming values strictly between zero and one, therefore (b) the property of MLR holds, and (c) for $K_0 = \{I_1, I_2\}$, $x_{+,1} = 1$, and $x_{+,2} = 2$ the SOL property is violated. Such a choice of values for the BLIM parameters is given by $p(K) = 1/5$ for any $K \in \mathcal{K}$ (uniform distribution), and $\beta_1 = 0.75$ / $\eta_1 = 0.10$, $\beta_2 = 0.82$ / $\eta_2 = 0.16$, and $\beta_3 = 0.71$ / $\eta_3 = 0.07$. This specification yields $P(K \supseteq K_0 | X_+ = x_{+,1}) \approx 0.532 > 0.477 \approx P(K \supseteq K_0 | X_+ = x_{+,2})$. Hence, this set of BLIM IRFs does not possess the SOL property.

In the example, the property of MLR does not imply the SOL property, even in case of this restrictive set of parametric BLIM IRFs satisfying the axioms of local independence and isotonicity. However, simulations demonstrate that violations of the SOL property occur only for extreme (unrealistic) values of some of the BLIM parameters; for non-extreme and thus practical parameter vectors the BLIM seems to satisfy the property of SOL. Therefore, it is necessary to check for the SOL property in any fitted BLIM, if at all of interest.

Statistical and probabilistic contributions to KST have been presented generalizing the theory of knowledge spaces in parametric (Stefanutti, 2006; Ünlü, 2006) and nonparametric (Ünlü, 2007, 2008) IRT related directions.

The proposed nonparametric Mokken-type formulation in KST is new. It must be further elaborated in research, as a necessary prerequisite for the development of a superior probabilistic test theory, with corresponding statistical inference methodology. Such a theory could include most of the existing IRT and KST models as special cases. For example, the elaboration of a Mokken-type scale analysis for the surmise relation or even surmise system model would be an important contribution. Especially, the nonparametric concepts discussed in this section will have to be applied in practice, and the practical implications of these properties and conditions should be studied using real data.

## 9.3 Methods for Empirical Construction and Validation of Knowledge Structures

**9.3.1 Data Driven Generation of Knowledge Structures.**
A knowledge structure can be constructed by several methods, for example querying experts, deriving the structure from psychological models or explorative data analysis (see later in this chapter, and Chapter 10).

There are several different data analytical methods available, which can be used to construct a knowledge structure. The central assumption of all these methods is that there is a true knowledge structure underlying the data and that the observed response patterns result from this true knowledge structure by lucky guesses and careless errors. The algorithms have to solve the challenge that both, the true knowledge structure and the probabilities for these random response errors are unknown.

We can separate the available methods into two groups. The first group of methods is related to Boolean analysis of questionnaires (see Flament, 1976). Algorithms of this group construct a surmise relation on the item set from the data (e.g., Sargin and Ünlü, 2009; Schrepp, 1999b; Theuns, 1994, 1998; van Buggenhaut and Degreef, 1987; van Leeuwe, 1974). This surmise relation determines the corresponding quasi ordinal knowledge space. The second group of methods constructs a knowledge structure directly from the data (e.g., Schrepp, 1999a).

All existing methods to construct knowledge structures from data share more or less the same structure. In a first step they construct a set of candidates for a surmise relation respectively knowledge structure. In the second step these candidates are evaluated against the data by using a specific fit measure. In the third step a best-fitting candidate is chosen.

Closely related to the empirical construction is the empirical evaluation of existing knowledge structures obtained, for example, by querying experts. In principle the fit measures used to select the best fitting candidate can be used also for this purpose, and a unified maximum likelihood significance testing procedure based on such a measure has been proposed by Ünlü and Sargin (2010).

We describe now concrete analysis methods to construct a knowledge structure from data. These are the currently most prominent methods of this group of data analytical algorithms in KST.[19]

Before we go into the detailed description of the methods we need some definitions. Let $Q$ be a set of $m$ questions. $D = \{d_1, ..., d_n\}$ is a set of $n$ observed response patterns to the questions in $Q$. Each $d_s$ is a mapping $d_s : Q \to \{0,1\}$ which assigns to each question $i$ the response $d_s(i) \in \{0,1\}$ of subject $s$. Let $p_i$ be the relative frequency of subjects who solve question $i$, i.e. $p_i =| \{d_s \in D \mid d_s(i) = 1\} \mid /n$. Let $b_{ij}$ be the number of response patterns which violate the dependency $i \leq j$ (a set of true dependencies, denoted by $\leq$, is assumed to underlie), i.e. $b_{ij} =| \{d_s \in D \mid d_s(i) = 0 \wedge d_s(j) = 1\} \mid$.[20]

[19] See Chapter 10 for applications.

[20] A dependency $i \leq j$ means that each person who is able to solve item $j$ is also able to solve item $i$. Thus, a data pattern $d_s$ violates such a dependency if $d_s(i) = 0$ and $d_s(j) = 1$. For an illustration, see also Figure 9.8.

Item tree analysis (ITA)

**Step 1:** A set of candidate relations, i.e. a set of quasi orders on $Q$, $\mathrm{PQO}(D) = \{\leq_L \mid L = 0, \ldots, n\}$ is constructed by defining $i \leq_L j \Leftrightarrow b_{ij} \leq L$. van Leeuwe (1974) showed that $\leq_0$ is transitive, but that this is not necessarily the case for relations $\leq_L$ with $L > 0$.[21]

**Step 2:** All relations $\leq_L$ are evaluated by the *correlational agreement coefficient* $\mathrm{CA}(\leq_L, D)$ defined by:

$$\mathrm{CA}(\leq_L, D) = 1 - \frac{2}{m(m-1)} \sum_{i<j} (r_{ij} - r^*_{ij})^2$$

where $r_{ij}$ is the Pearson correlation between i and j and $r^*_{ij}$ is defined by:

$$r^*_{ij} = \begin{cases} 1 & \text{if } i \leq_L j \wedge j \leq_L i \\ \sqrt{(1-p_i)p_j/(1-p_j)p_i} & \text{if } i \leq_L j \wedge j \not\leq_L i \\ \sqrt{(1-p_j)p_i/(1-p_i)p_j} & \text{if } i \not\leq_L j \wedge j \leq_L i \\ 0 & \text{otherwise.} \end{cases}$$

The coefficient $r^*_{ij}$ is interpreted as the expected correlation under the assumption that $\leq_L$ is the correct relation underlying the data. For a detailed explanation of the algorithm of ITA see van Leeuwe (1974) and Ünlü and Albert (2004). Potential problems of the correlational agreement coefficient are discussed in Ünlü and Albert (2004) and Schrepp (2006b). For a general discussion of the desirable properties of fit measures for quasi orders see Schrepp (2007).

**Step 3:** All relations that are not transitive and all relations that are inconsistent with more than a predefined percentage of the observed response patterns (i.e., the percentage of data patterns violating a dependency postulated by the binary relation exceeds a predefined threshold) are now eliminated from $\mathrm{PQO}(D)$. From the remaining relations the one with the highest $\mathrm{CA}(\leq_L, D)$ value is chosen.

Applications of ITA to data sets can be found in Bart and Airasian (1974), Bart and Krus (1973), Held and Korossy (1998), and van Leeuwe (1974), and in Chapter 10 of this volume.

Inductive item tree analysis (IITA)

**Step 1:** A set of candidate surmise relations $\mathrm{IPQO}(D)$ is defined by an inductive process. We start with the surmise relation $\leq_0$ defined by $i \leq_0 j \Leftrightarrow b_{ij} = 0$ for all $i, j \in Q$. Assume that the surmise relation $\leq_L$ is already constructed.

[21] Counterexamples in the latter case (for $L > 0$) can be easily constructed, and transitivity (and reflexivity) of the binary relation $\leq_0$ follows by contradiction.

In step $L+1$ of the process all item pairs $(i, j)$ which fulfill the condition $b_{ij} \leq L+1$ and do not cause an intransitivity to other dependencies already contained in $\leq_L$ are added to $\leq_L$ to construct $\leq_{L+1}$. For details of the construction process see Schrepp (1999b, 2003). Comments on this process and a first evaluation of the quality of the inductive construction procedure are presented by Sargin and Ünlü (2009).

**Step 2:** The probability $\gamma$ that a true dependency $i \leq_L j$ is violated due to random errors is estimated by:

$$\gamma = \frac{\sum\{b_{ij}/(p_j\, n) \mid i \leq_L j \wedge i \neq j\}}{(|\leq_L| - m)}$$

The probability $\gamma$ is now used to estimate for each pair of questions the expected number $t_{ij}$ of violations under the assumption that $\leq_L$ is the correct surmise relation underlying the data. We distinguish two cases:

1. $i \not\leq_L j$: In this case we assume that the items $i$ and $j$ are independent. Thus, $t_{ij}$ equals the expected number of response patterns $d$ with $d(i) = 0$ and $d(j) = 1$, so we have $t_{ij} = (1 - p_i)\, p_j\, n\, (1 - \gamma)$.
2. $i \leq_L j$ and $i \neq j$: In this case all violations of $i \leq_L j$ must result from random errors. Thus, $t_{ij} = \gamma\, p_j\, n$.

The fit between each $\leq_L$ and $D$ is measured by the diff$(\leq_L, D)$ coefficient:

$$\text{diff}(\leq_L, D) = \frac{\sum_{i \neq j}(b_{ij} - t_{ij})^2}{(m^2 - m)}$$

**Step 3:** The quasi order from IPQO$(D)$ which shows the smallest diff-value is chosen.

Applications of IITA can be found in Schrepp (2002, 2003). It was shown in a simulation study (Schrepp, 1999b) that IITA shows a better performance in reconstructing the correct surmise relation from data than ITA. An exception are cases where the underlying surmise relation is very close to a linear order; in these cases ITA shows the better performance. The study showed also that IITA is able to reconstruct the underlying structure of the items with high accuracy as long as the error probabilities are small (see Schrepp, 1999b, 2003).

An explanation for these observations is given by Sargin and Ünlü (2009). For instance, the above (original) IITA algorithm gives bad results when longer chains of items are present in the underlying quasi order, because in this case estimation of the expected numbers of counterexamples is methodologically inconsistent (for details, see Sargin and Ünlü, 2009, Section 2.2). Therefore, corrections and improvements to this 'original' algorithm have been made, resulting in two further data analytic procedures of the inductive type, the corrected and minimized corrected IITA algorithms. These algorithms use the corrected estimators, and for instance, therefore detect true dependencies of a chain structure more properly.

As proposed by Sargin and Ünlü (2009), a correct choice for $t_{ij}$ for $i \not\leq_L j$ depends on whether $j \not\leq_L i$ or $j \leq_L i$. If $i \not\leq_L j$ and $j \not\leq_L i$, set $t_{ij} = (1-p_i)p_j n$. Independence is assumed, and the additional factor $(1-\gamma)$ is omitted. If $i \not\leq_L j$ and $j \leq_L i$, set $t_{ij} = (p_j - (p_i - p_i\gamma))n$. This estimator is derived as follows. The observed number of people who solve item $i$ is $p_i n$. Hence the estimated number of people who solve item $i$ and item $j$ is $p_i n - t_{ji} = (p_i - p_i\gamma)n$. Note that $j \leq_L i$, and the estimator is $t_{ji} = p_i\gamma n$. Eventually this gives the estimate $t_{ij} = p_j n - (p_i - p_i\gamma)n = (p_j - (p_i - p_i\gamma))n$. This estimator is not only mathematically motivated, but is also interpretable. The corrected IITA algorithm is based on these estimators (using the same inductive construction procedure and rate $\gamma$ of the above original algorithm).

The minimized corrected IITA algorithm is obtained as follows. Let the diff coefficient be based on these corrected estimators. The diff coefficient can be viewed as a function of the error probability $\gamma$, and we minimize the function with respect to $\gamma$. The fit measure then favors quasi orders that lead to smallest minimum discrepancies, or equivalently, largest maximum matches, between the observed and expected summaries $b_{ij}$ and $t_{ij}$. This optimum error rate can be expressed in closed analytical form (Sargin and Ünlü, 2009, Section 3.2) and can now be used for an alternative IITA procedure (based on the estimators of the corrected algorithm and the inductive procedure of the original algorithm), in which a minimized diff value is computed for every quasi order.

The program ITA 2.0 implements both classical and inductive ITA. It can be downloaded free of charge under http://www.jstatsoft.org/v16/i10 (Journal of Statistical Software). The handling of the program is described in Schrepp (2006a).

The R package (R Development Core Team, 2012) DAKS for data analysis methods, procedures for simulating data and quasi orders, and for transforming different representations in KST has been developed by Ünlü and Sargin (2010); see also Sargin and Ünlü (2010). It is available from the Comprehensive R Archive Network[22]; the paper and package can also be downloaded from the Journal of Statistical Software's Web page[23]. In particular, this R package DAKS implements the original inductive ITA as well as the two corrected and minimized corrected IITA algorithms (including further simulation and statistical testing procedures).

Direct extraction of knowledge structures from data

**Step 1:** A set of subsets of of the power set $2^Q$ is defined by $\mathcal{K}_L = \{d \in D \mid f(d) \geq L\}$ where $f(d)$ is the frequency of $d$ in the data set.

**Step 2:** The central idea of the algorithm is to compare the observed frequencies of elements of $2^Q$ with an estimated frequency under the assumption

[22]See http://CRAN.R-project.org/package=DAKS

[23]See www.jstatsoft.org/v37/i02

that $\mathcal{K}_L$ is the correct knowledge structure. Therefore, careless error and lucky guess probabilities $\beta_L$ and $\eta_L$ are estimated from the data set and $\mathcal{K}_L$. Details of the estimation procedure can be found in Schrepp (1999a).

For each element $S \in 2^Q$ (i.e., a possible response pattern) and each knowledge state $K$ in $\mathcal{K}_L$ the probability that a subject in state $K$ shows the response pattern $S$ is calculated by:

$$P(S|K) = \beta_L^{|K \setminus S|} \eta_L^{|S \setminus K|} (1 - \beta_L)^{|K \cap S|} (1 - \eta_L)^{|Q \setminus (K \cup S)|}$$

Then a frequency distribution $F_L$ on $2^Q$ can be defined by:

$$F_L(S) = \frac{\sum_{K \in \mathcal{K}_L} P(S|K)}{| \mathcal{K}_L |}$$

The fit between $\mathcal{K}_L$ and the data set $D$ can now be measured by:

$$app(\mathcal{K}_L, D) = \frac{\sum_{S \in 2^Q} (f(S) - F_L(S))^2}{2^{|Q|}}$$

**Step 3:** The fit of each $\mathcal{K}_L$ is now evaluated by the app coefficient. The knowledge structure with the smallest value for $app(\mathcal{K}_L, D)$ is chosen. The ability of this algorithm to reconstruct the correct knowledge structure from data was tested in a simulation study, see Schrepp (1999a). The results showed that the method is able to reconstruct the underlying knowledge structure with high accuracy if enough response patterns are available in the data set.

SIZE/FIT TRADE-OFF EVALUATION CRITERION $\kappa$

Critical studies of the correlational agreement coefficient CA (see ITA in this section) pointed out that this measure should be used, if at all, only with special caution, to determine the optimal surmise relation for a given set of data. As an alternative (cf. also ITA in this section for the coefficient diff), Ünlü (2009) and Ünlü and Malik (2008b) proposed the coefficient $\kappa$, which is especially designed to trade-off the descriptive fit[24] of and the number of states in a knowledge structure.

The coefficient $\kappa$ is a 'size trading-off' fit measure. It has two constituents, $m_1$ and $m_2$, which account for the fit and size of a knowledge structure, respectively. The derivation of this coefficient is motivated by the unitary method of *proportional reduction in predictive error* (PRPE), which was used in a series of papers by Goodman and Kruskal (e.g., 1954, 1963).

Suppose an individual is randomly chosen from a population of reference, and we are asked to guess his/her response pattern, given, either

- 'no info': no further information (other than the multinomial sampling distribution on $2^Q$), or

[24] Cautionary note: In the sense that knowledge states manifest themselves (i.e., are observed) as response patterns in a data set.

- 'info': the knowledge structure $\mathcal{K}$ assumed to underlie the responses of the individual.

In the 'no info' case, we 'optimally' guess some response pattern $R_m \in 2^Q$ with the largest probability of occurrence $\rho(R_m) = \max_{R \in 2^Q} \rho(R)$. In the 'info' case, we 'proportionally' guess the knowledge states $K \in \mathcal{K}$ with their probabilities of occurrence $\rho(K)$. If $\mathcal{K} = \{K_1, K_2, \ldots, K_{|\mathcal{K}|}\}$, we guess $K_1$ with probability $\rho(K_1)$, $K_2$ with probability $\rho(K_2)$, ..., $K_{|\mathcal{K}|}$ with probability $\rho(K_{|\mathcal{K}|})$. To complete the prediction strategy, we abstain from guessing with probability $1 - \sum_{K \in \mathcal{K}} \rho(K)$, and in the sequel, view that as a prediction error.

In the 'no info' case, the probability of a prediction error is $1 - \rho(R_m)$; in the 'info' case, it is $1 - \sum_{K \in \mathcal{K}} \rho(K)^2$. The general probability formula of the method of PRPE (Goodman and Kruskal, 1954) quantifies the predictive utility, $\mathrm{PU}_{\mathrm{info}}$, of given information:

$$\mathrm{PU}_{\mathrm{info}} = \frac{\text{Prob. of error (no info)} - \text{Prob. of error (info)}}{\text{Prob. of error (no info)}}.$$

**9.3.2 Definition (First Constituent of $\kappa$: Measure of Fit).** The measure $m_1$ is defined as[25]

$$m_1 = \frac{(1 - \rho(R_m)) - (1 - \sum_{K \in \mathcal{K}} \rho(K)^2)}{1 - \rho(R_m)}.$$

The maximum likelihood estimate (MLE), $\widehat{m_1}$, for $m_1$ is

$$\widehat{m_1} = \frac{\sum_{K \in \mathcal{K}} N(K)^2 - N \cdot N(R'_m)}{N^2 - N \cdot N(R'_m)},$$

for a response pattern $R'_m \in 2^Q$ with absolute count $N(R'_m) = \max_{R \in 2^Q} N(R)$.

The definition of the second constituent of $\kappa$ is based on the following notion of a truncation of a knowledge structure.

**9.3.3 Definition (M-truncation).** Let $M \in \mathbb{N}$ be a truncation constant.[26]

An $M$-truncation of $\mathcal{K}$ is any subset, $\mathcal{K}_{\text{M-trunc}}$, of $\mathcal{K}$ which is derived as follows.

1. Order the knowledge states $K \in \mathcal{K}$ according to their probabilities of occurrence $\rho(K)$, say, from left to right, ascending from smaller $\rho$ values to larger ones. Knowledge states with equal probabilities of occurrence are ordered arbitrarily.

[25] Loosely speaking, the first constituent expresses the extent to which the manifest multinomial probability distribution on the response patterns is concentrated to the knowledge structure.

[26] A special choice for $M$ may be given in a model selection context, in particular penalizing for and evaluating the appropriate specification of selection sets of knowledge structures, for instance in ITA type data analysis methods. For details, see Ünlü (2009) or Ünlü and Malik (2008b).

2. Starting with the foremost right knowledge state, a knowledge state with the largest probability of occurrence, take the first $\min(|\mathcal{K}|, M)$ knowledge states, descending from right to left. The set of these knowledge states is called an $M$-truncation of $\mathcal{K}$, denoted by $\mathcal{K}_{\text{M-trunc}}$.

**9.3.4 Definition (Second Constituent of $\kappa$: Measure of Size).** The measure $m_2$ is defined as[27]

$$m_2 = \frac{\sum_{K \in \mathcal{K}} \rho(K)^2}{\sum_{K \in \mathcal{K}_{\text{M-trunc}}} \rho(K)^2},$$

where $M \in \mathbb{N}$ is a truncation constant, and $\mathcal{K}_{\text{M-trunc}}$ denotes an $M$-truncation of $\mathcal{K}$. The MLE, $\widehat{m_2}$, for $m_2$ is

$$\widehat{m_2} = \frac{\sum_{K \in \mathcal{K}} N(K)^2}{\sum_{K \in \widehat{\mathcal{K}_{\text{M-trunc}}}} N(K)^2},$$

where $\widehat{\mathcal{K}_{\text{M-trunc}}}$ is analogously defined as in the definition of an $M$-truncation (we have to replace $\rho(K)$ with its MLE $N(K)/N$ for any $K \in \mathcal{K}$).

**9.3.5 Definition (Coefficient $\kappa$: Size/Fit Trade-Off).** Let $M \in \mathbb{N}$ be a truncation constant. Let $C \in [0, 0.01]$ be a small, fixed non-negative shift constant.[28] The measure $\kappa$ is defined as

$$\kappa = m_2 \cdot (m_1 - C) = \frac{\sum_{K \in \mathcal{K}} \rho(K)^2}{\sum_{K \in \mathcal{K}_{\text{M-trunc}}} \rho(K)^2} \cdot \left( \frac{\sum_{K \in \mathcal{K}} \rho(K)^2 - \rho(R_m)}{1 - \rho(R_m)} - C \right).$$

The MLE, $\widehat{\kappa}$, for $\kappa$ is

$$\widehat{\kappa} = \frac{\sum_{K \in \mathcal{K}} N(K)^2}{\sum_{K \in \widehat{\mathcal{K}_{\text{M-trunc}}}} N(K)^2} \cdot \left( \frac{\sum_{K \in \mathcal{K}} N(K)^2 - N \cdot N(R'_m)}{N^2 - N \cdot N(R'_m)} - C \right).$$

The decision rule for applications of $\kappa$ (for the selection among competing knowledge structures) is as follows. The greater the (population) value of $\kappa$, the 'better' a knowledge structure 'performs' with respect to a trade-off between descriptive fit (cf. Footnote 24 on Page 176) and structure size.

[27] Loosely speaking, the second constituent expresses the extent to which the restricted multinomial probability distribution on the knowledge states is concentrated to a fraction of the knowledge structure. In a sense, both constituents vary 'on same scale,' which is generally not the case if fit is penalized, for instance, by the (large) number of states.

[28] The constant $C$ is introduced to compensate for a zero value of $m_1$ and can be omitted if that is not the case. Sensitivity analyses are necessary in order to control for this constant (and for $M$ as well), ideally to be estimated statistically.

The unknown ordering of the $\kappa$ values is estimated by the ordering of the corresponding MLEs.[29]

Properties of $\kappa$ have been discussed in the aforementioned references. An application of this criterion to empirical data obtained from dichotomously scored inductive reasoning test items is presented in the work by Ünlü and Malik (2008a). For a further application, see also Chapter 10 in this volume.

### Other construction methods

We have discussed four methods to construct knowledge structures respectively surmise relations from data in more detail. But there are of course other methods available, which we can only mention shortly. A potentially interesting approach is to consider knowledge structures as a special type of latent class models (see Martin and Wiley, 2000; Schrepp, 2005; Ünlü, 2006, and Section 9.2.8 of this chapter). This special view allows to use methods of latent class analysis for the empirical construction of a knowledge structure.

Other potentially interesting methods, which can be used in that context, are the extraction of association rules by data mining techniques (e.g., Toivonen, 1996) or the GUHA method (e.g., Hájek et al., 1966; Hájek and Havránek, 1977). Currently, however, there are no simulation studies available directly showing how accurate these methods are in reconstructing the underlying structure from simulated data.

Interactive graphical approaches, as complementary to numerical data analysis methods, using mosaic plots, glyphs, or barcharts and spineplots have been discussed by Ünlü and Malik (2011), Ünlü and Sargin (2009, 2011).

### Empirical comparison of existing knowledge structures

Sometimes researchers have to select between a number of given knowledge structures $\mathcal{K}_1, \ldots, \mathcal{K}_r$ (see Chapter 10 of this volume). These knowledge structures can result, for example, from querying $r$ experts or from $r$ competing psychological models of the problem solving behavior in the domain $Q$. It is of course possible to use the fit measures described above to solve this comparison task. If the knowledge structures are quasi ordinal knowledge spaces we can use the correlational agreement coefficient or the diff coefficient to compare the fit of the corresponding surmise relations to observed data. Concerning this task, detailed simulation studies (see Schrepp, 2007) show that the diff coefficient shows better results than the correlational agreement coefficient.

It is also possible to use the app coefficient to directly compare the knowledge structures. In a simulation study, Schrepp (2001) shows that this method is highly accurate in its evaluation of knowledge structures.

---

[29] So far, no latencies are assumed, and the idea underlying the coefficient can be utilized to incorporate such extensions of the approach. Note also that the measure $\kappa$ (as a size/fit trade-off evaluation) is designed to allow for a 'poorer' (descriptive) fit of a knowledge structure in favor of a smaller size (which may be important for efficient applications of adaptive knowledge assessment procedures).

### 9.3.6 Methods for Validation[30].

For using a specified knowledge structure, e.g., for adaptive teaching or adaptive assessment its validity is an essential and necessary pre-condition. Imagine for instance a system selecting the next to be presented item depending on a student's answer which is based on an empirically wrong prerequisite/surmise relation or knowledge space. As a result, the next item presented might not yield the expected information. Thus, empirical validity of the structure is a necessary pre-condition for adaptive assessment and teaching.

For validating a knowledge structure the whole set of items has to be presented to subjects of a heterogeneous sample, which has not necessarily be statistically representative for the whole population of students in question. However experts and novices or subjects who are specifically and isolated trained to perform a special subset of items are of minor or even no value for at least two methodological reasons. (a) The results of the investigation can be easily manipulated by selecting subjects of these types and (b) the data are of low diagnostic value, because the sets Q and the empty set belong by definition to the knowledge structure. On the other hand the question arises, what about items which are very easy or very hard to answer for all the subjects of that sample. In that case a ceiling or bottom effect would result for these items. They have little or no diagnostic value. The same is true of subjects. Those subjects who solve all or no items are useless for the validation. In praxis these items and subjects should not be taken into account for validation analysis.

Of course other usual validation methods have to be utilized. If for instance the hypothetical structure has been generated from data the method of cross validation has to be used for generating and validating the hypothetical structure.

In this section, we discuss different validation methods, based on the surmise relation as well as on the knowledge space. Strictly viewed, the goodness of fit of a given knowledge space (and its corresponding surmise relation) to a set of data can be defined by the number of response patterns which can be assigned to exactly one of the postulated states. However, working with a deterministic model we have to consider that the knowledge states describe the latent knowledge of persons, whereas the response patterns might include careless errors and/or lucky guesses. Furthermore, the fit of a knowledge structure to a set of data increases with an increasing size $|\mathcal{K}|$ (just as e.g., the fit of a surmise relation increases with a decreasing number of pairs). In the extreme case where all items are independent, that is the relation contains only the pairs $(q, q)$ resulting in $|\mathcal{K}| = 2^{|Q|}$, the data are always in accordance with the knowledge space.

In the following, we mention different validation methods, which take into account that space size may vary as well as that there may exist discrepancies between a person's true knowledge state and his or her response pattern.

[30]For illustrative empirical applications, see Chapter 10.

Because two approaches for validating specific KST-hypotheses—the validation via the surmise relation and the validation via the knowledge structure (or space)—are common, the question arises which one should be used in practice. Regarding the deterministic theory both structures are equivalent in view of a theorem of Birkhoff, generalized by Doignon and Falmagne (1985). However, the resulting structures are not necessarily the same when analyzing noisy data. Thus, usually both methodological approaches are used for validating specific hypotheses, because each of them captures different aspects.

## Validation of hypotheses via the surmise/prerequisite relation

Usually as a first and rough method to evaluate a surmise hypothesis is done by means of the solution frequency per item. After that more sophisticated methods are applied. In case of fitting the data to a hypothesis in terms of a surmise relation in principle the measures used to select the best fitting candidate see Section 9.3.6 can be used also for this purpose. Indices for the fit of a surmise relation are described below exemplarily by $VC$ and $\gamma$. In any case, to interpret the indices, we need to recognize that they are pragmatical approaches to test the fit of a surmise relation and that they are mainly used to compare different relations or their subsets.

*Percentage of correct solutions*

Imagine a relationship $ySx$, where item $x$ should only be solved in combination with a correct solution to $y$, while $y$ can also be solved by itself (see Figure 9.8).

| | | $x$ = 0 | $x$ = 1 |
|---|---|---|---|
| $y$ | 0 | $a_{xy}$ | $b_{xy}$ |
| | 1 | $c_{xy}$ | $d_{xy}$ |

**Figure 9.8.** Possible response patterns and assumed relationship for two items $x$ and $y$ ($ySx$).

*Indices for the fit of a surmise relation*

In order to take into account the number of pairs in the surmise relation in determining the goodness of fit, we suggest to use the following two indices ($VC$ and $\gamma$), which assess the fit of a hypothetical surmise relation to a set of data. Both indices account for the number of pairs in a relation and thus allow for a comparison of the different subsets.

The *violational coefficient* ($VC$) evaluates the fit of a surmise relation to a binary data matrix by relating the number of violations to the number of response vectors for each pair $ySx$, with $x \neq y$ in a surmise relation $S$ (Schrepp et al., 1999). Violations are defined as those cases, in which a person masters item $x$ but fails in mastering its prerequisite item $y$ (cell $b_{xy}$ in Figure 9.8). Denote the number of these cases by $v_{xy}$. Formally,

$$VC = \frac{1}{n(|S| - m)} \sum_{(y,x) \in S \setminus \Delta} v_{xy}, \tag{9.6}$$

with $n$ denoting the number of response patterns, $m$ the number of items, $|S|$ the number of pairs in the relation $S$, and $\sum v_{xy}$ the number of violations. Thus, $VC$ denotes the averaged number of violations for all item pairs $ySx$ contained in $S$ (with $x \neq y$). $VC$ varies within the limits $[0, 1]$, with 0 indicating a perfect fit (no violations at all).[31]

The second index used for validation is the *gamma-index* ($\gamma$), which was first proposed by Goodman and Kruskal (1954; see also Körner and Albert, 2001, and Scheiblechner, 2003). Assuming a surmise relationship $ySx$, the contradicting response vectors $\langle 1, 0 \rangle$ are called discordant (cell $b_{xy}$ in Figure 9.8), which corresponds to $VC's$ violations ($v_{xy}$). The vector $\langle 0, 1 \rangle$, on the other hand, confirms the hypothesis, because we expect some of the participants to solve item $y$ but not the more difficult item $x$. These cases are called concordant (cell $c_{xy}$ in Figure 9.8). The cases where either both or neither of the items are solved (cells $d_{xy}$ and $a_{xy}$ in Figure 9.8) are trivial and do not allow any conclusions on the hypothesis' validity.

For each item pair, $\gamma$ is calculated as follows:

$$\gamma = \frac{c_{xy} - b_{xy}}{c_{xy} + b_{xy}}, \tag{9.7}$$

with $c_{xy}$ denoting the number of concordant vectors and $b_{xy}$ the number of discordant vectors over all answer patterns. The $\gamma$-index varies within the limits $[-1, 1]$, with 1 indicating a perfect fit (no discordant vectors). In order to evaluate the overall fit of a model, a global index $\gamma_G$ is computed by accumulating the frequencies of concordant and discordant cases over all item pairs $(x, y)$ with $x \neq y$ in the relation. Item pairs which are independent with regard to a given hypothesis (in the Hasse diagram, there is no direct or indirect line connecting the two items) are not taken into account. Contrary to $VC$, the $\gamma$-index does not include the response vectors $\langle 0, 0 \rangle$ and $\langle 1, 1 \rangle$ (cells $a_{xy}$

---

[31] The coefficient $VC$ is an ad hoc measure utilized descriptively as a tool for purely exploratory purposes. So far, no sampling and inference are considered. In particular, intermediate values of the coefficient, as an average of numbers, must be interpreted with special caution. For instance, a larger value of $VC$ may result from few item pairs with many violations as opposed to a smaller value obtained from many item pairs with few violations. For a similar, but general discussion, see Ünlü and Albert (2004).

and $d_{xy}$ in Figure 9.8), which are always in accordance with a given surmise relation and are therefore diagnostically not relevant. Thus, $\gamma$ is a stricter test of a hypothesis than $VC$.

VALIDATION OF HYPOTHESES VIA THE KNOWLEDGE SPACE

The second approach to evaluate the hypothesized structures is to determine the agreement between a data set and a given knowledge space. For this purpose, symmetric distances and the *distance agreement coefficient* ($DA$) are calculated for the knowledge spaces between items $KSbI$ and their substructures (see above). Furthermore, the empirical distances are compared to those of simulated data sets.

*Mean minimal symmetric distance ($MMSD$)*

Generally, the distance $d$ between two sets $A$ and $B$ is defined as the number of elements contained in the symmetric set difference of $A$ and $B$ (Davey and Priestley, 1990). Formally,

$$d(A, B) = |(A \setminus B) \cup (B \setminus A)|, \tag{9.8}$$

To calculate the $MMSD$ between a binary data set and a knowledge structure, first the minimal distance $d$ between each response pattern and the knowledge states is computed and then averaged over all patterns (e.g., Albert and Held, 1994; Kambouri et al., 1994). The $MMSD$ has a theoretical minimum ($dmin$) and maximum ($dmax$). The theoretical minimum corresponds to perfect agreement ($dmin = 0$), the theoretical maximum denotes the greatest possible distance depending on the number of items ($dmax = m/2$ or $(m-1)/2$ for even or odd numbers respectively).

*Distance agreement coefficient (DA)*

We are mainly interested in the validity of the surmise relation between tests, *SRbT*. It is therefore necessary to check whether the $MMSD$ of the entire structure is primarily due to the $MMSD$s within or across tests. Considering the varying sizes of the postulated knowledge structures, a comparison of the substructures' fit has to account for the number of knowledge states within the respective power sets. Therefore, $DA$ is calculated, which relates the fit of a structure to a data set to the structure's size (e.g., Schrepp et al., 1999). Formally,

$$DA = \frac{ddat}{dpot}, \tag{9.9}$$

with $ddat$ the empirical $MMSD$ and $dpot$ the $MMSD$ between the $2^{|Q|}$ elements of the power set and the postulated knowledge structure. The lower $DA$, the better the fit of a knowledge structure to a given data set. $DA$'s minimum equals zero; no fit at all is characterized by $DA = 1$.

*Simulations*

For evaluating the goodness of fit of a KS-hypothesis a comparison between the fit for real data and for artificial, simulated data is useful. One way to compare the empirical *MMSDs* (*ddat*) with more appropriate values than the extremes (*dmin*, *dmax*) is to do simulations with various degrees of specificity. A comparison of the empirical and the simulated *MMSDs* shows, whether the fit of the knowledge structure to the empirical data set is better than chance. The least specific type is the computation of random response patterns with pure guessing for the respective number of items (random simulation). In this case, patterns are simulated with a constant probability of 0.5 to answer an item correctly (for each entry in the data matrix, the probability for a correct answer equals the probability for an incorrect answer, i.e. $p(1) = p(0) = 0.5$). For various sets of random data, the *MMSDs* ($dsim_r$) between the random data sets and a given knowledge structure are calculated. This simple and inappropriate method usually results in a much better fit of the KS-hypothesis for real data solely because of the incorrect choice of answer probabilities for the simulated data.

In further approaches, sets of data are simulated under consideration of the empirical solution frequencies (frequency simulations), that is the simulated data are based on the number of '1s' per row and/or column in the empirical data matrix. The following four types of frequency simulations seem to be useful: ($simf_1$) for each entry in the data matrix, the probability of a correct solution equals the mean solution frequency over all items and persons, ($simf_2$) for each row of entries, the probabilities are based on the marginal frequencies, i.e. the mean solution frequencies per person, ($simf_3$) for each column of entries, the probabilities are based on the marginal frequencies, i.e. the mean solution frequencies per item, and ($simf_4$) for each entry the probabilities are based on the marginal solution frequencies per item and person.

For ($simf_4$) the marginal frequencies of the simulated matrices equal those of the original data matrix, whereas the distributions of '0's and '1's differ. The used algorithm (Ponocny and Waldherr, 2002) randomly selects two rows and two columns of the given matrix. If the four selected entries show either the pattern $\begin{smallmatrix}10\\01\end{smallmatrix}$ or $\begin{smallmatrix}01\\10\end{smallmatrix}$, the '0's and '1's are exchanged, otherwise the algorithm selects another pair of rows and columns. The total number of potentially exchanged entries amounts to $n \times m + x$, with $n$ denoting the number of response patterns, $m$ the number of items, and $x$ an arbitrary number of additional selections. Most often for real data the fit is not significantly better than for ($simf_4$)-type artificial data. The reasons are: the chains of surmise relations are linear orders, and also real data have a random component, too.

Finally, response patterns can be simulated on the basis of the hypothesis (i.e. the postulated knowledge space), but under consideration of the probabilities for careless errors and lucky guesses (probability simulation). In the case of multiple-choice items, the probability for lucky guesses can be estimated from the number of answer alternatives (e.g. $1/8 = 0.125$ for 8 alternatives), whereas the probability for careless errors is unknown and has to be estimated from the real data. An example of applying simulated data for validating knowledge space hypotheses using the four types of frequency simulations is presented in Section 10.4.2 in this volume.

## 9.4 Applications

This sections covers procedures and software programs for the applications of performance-based Knowledge Space Theory in addition to ALEKS. We discuss three groups of such applications, (i) tools for obtaining, comparing, and validating knowledge spaces, (ii) adaptive assessment of knowledge, and (iii) adaptive training.

**9.4.1 Working with Knowledge Spaces.**
Over the time, a number of programs for working with knowledge spaces have been developed most of which are small command-line oriented UNIX-based tools. The following overview covers their whole area from obtaining knowledge spaces over comparing and combining them up to their validation. More exhaustive listings and documentation of these tools have been provided by Hockemeyer (2001) and Pötzi and Wesiak (2004)[32]. Finally, a web-based user interface for such tools is presented.

The elicitation of knowledge spaces has been a core part of related literature. Within performance-based knowledge spaces, the focus was on querying experts and on data-driven identification of knowledge spaces (see Section 9.3.1 above). For querying experts, several procedures for computer-based querying were developed reducing the number of questions to be asked to the experts (Dowling, 1993a; Koppen and Doignon, 1990; Müller, 1989; Stefanutti and Koppen, 2003). For the Müller/Dowling algorithm, the KQ_Win program for querying experts with graphical support was implemented (Dowling et al., 1996b; Freitag, 1999). An expert can construct the questions himself and can see the respective parts of the surmise system while doing this (see the screenshot in Fig. 9.9).

Since different experts tend to differ in their statements, Baumunk and Dowling (1997) have suggested to combine spaces elicited from different experts by taking into account only those prerequisite relationships in which the majority of experts agree. This would involve the computation of the intersection and union of knowledge spaces, a very tedious task. Dowling and Hockemeyer (1998) have suggested a more efficient procedure which works on the bases (see Section 8.2 in this volume) of the respective knowledge spaces. This work resulted in the programs `basis-section`, `basis-union`, and `surmise-closure` (Hockemeyer, 2001).

[32] For access to these tools, please send an email to `cord.hockemeyer@uni-graz.at`.

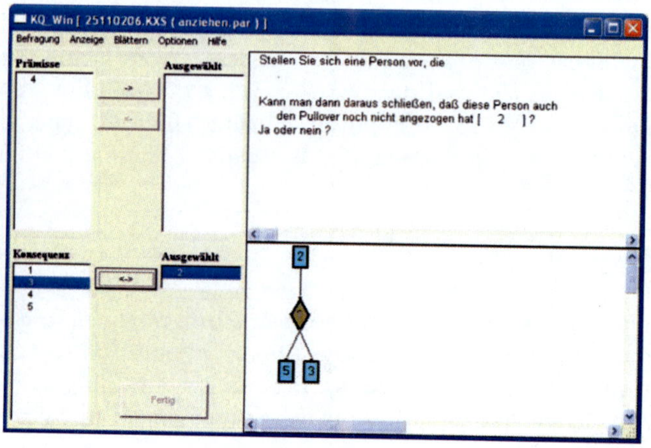


**Figure 9.9.** The KQ_Win program for querying experts

Numerous tools have been developed in the area of validation of knowledge spaces. The `di` program (Hockemeyer, 2001, based on ideas from an earlier program by Theo Held of the same name) computes the minimal symmetric distances between response patterns and knowledge states. It determines the distance agreement coefficient (see Section 9.3.6 above) and computes also several other indices, e.g., on item quality by identifying those test items which contribute strongly to invalidity of a given knowledge space (Baumunk et al., 1997).

A number of tools have been developed for simulating response patterns with changing degrees of randomness, e.g. for testing knowledge spaces against different null hypotheses. These simulation tools include completely random simulation, simulations with pre-specified solution probabilities for each item, or simulations based on given knowledge spaces with a pre-specified level of noise (careless errors and lucky guesses).

Several of these tools have also been used intensively for teaching at the Cognitive Science Section, University of Graz. Since students are hardly used to working with command line interfaces under UNIX nowadays, this frequently caused problems. Therefore, the ePsyt system[33] was developed. It allows to upload data files on knowledge spaces and to work with them through a Web interface. Figure 9.10 shows a screenshot of this system.

Besides the standalone tools (and their integration into the ePsyt web service environment) there are also implementations of similar functions as packages within mathematical software systems. Examples are a Mathematica™ package (Zaluski, 2001) and the KST (Stahl and Meyer, 2009) and DAKS (Sargin and Ünlü, 2010; Ünlü and Sargin, 2010) packages for R[34].

[33] See http://css.uni-graz.at/ePsyt/.

[34] See http://www.r-project.org.

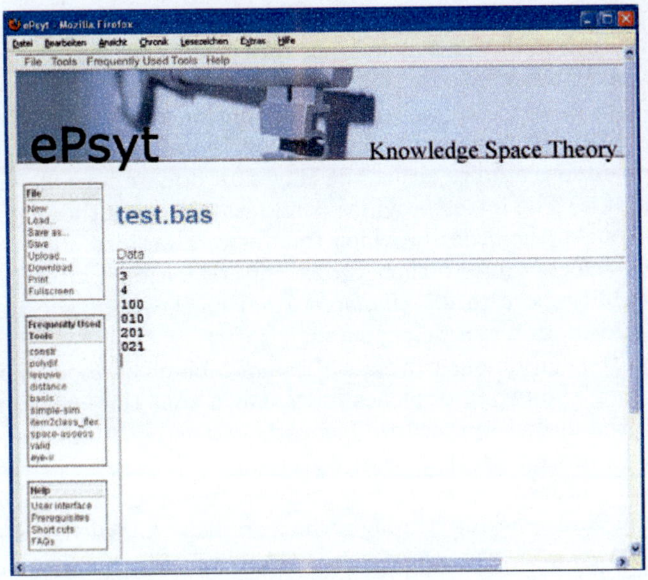


**Figure 9.10.** The ePsyt Web interface for KST tools

### 9.4.2 Adaptive Assessment of Knowledge.

The adaptive assessment of knowledge was the original motivation behind the development of Knowledge Space Theory (KST; Doignon and Falmagne, 1985). Several procedures for adaptively assessing a testee's knowledge space by posing a minimal number of test items were suggested already quite early. Degreef et al. (1986) suggested a deterministic procedure based on formal languages and a binary decision tree. Here, *deterministic* means that assessment is performed under the assumption that the testees' answers are determined only by their knowledge state without any noise like, e.g., lucky guesses or careless errors. As in classical adaptive testing, the basic idea of this procedure is to start with an item of medium difficulty and to continue with easier or with more difficult items depending on the correctness of the received answer. Items whose mastery (or non-mastery, respectively) can be concluded from earlier answers are not taken into consideration any more.

Soon afterwards, Falmagne and Doignon (1988a) suggested also two completely different non-deterministic procedures. A discrete non-deterministic procedure starts actually with a deterministic assessment. Assuming that this has led at least into the proximity of the correct knowledge state, they afterwards test items which may just have been learned or which could be learned right away for a person in that respective knowledge state (the *fringe*—see cf. Section 8.3 in this volume—of the respective knowledge state). If the preliminary result of the deterministic part of the procedure is close enough to the real knowledge state of the testee, this procedure will probably end up soon with the correct knowledge state.

A continuous non-deterministic procedure (Falmagne and Doignon, 1988b) estimates a probability distribution over the knowledge space. Starting, e.g., with a uniform distribution or with a distribution resembling the respective group of testees, the probability distribution is updated after each received answer. This update applies either the Bayesian update formula or a generalized version of it. The latter provides commutativity, i.e. the updated results are independent of the order in which the answers were received. Subsequent questions are selected due to their probability of mastery, either those which have a probabilty close to 0.5 (*halfsplit* rule) or those for which a maximal increase of information can be expected.

Although the continuous procedure seems to be much more exact, simulations comparing the two approaches have shown that the loss of information occurring in the discrete procedure has rather little effect on the accuracy of the assessment results (Hockemeyer, 2002).

All of these procedures share the problem that their implementation requires storing and working through the complete knowledge space. Since knowledge spaces can grow very large, the resulting computing times can be too long for usability in interactive systems. Therefore, new discrete procedures were developed which use the surmise system (or the base of the knowledge space) as their internal representation of the structural information. Dowling and Hockemeyer (2001) developed a deterministic procedure which interprets the surmise system as a set of rules defined by (i) an item is mastered if at least one of its clauses is mastered, and (ii) a clause is mastered if all of its items are mastered. They built a constraint propagation network where the variables stand for the mastery of the items and the clauses and where these variables are connected by logical AND and OR nodes according to the aforementioned rules. They proved that the update in this constraint propagation network is equivalent to the update in the procedure by Degreef et al. (1986) if also the distributive law for Boolean algebra is integrated into the network. Simulation results show that the procedure by Dowling and Hockemeyer (2001) works much faster than the classical one while it shows only small losses in efficiency. Hockemeyer (1997b) also developed a procedure for determining the fringe of a knowledge state using only the base of the knowledge space. Putting both procedures together allowed building an efficient discrete non-deterministic assessment procedure. Dowling et al. (1996a) took this up for building a system for the adaptive testing and training of knowledge in the area of fractions. Figure 9.11 shows a screenshot of the resulting AdAsTra system. AdAsTra starts with an adaptive knowledge test. Afterwards, the system suggests items from the fringe of the learner's knowledge state for training, i.e. items which the learner may just have learned (inner fringe) or is ready to learn (outer fringe).The former should be further trained to deepen their mastery. The latter items for which the learner has all necessary prerequisites may just be in the process of being learned.

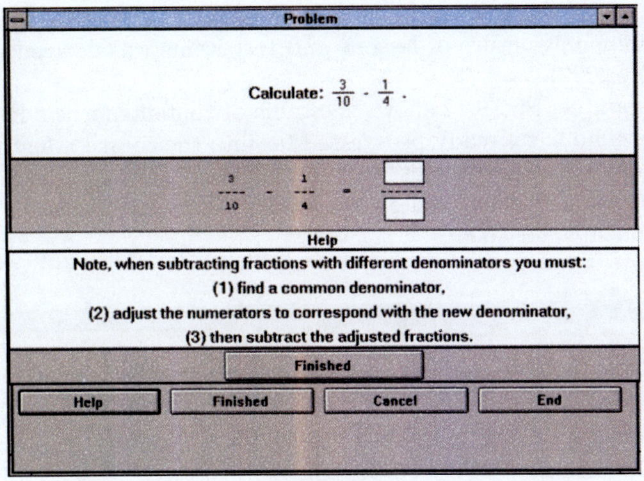


**Figure 9.11.** Screenshot of the AdAsTra system

### 9.4.3 Adaptive training of knowledge: the RATH prototype.

Starting with ALEKS and AdAsTra, the focus of KST applications shifted from the adaptive knowledge assessment to adaptive teaching and learning in the mid nineties. The RATH prototype (Hockemeyer, 1997a; Hockemeyer et al., 1998) is on the rim between performance- and competence-based KST.

RATH has two theoretical foundations. On the one side, it is based on Knowledge Space Theory. On the other side, it is based on a relational formulation of the Dexter Hypertext Reference Model (Halasz and Schwartz, 1990, 1994). Combining these two foundations, a surmise relation is mapped into prerequisite links between hypertext documents (Albert and Hockemeyer, 1997). These relational formalisms have the advantage that they can be easily and efficiently implemented using relational database technology. This led to the RATH prototype system which allows to scan a set (i.e. a course) of hypertext teaching documents equipped with prerequisite meta tags and to store them in the database. When a learner works with the documents, information about his/her knowledge is also stored there. On presenting a document to the learner, links pointing to other documents are hidden if this learner lacks any prerequisites for that linked document. Alternatively, such links could also be annotated as not recommendable.

As content, RATH uses lessons and test items on elementary probability theory based on a demand analysis of the test items by Held (1993, 1999). For each of these demands identified by Held, a corresponding lesson and some corresponding examples were created as documents. Held used the component-attribute approach (Albert and Held, 1994) for ordering the test items according to the demands they impose onto the learner. This procedure can also be used in the other direction, i.e., for ordering the identified demands. As a result, a combined structure of lessons and test items was derived (Albert and Hockemeyer, 2002).

As an example, Fig. 9.12 shows the table of contents for the RATH course for a learner who has already progressed far into the course. Most parts of the course are already open to this learner; however, some entries in that table of contents are not yet marked as hyperlinks because this learner still misses some prerequisites for them.

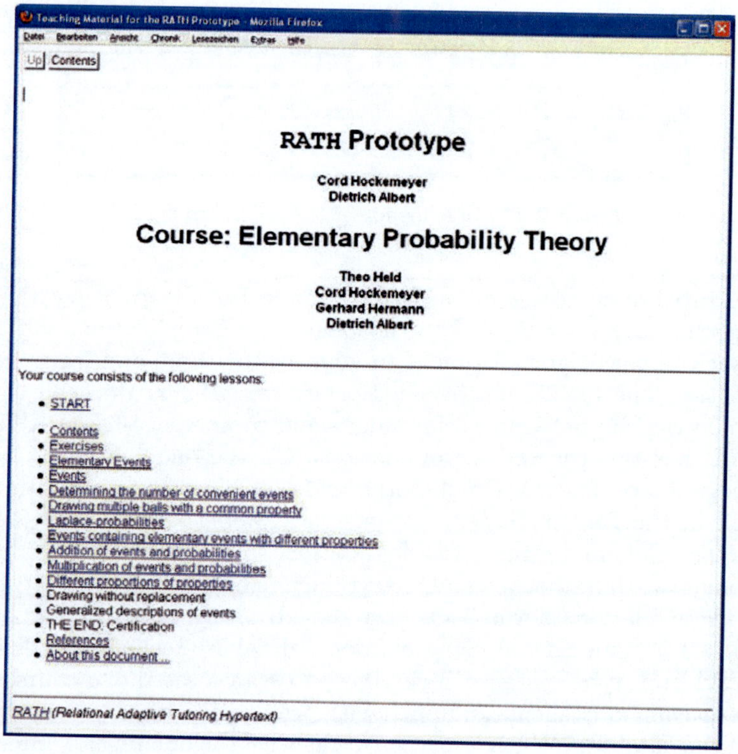


**Figure 9.12.** An experienced learner in RATH

More recently, the development of applications for the adaptive testing and learning were more oriented towards Competence-based Knowledge Space Theory. They are described in more detail in Chapters 11 and 12.

## 9.5 Discussion

The progress of Knowledge Space Theory outlined in this chapter is in theoretical extensions and methodological additions as well as in applications.

Furthermore, we focus on the relationships with some competitive and complementary theoretical conceptions. Although the progress is remarkable some critical questions have to be discussed with the aim to stimulate future research and development.

The theoretical results by taking into account multiple answer alternatives and by introducing the surmise relation between 'tests' are, at one hand, open for continuing theoretical research and going beyond the current state, e.g. by partially ordered answer alternatives or by surmise systems for tests/sets. On the other hand, the work has high potential for applications in fields of practical importance, like diagnosing and evaluating performances and abilities of persons, developing and restructuring curricula or web sites, organizing workflow, and so on. Initiatives to develop further and to exploit the results are necessary for solving problems of high theoretical and practical value.

The arguments for investigating the relationships between different theoretical approaches are of a different kind, e.g. to find common roots, to prepare some more general models and their hierarchy, or to use established methods from one field in another field. To abstract, the expectation is to foster synergies. Regarding the relationships of Knowledge Space Theory with Formal Concept Analysis, Latent Class Analysis and Item Response Theory, the expected synergies—according to our knowledge—are currently far behind of what seems possible or can be expected. Crossing the borders between the different fields, by e.g. collaboration, for solving theoretical and practical problems is still a program for future research and development. Of course, the reported results are of high value by there own, because of clarifying metatheoretical questions and problems, and by providing insight into the structure of a larger scientific field with common aims and goals.

The described methods for empirical generation and validation of knowledge structures are of high value and used quite often, no question. However, we have to keep in mind that almost all of these methods are of the ad hoc type, that they are (only) heuristics for solving methodological problems, and that they are not deeply investigated mathematically and therefore the properties of almost all of them are not very well understood. The methodology in this field has still to mature, developing more theoretically sound methods and algorithms derived from and based on Knowledge/Learning Space Theory by further research.

The mentioned applications represent important developments for demonstrating the usefulness of the theoretical structures, from a methodological as well as from an applied point of view. Of course the value of the developed software programs and systems is limited in time, due to theoretical and technical progress. Thus adaptations, re-constructions and new developments have to be performed in the future—and usability has to be improved!

Are the mentioned developments of Knowledge Space Theory still valuable in view of the skill- and competence-based extensions of Knowledge Space Theory (see Chapters 11 and 12 of this volume)? The answer is straightforward: yes, they are. The main reason is that Knowledge Space Theory, beyond having high value on its own, represents the link and the interface between the non-observable and the observable, between the structural models and the data.

In summary, the described recent contributions to Knowledge Space Theory are of high value for theoretical, methodological, and applicational reasons; they are significant steps and open the doors for further research, developments and applications, even in light of other important trends like Competence-based Knowledge Space Theory (Chapters 11 and 12 of this volume) and Learning Spaces (see Falmagne and Doignon, 2011, and Chapter 8 of this volume).

## Acknowledgements

Part of the research reported here was financially supported by two projects from the Austrian Science Foundation (FWF) and by HCM and TMR Marie-Curie-Fellowships of the European Commission.

# 10

# Heuristics for Generating and Validating Surmise Relations across, between and within Sets/Tests

Dietrich Albert[1, 2]

Gudrun Wesiak[1, 2]

Ali Ünlü[3]

## 10.1 Introduction

Generating and validating surmise relations across, between and within sets has been strongly stimulated for practical reasons. For instance, the number of psychological tests (sets of items) used simultaneously in diagnostic settings continues to grow. Therefore it is useful to develop new methods that economize assessment procedures and that allow for restructuring of tests. For this purpose, we replace the term 'sets' and apply in this chapter the ***S**urmise **R**elations across, **b**etween and within **T**ests (SRbT)* approach. Applying the SRbT approach is however not restricted to tests (sets of items); it can be used e.g. in Educational Science for developing and restructuring curricula in order to obtain among other things a meaningful sequence of courses (Hockemeyer, Albert, & Brandt, 1998). Another example in Computer Science is (re-)structuring hypertext documents respecting prerequisites (Albert and Hockemeyer, 1997).

The SRbT approach has been proposed by the first author and his co-workers (Albert, 1995; Brandt et al., 2003; Ünlü et al., 2004). It is based on the Knowledge Space Theory (KST) developed by Doignon and Falmagne (see Doignon and Falmagne, 1985, 1987; Falmagne, 1989a, 1993, and other publications by these authors), but in addition to a single set of elements or items, it relates several sets of items, that is, several tests in the current application. On the other hand, the approach in its current status is a simplification of Knowledge Space Theory: Surmise relations instead of surmise systems and quasi ordinal knowledge spaces instead of knowledge spaces are currently focused on as special cases.

[1]Department of Psychology, University of Graz, Austria
[2]Knowledge Management Institute, Graz University of Technology, Austria
[3]TUM School of Education, Technische Universität München, Germany

J.-C. Falmagne et al. (eds.), *Knowledge Spaces: Applications in Education*,
DOI 10.1007/978-3-642-35329-1_10, © Springer-Verlag Berlin Heidelberg 2013

Doignon and Falmagne (1999)[4] as well as others working with knowledge spaces (e.g. Albert and Lukas, 1999) referred to single tests only. However, tests assessing, for example, developmental or cognitive stages often require abilities that have prerequisites in some other test. Thus, the concept was extended to sets of tests. Knowledge about how, for example, various cognitive stages or tests are related facilitates the definition of developmental and learning paths as well as the selection of adequate tests to assess the impact of the applied training units. The definition of prerequisites between the items of different tests may also uncover, redundancies (equally informative items occurring in different tests) and help to restructure the tests. Looking at diagnostic settings, the implications established in SRbT allow an efficient assessment of a person's knowledge on various tests, because only those tests have to be presented for which the performance cannot be inferred from previously performed tests. In the case of a transitive relation on a set of tests, it might be possible to skip the presentation of entire tests.

This chapter illustrates methods for generating and validating hypotheses on how various tests are related. The theoretical foundations are introduced in Chapter 9 of this volume and are recapitulated here only as far as necessary for understanding this chapter.

One of the main challenges is to establish a valid SRbT or test knowledge space on a given set of items and tests. We will introduce two methods to generate a SRbT surmise relation between tests. The first approach is based on the principle of component-wise ordering of product sets (e.g., Albert and Held, 1999; Davey and Priestley, 1990). It is applied to generate theory-driven hypotheses. In a second approach we use van Leeuwe's (1974) method for generating data-driven hypotheses. Finally we present different validation methods, apply them and evaluate the obtained results with respect to their consistency, and compare the theory- and the data-driven models.

The discussed procedures are exemplified by the domain of inductive reasoning. In the final discussion, we summarize the results and briefly mention further research options in this area.

## 10.2 Basic Concepts of SRbT and Corresponding Test Knowledge Spaces

In the following, we briefly mention the basic concepts of SRbT and the corresponding test knowledge spaces. For more detailed descriptions we refer to Brandt et al. (2003), Ünlü et al. (2004), and Section 9.1.7 in this volume.

Doignon and Falmagne (1985, 1999; Falmagne et al., 1990; Falmagne and Doignon, 2011; and Chapter 8 in this volume) define the *knowledge state* of a person as the set of items in a specified domain $Q$ this person is able to master under ideal conditions. A *surmise* or *prerequisite relation* $S \subseteq Q \times Q$

[4] See also the much expanded edition Falmagne and Doignon (2011).

is a binary relation on a finite set $Q$ of items. Surmise relations are reflexive and transitive (quasi orders). An item set in accordance with a given surmise relation $S$ is called a *knowledge state*. The set of all knowledge states is called *knowledge structure* $\mathcal{K}$. A knowledge structure $\mathcal{K}$ which is closed under union and intersection is called a *quasi ordinal knowledge space*. This means that for any two knowledge states $K$ and $L$, their union $K \cup L$ and their intersection $K \cap L$ are also knowledge states.

Formally, a knowledge state $K$ is defined by

$$K \in \mathcal{K} \Longleftrightarrow (\forall (y,x) \in S : x \in K \Rightarrow y \in K) \tag{10.1}$$

A quasi ordinal knowledge space always includes the empty set $\emptyset$ and the complete set $Q$ of items. According to the *Birkhoff-Theorem* (Birkhoff, 1937), there exists a one-to-one correspondence between the family of all surmise relations and the family of all quasi ordinal knowledge spaces[5]. Organizing items according to a surmise relation reduces the number of possible item combinations, that is the power set $2^Q$ of all items, to a subset $\mathcal{K} \subseteq 2^Q$ of knowledge states.

So far, we have referred to single tests. However, in common psychological assessment procedures we often deal with a set $\mathcal{T} = \{A, B, C, \ldots\}$ of different tests that may be related. In some cases, these tests may be disjoint, that is, they contain different items. In the following, we call the elements of any partition of $Q$ a test. Of course, equivalently in the case of starting with several different tests their union results in $Q$. For a better understanding, we describe the approach for only two tests. However, the described concepts are applicable to any finite set of distinct tests.

In the following the basic concepts are recapitulated. For the formal definitions and illustrations, we refer to Section 9.1.7 of this volume.

For two tests $A, B \in \mathcal{T}$, we write $B \;\dot{\mathcal{S}}\; A$ to mean that we can surmise the correct solution of some non-empty subset of items in test $B$ from the correct solution of at least one item in test $A$. In other words, solving item(s) in test $B$ is a prerequisite for the solution of a given set of items in test $A$. We call $\dot{\mathcal{S}}$ a *surmise or prerequisite relationship between tests*. We have thus $\dot{\mathcal{S}} \subseteq \mathcal{T} \times \mathcal{T}$.

By definition, a surmise relation between tests is reflexive, but not necessarily transitive. However, there are special cases in which transitivity holds, namely 'left-', 'right-', and 'total-covering' surmise relations.

Two tests $A, B \in \mathcal{T}$ are in a *left-covering surmise relation*, $B \;\dot{\mathcal{S}}_l\; A$, from test $A$ to test $B$, if for each item $a \in A$ there exists a nonempty subset of prerequisites in test $B$. This means that a person who doesn't solve any item in $B$ will not be able to solve any item in $A$, either. There is no need to test this person on test $A$.

Two tests $A, B \in \mathcal{T}$ are in a *right-covering surmise relation*, $B \;\dot{\mathcal{S}}_r\; A$, from test $A$ to test $B$, if for each item $b \in B$, there exists at least one item $a \in A$ for

[5] $ySx \Leftrightarrow (\forall K \in \mathcal{K} : x \in K \Rightarrow y \in K)$, and $K \in \mathcal{K} \Leftrightarrow (\forall (y,x) \in S : x \in K \Rightarrow y \in K)$.

which $b$ is a prerequisite. Failing to solve any item in test $B$ implies a failure on a subset of items in test $A$. In other words, a person who solves all items in test $A$ is also able to solve all items in test $B$. Thus, there is no need to test this person on test $B$.

Finally, we speak of a *total-covering surmise relation* $B\ \dot{\mathcal{S}}_t\ A$ from test $A$ to test $B$ if the ordered pairs of tests are in a left- as well as right-covering relation.

The left-, right-, and total-covering surmise relations, and the surmise relation between tests satisfy the subset properties: $\dot{\mathcal{S}}_t \subseteq \dot{\mathcal{S}}_l$, $\dot{\mathcal{S}}_t \subseteq \dot{\mathcal{S}}_r$, $\dot{\mathcal{S}}_l \subseteq \dot{\mathcal{S}}$ and $\dot{\mathcal{S}}_r \subseteq \dot{\mathcal{S}}$.

Aside of the surmise relations between tests, we differentiate between various subsets of the surmise relation on the entire set $Q$ of items. $S_{QxQ}$ denotes the surmise relation on $Q$ and is referred to as *surmise relation between items (SRbI)*. The disjoint subsets of $S_{QxQ}$ for two tests $A$ and $B$ (with $A \cup B = Q$) are denoted $S_{AxA}$, $S_{BxB}$, $S_{AxB}$, and $S_{BxA}$. The subset $S_{AxA}$ united with $S_{BxB}$ is called *surmise relation within tests (SRwT)*. The union of the subsets $S_{AxB}$, $S_{BxA}$, and the set $\Delta = \{(q, q),\ \ q \in Q\}$ is called a surmise relation *across tests (SRxT)*. Each subset is defined as follows:

$$\begin{aligned}
\text{SRwT} &= S_{AxA} \cup S_{BxB}\\
\text{SRxT} &= S_{AxB} \cup S_{BxA} \cup \Delta\\
\text{SRbI} &= \text{SRwT} \cup \text{SRxT} = S_{QxQ}\\
&\text{where}\\
S_{AxA} &= \{(a_i, a_j)\,|\, a_i, a_j \in A \wedge a_i S a_j\}\\
S_{BxB} &= \{(b_i, b_j)\,|\, b_i, b_j \in B \wedge b_i S b_j\}\\
S_{AxB} &= \{(a, b)\,|\, a \in A, b \in B \wedge aSb\}\\
S_{BxA} &= \{(b, a)\,|\, a \in A, b \in B \wedge bSa\}\\
S_{QxQ} &= \{(y, x)\,|\, x, y \in Q \wedge ySx\}
\end{aligned} \tag{10.2}$$

If either the set $S_{AxB}$ or the set $S_{BxA}$ are not empty, then there exists a surmise relation between the tests $A$ and $B$ ($\dot{\mathcal{S}}_{\mathcal{T}x\mathcal{T}}$).

Extending the concepts of Doignon and Falmagne's KST approach, a *test knowledge state* $\dot{K}$ is defined as the $|\mathcal{T}|$-tuple of items per test a person in state $K$ is capable of mastering: $A \cap K, B \cap K, \ldots$ The collection of all test knowledge states, that are in agreement with the *SRbI*, is called the *test knowledge structure*, which is defined as the pair $(\mathcal{T}, \dot{\mathcal{K}})$. If a test knowledge structure is closed under union it is called a *test knowledge space*, if it is also closed under intersection we speak of a *quasi ordinal test knowledge space*. The surmise relation between tests can be derived from its corresponding test knowledge space. However, the reverse inference is not valid for a set of tests. As for the surmise relations we differentiate between the *knowledge spaces between items (KSbI)*, *within tests (KSwT)*, and *across tests (KSxT)*. These subsets are derived by applying the Birkhoff Theorem to the corresponding surmise relations, that is the *SRbI*, the *SRwT*, and the *SRxT*, respectively.

For given tests the above described structures have to be generated and validated before being applied. This is the subject of the next sections.

## 10.3 Generation of Hypotheses

There are two kinds of hypothesis specification, namely (a) theory based hypotheses on surmise relations between items and tests are established (top-down approach) and (b) sets of items (tests) and a body of data are given, and surmise relations are generated from the data (bottom-up approach). In this section we discuss (a) a theory-driven as well as (b) a data-driven approach to generate hypotheses. To exemplify the applied techniques, we use the sets of items and data of two inductive reasoning tests.

**10.3.1 Theory-driven Generation of Hypotheses.** Starting from psychological findings surmise relations between items and the corresponding knowledge spaces can be established (Albert and Lukas, 1999). In one approach common components, that is sets of cognitive demands or competencies, are assigned to the items and an order is established by way of set-inclusion. A generalization of this approach is based on the principle of component-wise ordering of product sets (see Davey and Priestley, 1990), which defines each component by a set of attributes of varying difficulty (e.g., Albert and Held, 1999).

### Component-wise ordering of items

The principle of component-wise ordering of product sets describes every item in a set $Q$ by a finite set of components $\mathcal{C} = \{A, B, C, \dots\}$. Each component consists of a set of attributes $A = \{a_1, a_2, \dots\}$, $B = \{b_1, b_2, \dots\}$, $C = \{c_1, c_2, \dots\}$. On each attribute set a partial order of difficulty is defined. Forming the Cartesian product of the components results in the set of all possible attribute combinations, which can be used for constructing items as well as for a demand analysis. Items that are described by the same attribute vector form an item class. By pairwise comparison of the attribute vectors, a surmise relation on the set of these item classes can be established. The ordering rule assumes, that an item class $y$ is prerequisite for an item class $x$, if the attributes of $y$ are equally or less difficult than the attributes of $x$. This rule is known as coordinatewise order, which corresponds to the dominance rule known from choice heuristics. The resulting surmise relation on the item classes is a partial order with the properties reflexivity, transitivity, and antisymmetry, which is a special case of a quasi order.

So far, all components are of equal importance. In a more specific case, a partial order can be defined on the components, meaning that some of the components are assumed to be of higher importance and some are independent with respect to importance. If one component $A$ is assumed to be more important than another component $B$, the attributes of component $A$ are compared first. Attributes of component $B$ are only considered for items equipped with the same attribute on $A$. The resulting surmise relation is also a partial order and contains more item pairs than the order for equally important components. As a special case, item classes can be ordered lexicographically, supposing a linear importance relation on the components. The resulting surmise relation becomes a linear order and contains for $m$ items $m(m+1)/2$ fully connected item pairs.

COMPONENT-WISE ORDERING OF TESTS

For a set of tests, the items of various tests are analyzed in order to specify components that are valid for the items in all tests under investigation. The $SRbI$, the $SRxT$, the $SRwT$ and their corresponding knowledge structures, as well as the surmise relation between tests are derived by a pairwise comparison of the attributes describing the item classes.

For an empirical demonstration, we chose the domain of inductive reasoning. To construct a test knowledge structure we defined components and attributes for inductive reasoning problems as they are usually used in aptitute tests (classifications, series continuations, analogies, and matrices). The theoretical analysis of the items is mainly based on Albert and Held (1999), Albert and Musch (1996), Bejar et al. (1991), Carpenter et al. (1990), Holzman et al. (1983), Klauer (1997), Schrepp (1999c), Sternberg and Gardner (1983), and Wesiak and Albert (2001). Based on these previous results the respective structures are derived in the sequel. For a detailed analysis of the items and the derivation of the theory-driven hypotheses see Wesiak (2003).

The components include *(A) number of answer alternatives*, *(B) operation difficulty*, *(C) number of operations* (rules for solving a problem), *(D) task ambiguity* (similarity of answer alternatives), *(E) salience* (easiness to detect the relevant rules), and *(F) material* (e.g. verbal or geometric-figural). Components $A$ and $F$ are assumed to be of less importance than components $B$, $C$, $D$, and $E$. Furthermore, components $A$ and $F$ as well as $B$ through $E$ are assumed to be independent from each other (see Table 10.1, first two columns). Thus, we assume a partial order on the set of components. The attributes of each component and the postulated difficulty orders of the attributes are presented in the last two columns of Table 10.1.

Using the method for the component-wise ordering of items, we formed the components' product to define item classes and established an order by a pairwise comparison of their attribute vectors. From the obtained $SRbI$, we derived its subsets (see Equation 10.2), including the surmise relation between tests, and the corresponding knowledge spaces (for details, see next section under *Hypotheses*).

**Table 10.1.** Components, attributes, and their prerequisites for inductive reasoning problems

| Components | Importance of components | Attributes | Prerequisites for attributes |
|---|---|---|---|
| $A$: Number of answer alternatives | $\{A\}$ | $a_i$: 2, 3, ..., $n$ | $\{a_{\leq i}\}^a$ |
| $B$: Operation difficulty | $\{A, B, F\}$ | $b_1$: O other (less difficult) | $\{b_1\}$ |
| | | $b_2$: D difficult | $\{b_1, b_2\}$ |
| $C$: Number of operations | $\{A, C, F\}$ | $c_i$: 1, 2, ..., $m$ | $\{c_{\leq i}\}^a$ |
| $D$: Task Ambiguity | $\{A, D, F\}$ | $d_1$: L low ambiguity | $\{d_1\}$ |
| | | $d_2$: H high ambiguity | $\{d_1, d_2\}$ |
| $E$: Salience | $\{A, E, F\}$ | $e_1$: L low demand | $\{e_1\}$ |
| | | $e_2$: H high demand | $\{e_1, e_2\}$ |
| $F$: Material | $\{F\}$ | $f_1$: P pictorial | $\{f_1\}$ |
| | | $f_2$: L letters | $\{f_1, f_2\}$ |
| | | $f_3$: V verbal | $\{f_1, f_3\}$ |
| | | $f_4$: N numerical | $\{f_1, f_2, f_4\}$ |
| | | $f_5$: G geometric-figural | $\{f_1, f_2, f_3, f_4, f_5\}$ |

*Note.* [a]The smaller the number of answer alternatives/operations the easier the attribute.

## Demonstration with two inductive reasoning tests

For a demonstration of the methods, we analyzed the items of two inductive reasoning tests developed by the "Heerespsychologischer Dienst des Bundesministeriums für Landesverteidigung" (Psychological Service of the Ministry of Defense) in Vienna, Austria. The supplied data set contained response patterns from 1221 male participants, who performed five intelligence tests, two of which consist of inductive reasoning problems (20 geometric matrices and 25 verbal analogies). Items were presented as speed-power tests in paper-pencil form (group sessions). The matrix test ($MT$) was always presented ahead of the analogy test ($AN$). Participants had to choose among eight answer alternatives for $MT$ and five alternatives for the $AN$. The given sequence of items within a test was not binding (participants had the possibility to skip items and to go back to previous items).

### *Pre-editing of data and tests*

The selected body of data is suboptimal for validation, however it is optimal for discussing important principles for handling data for validation. The validation of knowledge space hypotheses requires complete response patterns, because otherwise from not processing an item no conclusion about its solvability by that person can be drawn. That is, each participant has to process all of the items. In a speed-power test, the items presented towards the end of the test are generally not processed by all participants. In this body of data, 51 participants processed all 45 items. Of course, this sample of persons is highly selective and should not be used. Having only 51 answer patterns is not sufficient, although validating a knowledge space hypothesis does not need a representative sample. What is needed is a heterogeneous sample. The whole range in between novice and expert should be represented because in that case every knowledge state has a chance to be realized.

In order to get a heterogeneous sample large enough to handle the trade-off between the number of items and the number of complete response patterns, we iteratively removed items that were processed by the fewest number of participants and recounted the number of complete response patterns after each removal. The procedure was stopped as soon as the number of response patterns exceeded the number of postulated knowledge states in the $KSbI$. Thus, in principle, every state has a chance to become realized.

Furthermore, we removed the trivial response patterns from the data set, that is patterns with either all or none of the items solved. Schrepp (2006) commenting on Ünlü and Albert (2004) recommends to take the trivial response patterns into account for validation. From our point of view however, removing the trivial response patterns is absolutely necessary because of two reasons. (a) By definition the complete set $Q$ and the empty set $\emptyset$ are knowledge states. Thus, observing trivial answer patterns has no diagnostic value for validation. (b) Even more important seems, that including trivial response patterns allows to manipulate the results and to improve the fit. By choosing and increasing the number of novices and/or experts—who produce the trivial response patterns—the fit will be improved accordingly.

As a result of the procedure described above, we obtained a $732 \times 27$ data matrix (732 response patterns for 12 matrix and 15 analogy items) containing '1's and '0's for correct and incorrect answers respectively. The postulated $KSbI$ for 27 items contains 553 states, $Q$ and $\emptyset$ included (see below), out of $2^{27}$ elements of the power set.

*Hypotheses*

As a first step, $MT$ and $AN$ were analyzed item by item according to the components and attributes shown in Table 10.1. The resulting attribute combinations respectively item classes for the 27 items are shown in Table 10.2 (the second row lists $MT$ item classes, the first column $AN$ item classes). The item class 5O1LLV, for example, has **5** answer alternatives ($A$), **O**ther operation difficulty ($B$), **1** operation ($C$), **L**ow ambiguity ($D$), **L**ow demand on salience ($E$), and **V**erbal material ($F$).

The prerequisites for each item class were defined by applying the principle of component-wise ordering as described above (see Table 10.1 for the postulated orders on the components and attributes). As a result, we obtain the $SRbI$ (for the item classes of both tests) and its corresponding quasi ordinal test knowledge space. According to Definition 10.2, the $SRbI$ consists of the subsets $SRwT$ and $SRxT$.

If we consider only those pairs of item classes, which belong to the same test, we receive the two $SRwT$. The two subsets of the $SRwT$ are referred to as $SRwMT$ and $SRwAN$ for item classes of the matrix and analogy test respectively. The Hasse diagram in Figure 10.1 illustrates the orders of the $SRwMT$ (left ellipse) and the $SRwAN$ (right ellipse). On the other hand, we consider pairs of item classes which belong to different tests. This subset of pairs results in the $SRxT$, which is depicted in Table 10.2. Generally, an item pair is in a relationship if the item class in test $A$ has a prerequisite item class in test $B$. For example, in Table 10.2, the 'x' in column 8O1LLG and row 5O1LHV indicates that the $MT$ class 8O1LLG is prerequisite for the $AN$ class 5O1LHV: With respect to the components' importance we first compare components $B$ through $E$. The attributes of components $B$, $C$, and $D$ are of equal difficulty (O1L), whereas the attributes of component $E$ differ. The $MT$ class shows attribute L ($e_1$, low demand), which is easier than the $AN$ class's attribute H ($e_2$, high demand). Thus, the $MT$ class is prerequisite for the $AN$ class, and the attributes of the less important components $A$ and $F$ are not taken into account. A graphical illustration of the order in Table 10.2 is given in Figure 10.1 (arrows going from one ellipse to the other).

The $SRxT$ implies the following hypotheses on the surmise relation between tests: There exists a right-covering surmise relation from $AN$ to $MT$ ($MT\ \dot{\mathcal{S}}_r\ AN$—each matrix item is prerequisite for an analogy item, indicated by 'x' in Table 10.2) and a left-covering surmise relation from $MT$ to $AN$ ($AN\ \dot{\mathcal{S}}_l\ MT$—each matrix item has a prerequisite in $AN$, indicated by '$\diamond$' in Table 10.2). Thus, $(MT, AN) \in \dot{\mathcal{S}}_r$ and $(AN, MT) \in \dot{\mathcal{S}}_l$, whereas $\dot{\mathcal{S}}_t = \emptyset$ (not counting $\Delta$).

The validation of the theory-driven hypothesis is presented below together with the validation of the data-driven hypothesis.

**10.3.2 Data-driven Generation of Hypotheses.** There exist a variety of data-analytic techniques for detecting dependencies between items, as for example Guttman's deterministic scalogram analysis (Guttman, 1944; Ducamp and Falmagne, 1969), feature pattern analysis (Feger, 1994, chap. 2, 2000), or Boolean analysis (e.g., van Buggenhaut and Degreef, 1987; Schrepp, 2003; Theuns, 1998), see also Chapter 9 in this volume. Whereas feature pattern analysis and Guttman scaling try to represent the data in such a way that all items are connected, Boolean analysis also allows independencies. One application of Boolean analysis which generates a quasi ordinal knowledge space from dichotomous data sets is *Item Tree Analysis* (ITA). ITA was proposed by Bart and Krus (1973), further developed by van Leeuwe (1974; see also Section 9.3.1 in this volume), and applied in the context of knowledge space theory by Held and Korossy (1998).

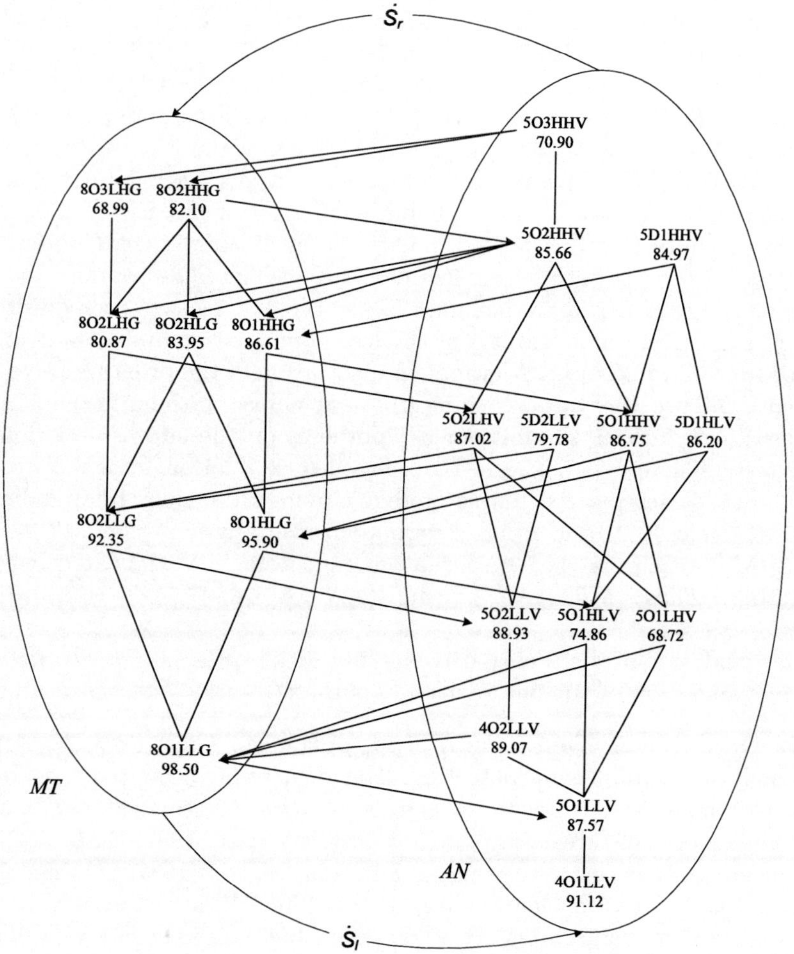


**Figure 10.1.** Hasse diagram for the theory-driven surmise relations between items and tests and percentage of correct solutions for item classes.

**Table 10.2.** Surmise relation across tests[a]

| | | *Matrices* | | | | | | | |
|---|---|---|---|---|---|---|---|---|---|
| | | 8O1LLG | 8O1HLG | 8O1HHG | 8O2LLG | 8O2HLG | 8O2LHG | 8O2HHG | 8O3LHG |
| *Analogies* | | 3,5 | 1,7 | 6 | 2,10 | 4,11 | 8 | 9 | 12 |
| 4O1LLV | 24 | ◇ | ◇ | ◇ | ◇ | ◇ | ◇ | ◇ | ◇ |
| 5O1LLV | 13 | ◇ | ◇ | ◇ | ◇ | ◇ | ◇ | ◇ | ◇ |
| 5O1LHV | 18 | x | | ◇ | | | ◇ | ◇ | ◇ |
| 5O1HLV | 15 | x | ◇ | ◇ | | ◇ | | ◇ | |
| 5O1HHV | 27 | x | x | ◇ | | | | ◇ | |
| 4O2LLV | 16 | x | | | ◇ | ◇ | ◇ | ◇ | ◇ |
| 5O2LLV | 19 | x | | | ◇ | ◇ | ◇ | ◇ | ◇ |
| 5O2LHV | 20,22 | x | | | x | | ◇ | ◇ | ◇ |
| 5O2HHV | 23,25 | x | x | x | x | x | x | ◇ | |
| 5O3HHV | 26 | x | x | x | x | x | x | x | x |
| 5D1HLV | 17 | x | x | | | | | | |
| 5D1HHV | 14 | x | x | x | | | | | |
| 5D2LLV | 21 | x | | | x | | | | |

*Note.* An 'x' indicates that the matrix class in column $i$ is prerequisite for the analogy class in row $j$, a '◇' indicates that the analogy class in row $j$ is prerequisite for the matrix class in column $i$.

[a] Pairs $(q, q)$ are not shown.

| | | $x$ | |
|---|---|---|---|
| | | 0 | 1 |
| $y$ | 0 | $a_{xy}$ | $b_{xy}$ |
| | 1 | $c_{xy}$ | $d_{xy}$ |

**Figure 10.2.** ITA: possible response patterns and assumed relationship for two items $x$ and $y$ ($ySx$).

### Generation of order relations

ITA can be applied to dichotomous data, where an analysis of the empirical responses yields four possible correct/incorrect response patterns for each item pair. The contingency table in Figure 10.2 shows the possible response patterns for two items $x$ and $y$, namely that both items are answered correctly, which is denoted $\langle 1, 1\rangle$, that only item $x$ is answered correctly $\langle 1, 0\rangle$, only item $y$ is answered correctly $\langle 0, 1\rangle$, or that none of the items is answered correctly $\langle 0, 0\rangle$. A surmise relationship $ySx$ between two items $x$ and $y$ ($y$ is surmisable from $x$) is assumed, whenever the frequency of the pattern $\langle 1, 0\rangle$ equals zero (see Figure 10.2 , cell $b_{xy}$). By analyzing all possible item pairs an order relation is established.

Taking into account that empirical data are usually noisy, that is, participants make mistakes in the form of lucky guesses and careless errors, the criterion of a zero in cell $b_{xy}$ is usually too strong. Therefore, a tolerance level $L$ is set to specify the percentage of allowed contradictions ($b_{xy} \leq L$), that is participants who show the response pattern $\langle 1, 0\rangle$ for a certain item pair.

Van Leeuwe (1974) proved that transitivity holds for $L = 0$, whereas intransitivities can occur for tolerance levels $L > 0$. Van Leeuwe suggests to omit all not transitive ITA solutions. However, sometimes the set of transitive solutions only consists of an order relation, where all items are independent. That means that in those cases the derived hypothesis cannot be falsified at all because there are no prerequisites and the space equals the power set. Therefore, we suggest to search for the smallest transitive closure of a not transitive relation by adding pairs to the relation[6]. As a consequence, the $SRbI$ becomes more specific (there are more pairs in the relation) and the knowledge space

[6]The (reflexive and) transitive closure of a binary relation $\mathcal{R}$ on a finite set $X$ can be implemented and computed based on the fact that it equals the relation $\bigcup_{i=0}^{I}\mathcal{R}^i$ for some positive integer $I$ such that $\bigcup_{i=0}^{I'}\mathcal{R}^i = \bigcup_{i=0}^{I}\mathcal{R}^i$ for all naturals $I' > I$, where $\mathcal{R}^i$ denotes the $i$th (relative) power of $\mathcal{R}$ and $\mathcal{R}^0$ is the identity relation (i.e., diagonal $\Delta$) on the ground set $X$. In other words, except for $\mathcal{R}^0 = \Delta$ and $\mathcal{R}^1 = \mathcal{R}$, pairwise (relative) products $\mathcal{R}^2 = \mathcal{R}\mathcal{R}, \mathcal{R}^3 = \mathcal{R}^2\mathcal{R}, \ldots, \mathcal{R}^I = \mathcal{R}^{I-1}\mathcal{R}$ are formed successively, until stagnation.

contains fewer states. The derived hypothesis will be falsified more easily than without the added pairs—that means the method is conservative.

*Application of ITA to a set of tests*

To check whether ITA is able to detect known underlying (sub)structures of data for two tests, we conducted a simulation. Schrepp (1999b) showed in a simulation study for a set of eight items that ITA performs better, the larger the data set, the lower the noise probabilities in the data, and the fewer non-connected (incomparable) item pairs are in the relation.

Schrepp based his analysis on different structures within a single test and used the correlational agreement coefficient $CA$ to select a relation for each set of simulated data. For a formal definition of the $CA$ see Section 9.3.1 of this volume.

To check the applicability of ITA for a set of tests, we constructed a hypothetical $SRbI$, which we partitioned in two tests $A$ and $B$ with five items each. To ascertain that ITA is able to differentiate between different kinds of order relations, we defined a partial order on test $A$ and a linear order on test $B$ (see Figure 10.3). For test $A$ 7 item pairs and 9 knowledge states, for test $B$ 10 pairs and 6 states have been obtained. $SRbI$ with 10 items has 35 pairs and 23 states. The relationships across the two tests imply a left-covering surmise relation from test $A$ to test $B$ ($B\ \dot{\mathbb{S}}_l\ A$) and a right-covering surmise relation from test $B$ to test $A$ ($A\ \dot{\mathbb{S}}_r\ B$).

In a second step, we simulated 1000 response patterns based on the corresponding knowledge space of the $SRbI$. Each state has been sampled with replacement with the same probability (1/23). To receive the simulated patterns, a certain amount of noise was applied to each sampled state. To keep it simple, we applied the same probability of .10 for lucky guesses and careless errors once to each of the 10 entries of each sampled state vector. To re-establish the $SRbI$ from the simulated data set, we used ITA to compute several order relations with tolerance levels at $0\% \leq L \leq 15\%$. Then, the set of items was repartitioned into the two tests in order to recover the $SRwT$, the $SRxT$, and the surmise relation between tests.

The original $SRbI$ as well as the left- and right-covering surmise relations between the tests could accurately be recovered at a tolerance level of 10%, that is exactly the amount of noise put into the data set. For $L = 9\%$ the number of item pairs in the generated relation is too low (34 pairs), for $L = 11\%$ it is too high (36 pairs) compared to the 35 pairs in the original relation. Thus, by matching the tolerance level with the amount of noise in the data, ITA is able to recover the relations for more than one test and different kinds of orders (partial and linear orders), at least for our simulated data set.

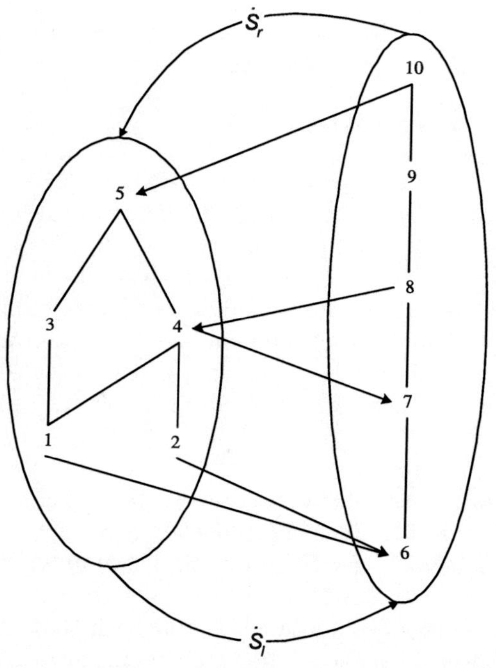


**Figure 10.3.** Example for a surmise relation between two tests.

SELECTING AN OPTIMAL TOLERANCE LEVEL AND ORDER RELATION

However, the percentage of noise in the data is mostly not known. Thus, the question remains, which of the order relations generated with various tolerance levels should be chosen as the best or optimal ITA solution.

To estimate the optimal tolerance level, van Leeuwe (1974) and Held and Korossy (1998) applied the correlational agreement coefficient ($CA$), which evaluates the relative fit of alternative surmise relations $S$ (generated with various tolerance levels) to a set of data. The $CA$ was introduced by van Leeuwe (1974) in order to account for the trade-off between the absolute goodness of fit (regarding the answer patterns in contradiction with the assumed relation) and the number of pairs in the inferred relation. A low $CA$ value should not only result in the case of too many generated pairs, but also in the case of too few pairs. The agreement of a surmise relation and a set of data is better, the higher the value of $CA$.

For selecting a relation, we only consider solutions, which contain at least one item pair (for solutions where all items are incomparable, a falsification of the generated model is not possible).

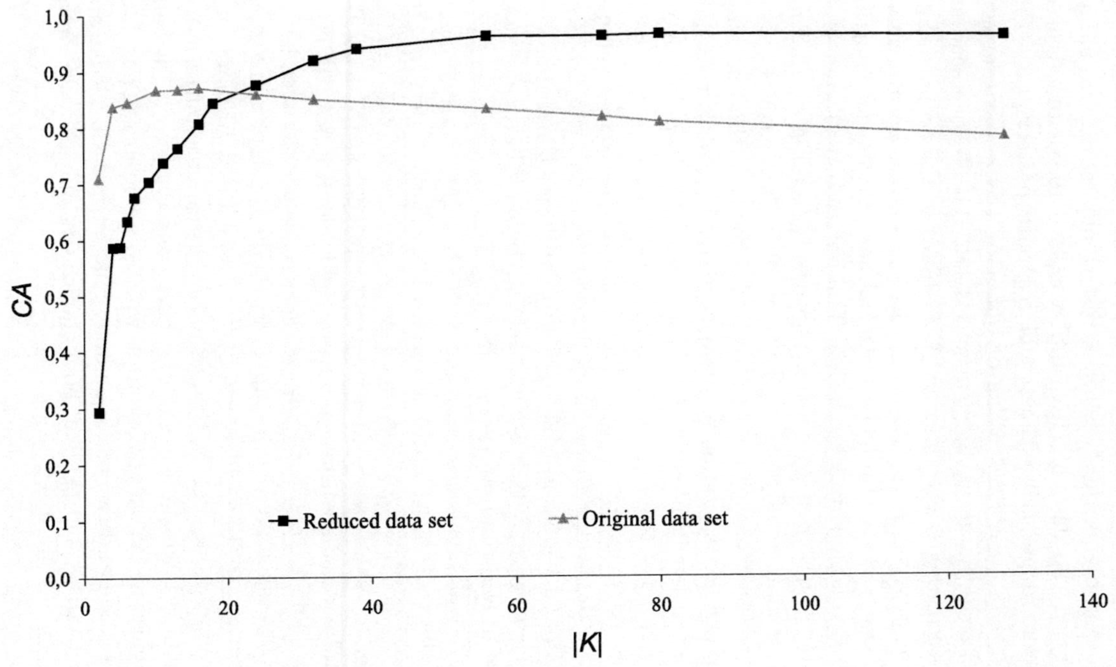


**Figure 10.4.** $CA$ as a function of space size $|\mathcal{K}|$ for ITA solutions with (original data set) and without (reduced data set) trivial response patterns (data by Held and Korossy, 1998).

Regarding the obtained $CA$ values for the relation depicted in Figure 10.3, we found that $CA$ decreases with increasing $L$ (between 0.96 for $L = 2\%$ and 0.84 for $L = 15\%$; see Table 10.3). For the correct solution at $L = 10\%$, $CA$ equals 0.91. We obtained the highest $CA$ value for a tolerance level of only 2% at which the resulting relation contains only one pair, that is almost all items are incomparable which is obviously not true (see Figure 10.3). Thus the question is, why Held and Korossy (1998) found a reasonable structure in their data applying the $CA$ in contrast to our result with simulated data. We reanalysed the data of Held and Korossy (1998) and figured out that the $CA$-result depends on the trivial response patterns see (Figure 10.4). For equal tolerance levels, adding or removing trivial patterns to a data set does not change the resulting surmise relation, but $CA$ decreases or increases in value. Accordingly, the obtained $CA$ values of the reduced data set from Held and Korossy (1998) decreased with increasing $L$ with only two pairs for the highest $CA$, cf. Table 10.3. Trivial patterns do not contribute to the establishment of the surmise relation. The patterns $\langle 0,0 \rangle$ and $\langle 1,1 \rangle$ increase the frequencies in cells $a_{xy}$ and $d_{xy}$ respectively (see Figure 10.2 ) but do not influence the number of pairs in the relation. ITA considers only cell $b_{xy}$ to determine whether or not an item pair is element of the relation. Thus, for equal tolerance levels, adding or removing trivial patterns to a data set does not change the resulting surmise relation, but $CA$ decreases or increases in value. For details of the $CA$ see Chapter 9.

Because the investigation of $CA$ showed that the largest value should not be applied to determine the optimal surmise relation for a given set of data, Ünlü (2009), and Ünlü and Malik (2008b) developed the Kappa coefficient $\kappa$. This coefficient takes better account of the trade-off between the descriptive goodness of fit and the number of pairs in a relation. $\kappa$ varies $(-\infty, -C]$ with an arbitrary small fixed correction constant $C$. Higher values indicate a better trade-off between the size and the fit of a knowledge structure. For a formal definition and computing of the coefficient $\kappa$, see Section 9.3.1 in this volume.

Applying $\kappa$ to the data set for the test knowledge space derived from Figure 10.3 and the reduced data set from Held and Korossy (1998), we found the optimal knowledge spaces (highest $\kappa$ values) for tolerance levels $L = 8\%$ and $L = 15\%$ respectively (see Table 10.3). The corresponding surmise relations contain 30 pairs for the space in Figure 10.3 and 13 pairs for Held and Korossy's reduced data set (without the trivial response patterns). The solution for the simulated data is slightly more conservative (i.e. it contains less pairs), but close to the original relation with 35 item pairs. Thus, for selecting a surmise relation and its corresponding quasi ordinal knowledge space using the $\kappa$ coefficient is recommended. For comparative purposes we computed also the $CA$-values in the following analysis.

**Table 10.3.** Properties of *SRbIs* generated by ITA with different tolerance levels $L$ for simulated and real data

| | Simulated data $N = 1000$ for 10 items | | | Reduced data from Held & Korossy (1998) $N = 679$ for 7 items | | |
|---|---|---|---|---|---|---|
| $L$ | $CA$ | No. of pairs\\$\Delta$ | $\kappa$ | $CA$ | No. of pairs\\$\Delta$ | $\kappa$ |
| 2% | .9566 | 1 | -.03608 | - | - | - |
| 3% | .9551 | 3 | -.03590 | - | - | - |
| 4% | .9519 | 6 | -.03558 | - | - | - |
| 5% | .9474 | 13 | -.03482 | - | - | - |
| 6% | .9456 | 18 | -.03427 | .9633 | 2 | -.06384 |
| 7% | .9437 | 22 | -.03393 | .9633 | 2 | -.06384 |
| **8%** | .9328 | **30** | **-.03361** | .9603 | 3 | -.06342 |
| 9% | .9152 | 34 | -.03374 | .9611 | 4 | -.06304 |
| 10% | .9074 | 35 | -.03378 | .9611 | 4 | -.06304 |
| 11% | .9020 | 36 | -.03411 | .9611 | 4 | -.06304 |
| 12% | .8954 | 37 | -.03441 | .9397 | 6 | -.06484 |
| 13% | .8893 | 38 | -.03604 | .9192 | 8 | -.06382 |
| 14% | .8730 | 40 | -.03724 | .8767 | 11 | -.06235 |
| **15%** | .8444 | 43 | -.03845 | .8444 | **13** | **-.06044** |
| 16% | - | - | - | .8080 | 15 | -.06115 |
| | | | | ⋮ | ⋮ | ⋮ |

*Note.* For tolerance levels < 2% (simulated data) and < 6% (reduced data) the number of pairs = 0 and the resulting relations are therefore not considered. Bold printed numbers indicate the best solution according to $\kappa$.

### DEMONSTRATION WITH A DATA SET FOR TWO INDUCTIVE REASONING TESTS

Participants, materials, procedure, and the pre-editing of data are identical to that used for the theory-driven hypotheses. In order to generate and validate the data-driven hypotheses, the set of response patterns was split in halves, leaving 366 response patterns each for the generation and the validation process.

Using ITA, we generated surmise relations for tolerance levels between 5% and 15%. We started by increasing $L$ in steps of 1%, but approaching the optimal $\kappa$ we reduced the intervall to .5% and finally .1%. The respective $\kappa$ indices (see Table 10.4) indicate the best solution for $L$ = 9.9%, 10.0% and 10.1% or $L_a = 36$ patterns ($\kappa$ = -0.02999). Since the absolute frequencies ($L_a$) do not change for each .1% step, the relative tolerance levels which refer to the same absolute frequency are presented in same rows. Furthermore, Table

**Table 10.4.** Properties of *SRbIs* generated by ITA with different tolerance levels $L$ for two inductive reasoning tests ($N = 366$ response patterns for 27 items)

| $L$ | $L_a$ | $CA$ | No. of intransitive triplets | No. of added pairs | No. of pairs\$\Delta$ | $\lvert\mathcal{K}\rvert$ | $\kappa$ |
|---|---|---|---|---|---|---|---|
| 0.0% | 0 | .9952 | 0 | 0 | 1 | –[a] | –[a] |
| 5.0% | 18 | .8413 | 16 | 16 | 193 | 280,833 | -.03670 |
| 6.0% | 21 | .8047 | 16 | 15 | 225 | 64,769 | -.03537 |
| 7.0% | 25 | .7622 | 18 | 6 | 261 | 10,113 | -.03308 |
| 8.0% | 29 | .6218 | 56 | 49 | 350 | 3,393 | -.03218 |
| 9.0% | 32 | .6068 | 70 | 23 | 362 | 1,155 | -.03093 |
| 9.5% | 34 | .5431 | 85 | 29 | 405 | 581 | -.03026 |
| 9.6%, 9.7%, 9.8% | 35 | .5171 | 59 | 35 | 425 | 417 | -.03008 |
| **9.9%, 10.0%, 10.1%** | **36** | .5115 | **87** | **26** | **429** | **329** | **-.02999** |
| 10.2%, 10.3% | 37 | .5095 | 104 | 19 | 431 | 233 | -.03005 |
| 10.4%, 10.5% | 38 | .5074 | 96 | 13 | 432 | 225 | -.03005 |
| 11.0% | 40 | .4087 | 114 | 60 | 496 | 161 | -.03005 |
| 12.0% | 43 | .3619 | 193 | 59 | 525 | 53 | -.03046 |
| 13.0% | 47 | .3270 | 248 | 50 | 549 | 37 | -.03055 |
| 14.0% | 51 | .2887 | 244 | 47 | 574 | 21 | -.03083 |
| 15.0% | 54 | .2887 | 183 | 28 | 574 | 21 | -.03083 |

*Note.* $L_a$ = absolute frequency of contradicting answer patterns.
[a]Due to too many states $\mathcal{K}$ and, therefore, $\kappa$ could not be computed.

10.4 shows for each generated surmise relation the $CA$ (which again shows a steady decrease with increasing tolerance levels), the number of intransitive triplets, the number of pairs in the resulting surmise relation after transitive closure (without $\Delta$), the cardinality of the knowledge space $\mathcal{K}$. In the case of transitivity violations, the transitive closure for the empirical relation was computed by adding pairs to the relation until transitivity holds ('No. of added pairs' in Table 10.4).

The derived $SRbI$ contains 429 out of 702 ($27 \times 27$ without $\Delta$) possible pairs, the corresponding knowledge space 329 out of $2^{27}$ possible knowledge states. Partitioning the set of items into $MT$ and $AN$, we obtained 5 item classes in $MT$ (see Figure 10.5, left ellipse) and 8 item classes in $AN$ (see Figure 10.5, right ellipse). The matrix items 1-7 and 10 as well as the analogy items 13, 16, 17, 19, 22-24, and 27 are equivalent (each item is prerequisite for the other items).

Regarding the obtained $SRxT$ (see Figure 10.5), we found that both relationships between the two tests are elements of the left-covering surmise relation ($MT\ \dot{\mathcal{S}}_l\ AN$ and $AN\ \dot{\mathcal{S}}_l\ MT$). This means that a person who doesn't solve any item in $MT$ will not be able to solve any item in $AN$ and vice versa.

A comparison of the theory- and the data-driven hypotheses shows that in both cases a left-covering surmise relation from $MT$ to $AN$ ($AN\ \dot{\mathcal{S}}_l\ MT$) is assumed. However, instead of the theoretically derived right-covering surmise relation from $AN$ to $MT$ ($MT\ \dot{\mathcal{S}}_r\ AN$), a left-covering surmise relation from $AN$ to $MT$ is obtained from the data. A detailed comparison of the two hypotheses is given at the end of the next section.

## 10.4 Validation of Hypotheses

In this section, we discuss different validation methods, based on the surmise relation as well as on the knowledge space, and exemplify their application on the theory- and data-driven hypotheses.

Strictly viewed, the goodness of fit of a given knowledge space (and its corresponding surmise relation) to a set of data can be defined by the number of response patterns which can be assigned to exactly one of the postulated states. However, working with a deterministic model we have to consider that the knowledge states describe the latent knowledge of persons, whereas the response patterns might include careless errors and/or lucky guesses. Furthermore, the fit of a knowledge structure to a set of data increases with an increasing size $|\mathcal{K}|$ (just as the fit of the corresponding surmise relation increases with a decreasing number of pairs). In the extreme case all items are independent, that is the relation contains only the pairs $(q, q)$, hence $|\mathcal{K}| = 2^Q$; the data are always in accordance with the knowledge space.

In the following, we apply different validation methods[7], which take into account that the space size may vary as well as that there may exist discrepancies between a person's true knowledge state and his or her response pattern. All of the discussed procedures can be applied to surmise relations and

[7]The programs used for validation procedures were developed in C/C++ by Cord Hockemeyer and Susanne Poetzi and are described in Hockemeyer (2001) and Pötzi and Wesiak (2004). They can be used by registering at http://css.uni-graz.at/ePsyt.

knowledge spaces between items ($SRbI/KSbI$), within tests ($SRwT/KSwT$), and across tests ($SRxT/KSxT$). In the latter two cases, the postulated pairs in the surmise relation (or the states in the knowledge space) are reduced to the respective subsets of the $SRbI/KSbI$ (see definition 10.2). For the $SRwT/KSwT$ additionally the answer patterns are reduced to the set of items in the respective test. Furthermore, we demonstrate how a right-, left-, or total-covering surmise relation is derived empirically. Theory- and data-driven hypotheses are validated separately, followed by a comparison of the obtained results.

**10.4.1 Validation of Hypotheses via the Surmise Relation.** The fit of a surmise relation to a data set is evaluated by means of the solution frequencies per item class and by two indices ($VC$ and $\gamma$).

PERCENTAGE OF CORRECT SOLUTIONS

Imagine a relationship $ySx$, where item $x$ should only be solved in combination with a correct solution to $y$, whereas $y$ can also be solved by itself (see Figure 10.2 ). In this case, we expect that the solution frequency for item $y$ is equal to or higher than the solution frequency for item $x$.

Figures 10.1 and 10.5 illustrate this validation method for the theory- and data-driven $SRbI$ by means of the relative solution frequencies (in percent). For item classes with more than one item (see Table 10.2), the solution frequencies are averaged over all items in the respective class.

Regarding the theory-driven hypotheses, the results for the $SRwMT$ (left ellipse in Figure 10.1) show that 17 out of 18 pairs of item classes confirm the hypothesis, that is, the relation is verified by all but one pairs (item classes 8O2HHG and 8O2LHG). The results for the $SRwAN$ (right ellipse in Figure 10.1) show deviations primarily at the bottom of the structure. More precisely, 37 out of 49 pairs of item classes confirm the hypothetical surmise relation.

For the $SRxT$ (see Table 10.2), that is relationships between item classes of different tests (lines going from one ellipse to the other in Figure 10.1), 26 out of 29 pairs confirm the surmise relation from $AN$ to $MT$ and only 25 out of 40 pairs the surmise relation from $MT$ to $AN$. A point worth noting is that most deviations occur for assumptions regarding analogy items being prerequisite for a matrix item. Thus, the assumptions regarding the components *material* and answer alternatives may be too strong. It states that the matrix items (geometric-figural material, eight alternatives) are more difficult than the analogy items (verbal material, five alternatives).

Regarding the hypotheses of a right-covering surmise relation from $AN$ to $MT$ and a left-covering surmise relation from $MT$ to $AN$, we expected that each of the item classes in $MT$ is prerequisite for at least one item class in $AN$ ($MT\ \dot{\mathbb{S}}_r\ AN$) and has at least one prerequisite in $AN$ ($AN\ \dot{\mathbb{S}}_l\ MT$; see also Table 10.2). Whereas the right-covering relation is confirmed by all but one item class in $MT$ (8O3LHG has the lowest solution frequency and thus

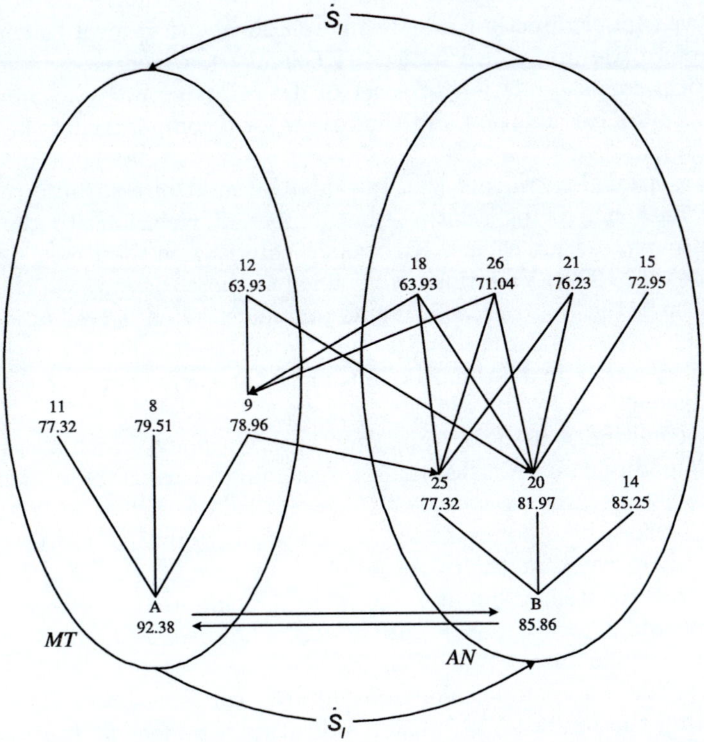


**Figure 10.5.** Hasse diagram for the data-driven surmise relations between items and tests and percentage of correct solutions for item classes (class A includes items 1-7, and 10; class B items 13,16,17,19,22-24, and 27).

is not prerequisite for an analogy item class), the left-covering relation is contradicted by three out of eight item classes (8O1LLG, 8O1HLG, and 8O2LL have the highest solution frequencies and thus do not have a prerequisite in *AN*).

For the data-driven hypotheses (derived from one half of the data), the solution frequencies (computed from the other half) confirm all but two of the 36 generated pairs in the *SRbI* (see Figure 10.5). Exceptions are the two pairs $(B, A)$ and (25,9) in the *SRxT*. Because of the postulated left-covering surmise relation for both tests, each item class of a test should have at least one prerequisite class in the other test. This is true of 12 out of the 13 item classes. Considering the relative solution frequencies, only item class {A} in *MT* does not have a prerequisite in *AN*.

### Indices for the fit of a surmise relation

Because of the different numbers of pairs in the two postulated *SRbI* and their subsets, we cannot directly compare the relations with respect to goodness of fit. Therefore, we apply two indices ($VC$ and $\gamma$), which assess the fit of a hypothetical surmise relation to a set of data. Both indices account for the number of pairs in a relation and thus allow for a comparison of the different subsets.

The *violational coefficient* ($VC$) evaluates the fit of a surmise relation to a binary data matrix by relating the number of violations to the number of response vectors for each non reflexive pair $ySx$ in a surmise relation $S$ (Schrepp et al., 1999). Violations are defined as those cases, in which a person masters item $x$ but fails in mastering its prerequisite item $y$ (cell $b_{xy}$ in Figure 10.2 ). $VC$ varies within the limits [0,1], with 0 indicating a perfect fit (no violations at all). For a formal definition of the coefficient $VC$ see Section 9.3.6 of this volume.

The second index used for validation is the $\gamma$-*index*, which was first proposed by Goodman and Kruskal (1954). Assuming a surmise relationship $ySx$, the contradicting response vectors $\langle 1, 0\rangle$ are called discordant (cell $b_{xy}$ in Figure 10.2), which corresponds to $VC's$ violations ($v_{xy}$). The vector $\langle 0, 1\rangle$, on the other hand, confirms the hypothesis, because we expect some of the participants to solve item $y$ but not the more difficult item $x$. These cases are called concordant (cell $c_{xy}$ in Figure 10.2 ). The cases where either both or neither of the items are solved (cells $d_{xy}$ and $a_{xy}$ in Figure 10.2 ) are trivial and do not allow any conclusions on the hypothesis' validity. The $\gamma$-index varies within the limits [-1,1], with 1 indicating a perfect fit (no discordant vectors). The $\gamma$-index can also be used to evaluate the overall fit of a model by accumulating the frequencies of concordant and discordant cases over all item pairs. This global index is referred to as $\gamma_G$. $\gamma$ is a stricter test of a hypothesis than $VC$. For a formal definition of the $\gamma$-index, see Section 9.3.6 of this volume.

*Results*

To find out how much the various subsets of the *SRbI* (i.e., the $SRxT$, the $SRwMT$, and the $SRwAN$) contribute to the overall result, we specify their respective pairs and calculate $VC$ and $\gamma_G$ for both the theory- and data-driven hypotheses (see Table 10.5). Note that the indices refer to relationships between single items as compared to item classes.

Regarding the theory-driven hypothesis, both indices indicate that the $SRwMT$ fits the data set best ($VC = .04$, $\gamma_G = .61$), while the $SRwAN$ deviates most from the data ($VC = .11$, $\gamma_G = .11$). Table 10.5 also shows that between 66.15% and 95.35% of the pairs have positive $\gamma$ indices. McNemar $\chi^2$ tests show for all subsets that the number of concordant vectors ($c_{xy}$) is significantly higher than the number of discordant vectors ($b_{xy}$).

**Table 10.5.** $VC$ and $\gamma$ for the theory- and data-driven surmise relations between items and their subsets

| | $m$ | No.of pairs\ $\Delta$ | $VC^a$ | $\gamma_G{}^a$ | % of pairs with $\gamma > 0$ | No. of $c_{xy}$ | No. of $b_{xy}$ | $\chi^2$ |
|---|---|---|---|---|---|---|---|---|
| theory-driven surmise relation ($N = 732$ response patterns) | | | | | | | | |
| *SRbI* | 27 | 231 | .08 | .28 | 74.03 | 24079 | 13563 | 2937.84 |
| *SRxT* | 27 | 123 | .08 | .29 | 70.73 | 12751 | 7092 | 1613.88 |
| *SRwMT* | 12 | 43 | .04 | .61 | 95.35 | 4655 | 1137 | 2136.80 |
| *SRwAN* | 15 | 65 | .11 | .11 | 66.15 | 6673 | 5334 | 149.32 |
| data-driven surmise relation ($N = 366$ response patterns) | | | | | | | | |
| *SRbI* | 27 | 429 | .08 | .27 | 70.16 | 22292 | 12911 | 2499.88 |
| *SRxT* | 27 | 333 | .08 | .20 | 62.76 | 15059 | 10131 | 964.08 |
| *SRwMT* | 12 | 89 | .06 | .35 | 68.54 | 4016 | 1917 | 742.59 |
| *SRwAN* | 15 | 119 | .10 | .21 | 74.79 | 6762 | 4408 | 496.09 |

*Note.* $m$ = number of items, $c_{xy}$ = concordant vectors, $b_{xy}$ = discordant vectors, $\chi^2$ are McNemar $\chi^2$ tests for $c_{xy}$ and $b_{xy}$ (for $df = 1$ and $p = .001$, $\chi^2_{crit} = 10.83$).
[a]For $VC$ lower values, for $\gamma_G$ higher values indicate a better fit.

For the data-driven hypothesis the best results are also found for the *SRwMT* ($VC = .06$, $\gamma_G = .35$). The percentage of pairs with positive $\gamma$ indices ranges between 62.76% and 74.79% and the four $\chi^2$ values are highly significant. Except for the *SRwAN*, the indices show slightly better results for the theory-driven than the data-driven hypotheses.

Regarding the postulated surmise relations between tests, the indices do not permit direct conclusions about their validity, because we cannot infer from the indices, which and how many items in test $A$ have a prerequisite in test $B$.

Summarized, the indices which were developed for validating single tests are also adequate for assessing to which extent the different subsets defined in Equation 10.2 contribute to the validity of the *SRbI*.

### 10.4.2 Validation of Hypotheses via the Knowledge Space.

For validating and evaluating the hypothesized structures via the knowledge space the agreement between a data set and a given knowledge structure is determined. The *Mean Minimal Symmetric Distance (MMSD)*, and the *Distance Agreement coefficient* (*DA*) are calculated for the *KSbI* and its substructures. Furthermore, the empirical distances are compared to those of simulated data sets.

The *MMSD* compares each response pattern with all the knowledge states and takes the minimal symmetric set difference. The theoretical minimum corresponds to perfect agreement ($dmin = 0$), the theoretical maximum denotes the greatest possible distance depending on the number of items ($dmax = m/2$ or $(m-1)/2$ for even or odd numbers respectively). For details see Section 9.3.6 of this volume.

Table 10.6 (4th column) shows the $MMSDs$ ($ddat$) between the data sets and the theory- as well as the data-driven $KSbI$ and their substructures. The values for all of the postulated knowledge spaces are far below their theoretical maxima ($ddat$ varies between 0.66 and 2.70 as compared to $dmax$ ranging between 6 and 13; see the corresponding $m$-values in the 2nd column). A direct comparison of the various $ddat$ values is not reasonable, because the number of items as well as the number of knowledge states influence $ddat$.

Focusing on the validity of the $SRbT$ means to check whether the $MMSD$ of the entire structure is primarily due to the $MMSDs$ within or across tests. Regarding the varying sizes of the hypothesized knowledge structures, a comparison of the substructures' fit has to account for the number of knowledge states within the respective power sets.

The *Distance agreement coefficient* $DA$ relates the fit of a structure to a data set to the structure's size. The lower $DA$, the better the fit of a knowledge structure to a given data set. $DA$'s minimum equals zero, a random fit is characterized by $DA = 1$. For a formal definition see Section 9.9.3.6 in this volume.

The $DA$ values for the theory- and data-driven $KSbI$ and their substructures are shown in Table 10.6 (5th column). For both approaches, the lowest $DA$ results for the $KSwMT$ and the highest $DA$ for the $KSwAN$. This corresponds to the results derived via the surmise relation.

SIMULATIONS

For evaluating the goodness of fit for knowledge structures simulation methods are appropriate. A detailed description can be found in Section 9.3.6 of this volume.

Simulations with various degrees of specificity are performed to compare the empirical $MMSDs$ ($ddat$) with more appropriate values than the extremes ($dmin$, $dmax$) For various sets of random data, the $MMSDs$ ($dsim_r$) between the random data sets and a given knowledge structure are calculated. Furthermore, we applied the following four types of frequency simulations: ($simf_1$) for each entry in the data matrix, the probability of a correct solution equals the mean solution frequency over all items and persons, ($simf_2$) for each row of entries, the probabilities are based on the marginal frequencies, i.e., the mean solution frequencies per person, ($simf_3$) for each column of entries, the probabilities are based on the marginal frequencies, i.e., the mean solution frequencies per item, and ($simf_4$) for each entry the probabilities are based on the solution frequencies per item and person.

For ($simf_4$) the marginal frequencies of the simulated matrices equal those of the original data matrix, whereas the distributions of '0's and '1's differ. The algorithm of (Ponocny and Waldherr, 2002) is used with the total number

**Table 10.6.** *MMSDs* between the empirical and simulated data sets and the postulated knowledge spaces

| | $m$ | $\vert\mathcal{K}\vert$ | $ddat$ $(SD)$ | $DA$ | $dsim_r$ $(SD)$ | $dsim_{f1}$ $(SD)$ | $dsim_{f2}$ $(SD)$ | $dsim_{f3}$ $(SD)$ | $dsim_{f4}$ $(SD)$ | $dsim_p$ $(SD)$ |
|---|---|---|---|---|---|---|---|---|---|---|
| theory-driven knowledge space ($N = 732$) | | | | | | | | | | |
| *KSbI* | 27 | 553 | 2.70 (1.87) | .32 | 8.43 (1.83) | 3.34 (1.60) | 3.27 (2.30) | 2.96 (1.48) | 2.75 (1.86) | 2.47 (1.46) |
| *KSxT* | 27 | 39348 | 1.21 (1.12) | .24 | 5.12 (1.36) | 1.72 (1.13) | 1.72 (1.48) | 1.35 (0.93) | 1.28 (1.10) | 1.57 (1.15) |
| *KSwMT* | 12 | 53 | 0.66 (0.78) | .22 | 3.04 (1.17) | 1.40 (0.85) | 0.97 (1.21) | 0.95 (0.74) | 0.65 (0.79) | 0.85 (0.85) |
| *KSwAN* | 15 | 99 | 1.56 (1.27) | .39 | 4.00 (1.32) | 1.97 (1.14) | 1.65 (1.52) | 1.80 (1.11) | 1.57 (1.28) | 1.38 (1.08) |
| data-driven knowledge space ($N = 366$) | | | | | | | | | | |
| *KSbI* | 27 | 329 | 2.24 (2.05) | .25 | 8.90 (1.76) | 3.16 (1.60) | 3.16 (2.47) | 2.46 (1.42) | 2.25 (1.96) | 2.06 (1.34) |
| *KSxT* | 27 | 879 | 2.02 (1.83) | .26 | 7.82 (1.62) | 3.08 (1.57) | 3.02 (2.28) | 2.29 (1.38) | 2.05 (1.75) | 2.09 (1.38) |
| *KSwMT* | 12 | 13 | 0.67 (0.93) | .18 | 3.73 (1.10) | 1.44 (0.94) | 1.13 (1.36) | 0.99 (0.85) | 0.70 (0.90) | 1.00 (0.93) |
| *KSwAN* | 15 | 41 | 1.44 (1.40) | .32 | 4.48 (1.26) | 1.98 (1.19) | 1.77 (1.65) | 1.59 (1.11) | 1.44 (1.36) | 1.16 (1.03) |

*Note.* $m$ denotes the number of items, $|\mathcal{K}|$ the number of knowledge states; means and $SDs$ for simulated data $dsim_i$ are the averaged values from 1000 data sets each; $dsim_{f1}$ is based on the mean solution frequency $f$, $dsim_{f2}$ on $f$ per person, $dsim_{f3}$ on $f$ per item, $dsim_{f4}$ on $f$ per person and item.

of potentially exchanged entries amounting to $N \times m + x$, where $n$ denotes the number of response patterns, $m$ the number of items, and $x = 3000$ an arbitrary number of additional selections.

Finally, response patterns are simulated on the basis of the hypothesis (i.e. the postulated knowledge space), but under consideration of the probabilities for careless errors and lucky guesses (probability simulation). In the case of multiple-choice items, the probability for lucky guesses is estimated from the number of answer alternatives (e.g., $1/8 = 0.125$ for 8 alternatives), whereas the probability for careless errors is unknown and will therefore be varied (see below).

We expect the postulated knowledge spaces (theory- and data-driven) to fit the empirical data better than the random simulation ($dsim_r$) and the four frequency simulations ($dsim_f$). Such a result would indicate that the agreement of our model with the empirical data set is not put forth by random and that the specific patterns of the responses are relevant for the prediction of the testees' solution behavior. Furthermore, we expect the postulated knowledge spaces to fit the empirical data equally well as data from the probability simulation ($dsim_p$). That is, because the response patterns are simulated under the assumption of a correct model, although permitting a certain amount of noise in the data.

*Results*

Table 10.6 depicts not only the $MMSDs$ and standard deviations ($SD$) for the empirical data ($ddat$), but also for the simulated data sets. The values for $dsim_r$, $dsim_{fi}$, and $dsim_p$ are the averaged $MMSDs$ and $SDs$ from 1000 data sets each. The number of response patterns and the number of items per data set correspond to the numbers in the respective empirical data sets, e.g. 732 patterns and 12 items for $KSwMT$. For the probability simulation the probability for careless errors ($\beta$) was varied in ten .01 steps with $.05 < \beta \leq .15$ and 100 data sets per step.

Figure 10.6 shows the distance distributions of the empirical and the simulated data sets for the two $KSbIs$. It can be seen that the distances obtained from $dsim_r$ are much higher than those of all other data sets (empirical and simulated). Thus, $dsim_r$ turns out to be absolutely inadequate to judge the validity of the postulated models. We will therefore no longer consider this type of simulation.

The results in Table 10.6 show for the theory- as well as the data-driven hypotheses, that the empirical $MMSDs$ ($ddat$) are lower (better) than those for $dsim_{f1,f2,f3}$ and about the same as those for $dsim_{f4}$. For $dsim_p$, both the theory- and the data-driven hypotheses on the $KSxT$ and the $KSwMT$ result in lower values for the empirical data, whereas on the $KSbI$ and the $KSwAN$ they result in higher values for $ddat$. Figure 10.7 visualizes the results ($MMSDs$ and standard errors $SE$) for the empirical data, the frequency, and the probability simulations. The four types of frequency simulations show a

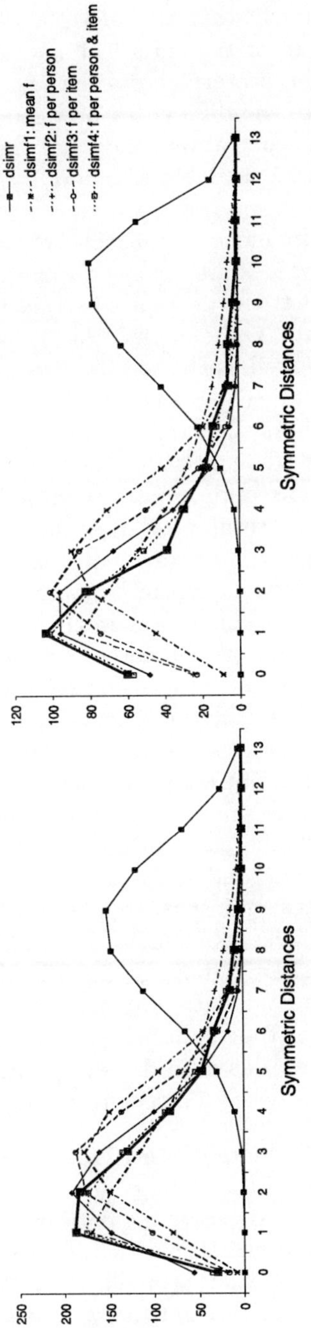


**Figure 10.6.** Distributions for empirical and simulated minimal symmetric distances (averaged over 1000 data sets each).

trend as expected, namely that the $MMSDs$ generally decrease, the more information is gained from the empirical solution frequencies. $MMSDs$ for $dsim_p$ show no common pattern across the various substructures.

In order to evaluate the differences between the empirical and simulated data sets regarding the central tendency and the distribution of the minimal symmetric distances, we calculated Mann-Whitney $U$ and $\chi^2$ tests (see Table 10.7).

The results for the frequency simulations $dsim_{f1,f2,f3}$ are almost all highly significant. The $ddat$ values are always lower than those for simulated data sets, which indicates that the postulated structures are an important predictor for the empirical data. Only for $AN$, the theory-driven hypothesis explains the data about equally well as the solution frequencies per person do. In this case, the linear order induced by the solution frequencies explains the empirical data as good as the postulated partial order does. All in all, up to the specification of solution frequencies per item, the postulated knowledge states are for the most part more appropriate to explain the empirical data than the solution frequencies. The results for the distance distributions derived from frequency simulations per item and person ($dsim_{f4}$) reveal no significant difference between the empirical and the simulated data sets. Thus, the marginal frequencies for rows and columns explain the data equally well as the postulated models. One possible reason for the obtained results for $dsim_{f4}$ are the high solution frequencies in the empirical data. With 85.71% and 83.5% of '1's in the data matrices for the theory- and the data-driven hypotheses, respectively, the number of changed entries is extremely restricted.

Comparing the distributions of the empirical minimal symmetric distances ($ddat$) to those of data sets simulated on the basis of the hypotheses ($dsim_p$), $U$-tests show significant differences in favor of the hypotheses for the theory- and the data-driven $KSxT$ and $KSwMT$ ($-6.26 \leq U(z) \leq -2.00$, $p < .05$) and significant differences against the hypotheses for both $KSwAN$ ($U(z) =$ 2.12 and 2.09, $p < .05$). The two values for the $KSbI$ are not significant. Furthermore, the $\chi^2$-tests show significant differences between the distributions of the empirical data and the data obtained by the probability simulation.

Overall, the results of the validation procedures via the knowledge space confirm the theory- as well as data-driven hypotheses on the $KSwMT$ and the $KSxT$, whereas the $KSwAN$ needs to be improved.

### 10.4.3 Comparison of Theory- and Data-driven Hypotheses.

To check the agreement of the theory- and the data-driven hypotheses, we used the whole set of data to generate a data-driven model. For this purpose, we first performed a cross-validation by exchanging the two data sets used for the generation and validation of hypotheses. In the following, we refer to the whole set relation as $SRbI_I$ and to the relation generated for the cross-validation as $SRbI_{II}$.

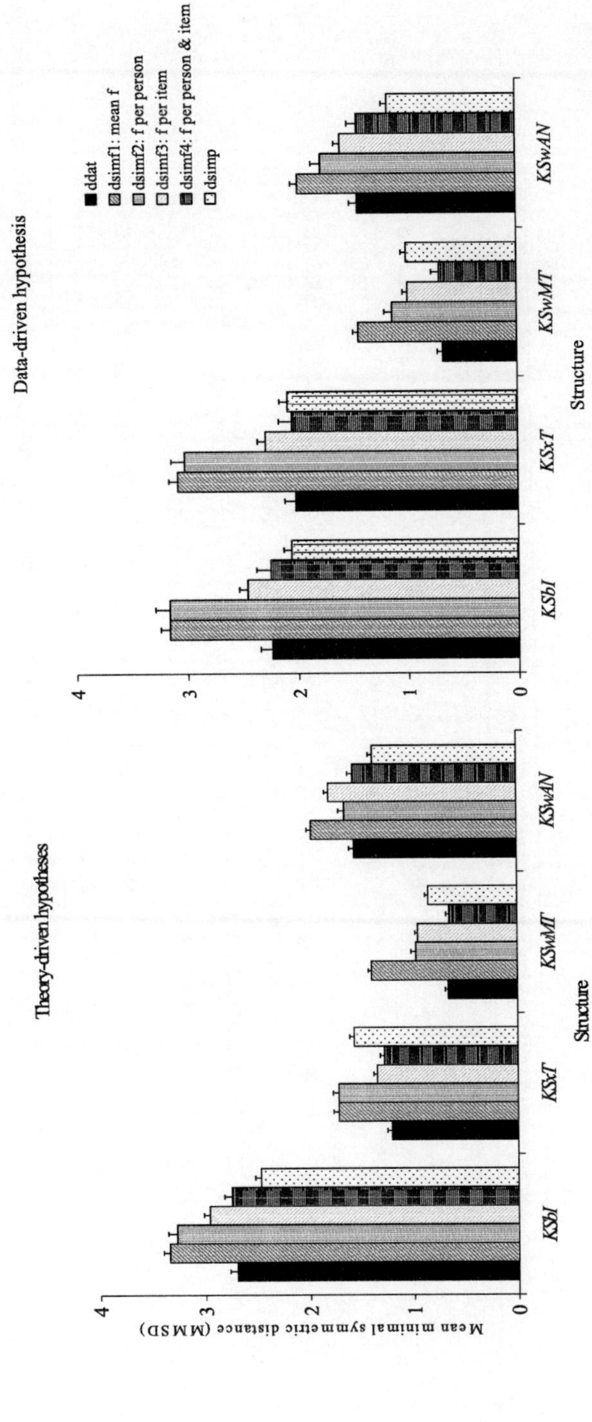


**Figure 10.7.** $MMSDs$ $(+SE)$ for the empirical data ($ddat$) and simulated data sets (with $f$ denoting solution frequency).

**Table 10.7.** Comparison of the empirical (*ddat*) and simulated *MMSDs* for theory- and data-driven knowledge spaces

| | $dsim_{f1}$ | | $dsim_{f2}$ | | $dsim_{f3}$ | | $dsim_{f4}$ | | $dsim_p$ | |
|---|---|---|---|---|---|---|---|---|---|---|
| | $U(z)$ | $\chi^2$ (*df*) | $U(z)$ | $\chi^2$ (*df*) | $U(z)$ | $\chi^2$ (*df*) | $U(z)$ | $\chi^2$ (*df*) | $U(z)$ | $\chi^2$ (*df*) |
| theory-driven knowledge space ($N_{1=2} = 732$) | | | | | | | | | | |
| *KSbI* | -8.81** | 315.97(8)** | -4.21** | 51.30(11)** | -5.12** | 184.74(7)** | -0.69 | 4.96(9) | 0.92 | 143.38(7)** |
| *KSxT* | -9.32** | 223.68(5)** | -6.54** | 88.99(6)** | -3.99** | 98.39(4)** | -1.35 | 8.46(5) | -6.26** | 81.77(5)** |
| *KSwMT* | -16.87** | 1303.42(4)** | -3.27** | 145.91(5)** | -8.32** | 202.68(3)** | 0.28 | 7.99(4) | -4.61** | 42.05(3)** |
| *KSwAN* | -7.51** | 247.21(5)** | -0.17 | 82.67(6)** | -4.94** | 120.15(5)** | -0.16 | 2.87(5) | 2.12* | 68.30(5)** |
| data-driven knowledge space ($N_{1=2} = 366$) | | | | | | | | | | |
| *KSbI* | -8.57** | 450.42(7)** | -5.43** | 65.44(10)** | -4.07** | 165.39(6)** | -0.45 | 6.10(8) | 0.46 | 100.95(6)** |
| *KSxT* | -9.46** | 488.68(7)** | -6.38** | 110.84(9)** | -4.03** | 123.47(6)** | -0.57 | 3.71(7) | -2.00* | 50.75(6)** |
| *KSwMT* | -11.75** | 675.89(4)** | -4.23** | 46.88(5)** | -6.14** | 122.15(3)** | -0.76 | 3.69(4) | -5.54** | 75.76(4)** |
| *KSwAN* | -6.74** | 237.55(5)** | -2.27* | 23.70(6)** | -3.13** | 69.21(4)** | -0.16 | 1.81(5) | 2.09* | 48.86(4)** |

*Note.* $dsim_{f1}$ is based on the mean solution frequency $f$, $dsim_{f2}$ on $f$ per person, $dsim_{f3}$ on $f$ per item, $dsim_{f4}$ on $f$ per person and item; $U(z)$ is the standardized $U$-score for large samples with tied ranks; for $\chi^2$ tests, frequencies of less than 5 where pulled together. *$p$ ¡ .05. **$p$ ¡ .01.

*Cross-validation of the data-driven hypotheses*

Applying ITA and the $\kappa$ coefficient, the optimal $SRbI_{II}$ resulted for a tolerance level $L = 12\%$. The $SRbI_{II}$ contains 452 pairs (not counting $\Delta$), the corresponding knowledge space 233 states. The indices used for validation ($VC_{II} = .06, \gamma_{G_{II}} = .29, DA_{II} = .19$) show similar results as those for the $SRbI_I$ ($VC_I = .08, \gamma_{G_I} = .27, DA_I = .25$; cf. Tables 10.5 and 10.6). A direct comparison of the two generated relations reveals in both cases a left-covering surmise relation between the two tests with $AN\ \dot{\mathrm{S}}_l\ MT$ and $MT\ \dot{\mathrm{S}}_l\ AN$. A closer look at the item pairs shows that the 702 entries in the 27x27 item by item matrix ($\Delta$ is not considered) are concordant with each other in 95.01% (423 common item pairs and 244 common independencies). The 35 discordant entries result from 6 item pairs derived only for $SRbI_I$ and 29 item pairs derived only for $SRbI_{II}$. Calculating the $\gamma_G$-index for concordant and discordant entries results in $\gamma_G = .90$. Because of this high agreement between the two relations, we generated the final data-driven hypothesis from the whole data set.

*Theory- vs. data-driven surmise relations*

The 732 response patterns used for the data-driven hypotheses ($SRbI_d$) lead to an optimal ITA solution at a tolerance level of $L = 11\%$. The $SRbI_d$ contains 430 non reflexive item pairs, the corresponding knowledge space 273 states as compared to the theory-driven model ($SRbI_t$) with 231 non reflexive pairs and 553 states (cf. Tables 10.5 and 10.6).

One reason for the high number of pairs in the $SRbI_d$ are the high solution frequencies with a mean of $M = 85.71\%$ ($SD = 0.083$) correct solutions. With a tolerance level of $L = 11\%$, each item that was solved by at least 89% of the participants is automatically prerequisite for all other items. ITA assumes a prerequisite relationship $ySx$, whenever the percentage of response vectors $\langle 1,0 \rangle \leq L$; with at least 89% correct solutions for item $y$, $\langle 1,0 \rangle$ can maximally amount to 11%. Out of the 27 presented items, 11 items fulfill this criterion (at least 89% correct solutions), which corresponds to 286 automatically generated pairs in the relation. To account for this ceiling effect, we removed all pairs $ySx$ in the $SRbI_d$ for which item $y$ was solved by at least 89% of the participants, and which are not contained in the theory-driven $SRbI_t$. This procedure left 283 item pairs in the $SRbI_d$, 192 of which are also contained in the $SRbI_t$.

A direct comparison of the theory- and the data-driven hypotheses reveals a left-covering surmise relation from $MT$ to $AN$ for both models, whereas the relation from $AN$ to $MT$ is right-covering in the theory-driven model and left-covering in the data-driven model. Considering that adding a single item pair can suffice to make a relation left- or right-covering, we have to take a closer look at the relations' subsets. The number of common entries (pairs or non-pairs occurring in both relations) amounts to 81.48% (572 of 702) for the $SRbI$, 82.78% (298 of 360) for the $SRxT$, 90.15%

(119 of 132) for the $SRwMT$, and 73.81% (155 of 210) for the $SRwAN$ (not counting $\Delta$). The $SRbI_t$ contains 39, the $SRbI_d$ 91 additional pairs, which subdivide into 21/2/16 and 41/11/39 for the theory- and data-driven $SRxT/SRwMT/SRwAN$ respectively. The resulting $\gamma_G$-indices for concordant and discordant entries are .63, .66, .80, and .48 for the $SRbI$, the $SRxT$, the $SRwMT$, and the $SRwAN$ respectively. The $\gamma_G$ indices show that the agreement of the single subsets corresponds to the results found by means of the various validation procedures: The hypotheses on $MT$ show the best agreement of the two models and also proved to be the best predictor for the empirical data. On the other hand, the two models for $AN$ differ most from each other, which also corresponds to the least fitting results for the theory- as well as data-driven relation within $AN$. Thus, an improvement of the model is primarily necessary for the postulated order on the analogy items.

Overall, the results show that in this case both the theory- and the data-driven approach lead to comparable models. Which of the two generation methods should be preferred might also depend on individual needs, the domain of information, and the available data set.

## 10.5 Discussion

This work describes how different heuristic methods used for the generation and validation of surmise relations and knowledge spaces can be applied to a set of tests. One advantage of the presented methods is the possibility of adaptive testing procedures, once a surmise relation is established. The concepts surmise relation across, between and within tests allow adaptive applications for large test batteries. Promising other fields different from psychological diagnostics are applications in, among others, knowledge testing, curriculum development and structuring hypertexts.

Here, we focused on which methods exist and how to apply them to tests in practice, but not on the mathematical investigation of their properties. The indices $\gamma$, $VC$, and $DA$ have already proved to be of practical value in earlier studies (e.g., Albert and Lukas, 1999), but mathematical analyses are still open to further research. For an extensive analysis of the mathematical properties of $\kappa$ and $CA$, we refer to Ünlü (2009), and Ünlü and Albert (2004), and Ünlü and Malik (2008b). The order-theoretical and metrical properties of a set of knowledge spaces is treated by Suck (1999), who also deals with the symmetric distance and alternative metrics to assess the distance between two knowledge spaces. Moreover, once again we refer to Section 9.3.6 of this volume.

We presented two methods for generating either theory- or data-driven hypotheses on surmise relations between tests. The theory-driven approach is based on the analysis of problem demands. We showed how to extract and specify general components for various inductive reasoning problems, that is how items can be ordered with respect to their specific attribute combinations.

Applying a data-driven approach, hypotheses for surmise relations between two tests were established from a set of data by means of ITA. ITA can be applied to establish surmise relations between tests, when combined with the $\kappa$ coefficient to choose among the generated relations, whereas $CA$ turned out to be an inadequate selection criterion. Furthermore, ITA seems to be able to differentiate between partial and linear orders. The successful discrimination between different order relations implies that for some sets of items a partial order is more appropriate than a linear order to describe a data set, and vice versa. However, more systematic research on this issue is called for.

A comparison of the theory- and the data-driven models resulted in 192 item pairs generated by both methods. The additional 91 pairs contained in the $SRbI_d$ are for the main part attributable to the high solution frequencies, whereas the additional 39 pairs contained in the $SRbI_t$ indicate that an improvement of the theory-driven hypothesis is necessary for this part of the relation. Regarding the investigated sample, it seems that the mentioned additional pairs are not elements of the relation. In real diagnostic settings, the generation of theory-driven hypotheses is generally preferable, because it is derived independent of the sample and it allows an exact specification of the cognitive demands the testees are able to meet. However, the additional generation of a data-driven model is an efficient tool to uncover inadequate assumptions and to improve the original model. Furthermore, the application of the data-driven approach is useful for domains, in which theoretically founded hypotheses are not yet available. This, of course, requires a large data set and the application of the results is limited to the investigated population.

The empirical validation of surmise relations between tests requires a separate analysis of the whole $SRbI$ and its subsets. Otherwise it is not possible to determine how different parts of the relation contribute to the results. Therefore, the validation procedures include an evaluation of the surmise relation on the entire set of items, within the single tests, across the tests and between the tests.

Validating a structure by means of two different methods, namely by way of the knowledge space and the surmise relation allows a comparison of the obtained results and offers a possibility to check whether the applied methods lead to the same results. In KST there exists a one-to-one correspondence between quasi ordinal knowledge spaces and surmise relations on an item set. Thus, the application of methods based on the knowledge space should in principle yield the same results as methods based on the surmise relation. Indeed, both the indices for the adequacy of a surmise relation ($VC$ and $\gamma$) and the index for the adequacy of a knowledge structure ($DA$) indicate the best fit for $MT$ and the least fitting structure for $AN$.

Table 10.8 shows the results of the validation via the surmise relation and via the knowledge space for the example depicted in Figure 10.3. We simulated a set of 1000 response patterns based on the $SRbI$. In order to obtain similar error rates for all subsets of the relation, we fixed the amount of deviations by setting the percentages for careless errors and lucky guesses to

**Table 10.8.** Validation of the surmise relation and the knowledge space from the example in Figure 10.3 ($N = 1000$)

| | $m$ | No. of pairs\\$\Delta$ | $\lvert\mathcal{K}\rvert$ | *ddat* (*SD*) | *dpot* (*SD*) | *DA* | *VC* | $\gamma_G$ |
|---|---|---|---|---|---|---|---|---|
| *SRbI* | 10 | 35 | 23 | .64 (.76) | 2.58 (1.06) | .25 | .055 | .76 |
| *SRxT* | 10 | 18 | 82 | .44 (.63) | 1.68 (0.86) | .26 | .057 | .76 |
| *SRwA* | 5 | 7 | 9 | .24 (.47) | 0.94 (0.70) | .26 | .058 | .73 |
| *SRwB* | 5 | 10 | 6 | .27 (.49) | 1.13 (0.70) | .24 | .051 | .79 |

*Note.* $m$ denotes the number of items, $\lvert\mathcal{K}\rvert$ the number of knowledge states.

10% each. Space sizes vary between 7 and 82 states and the *ddat* values range from .24 to .64. However, the three indices $DA$, $VC$, and $\gamma_G$ reveal maximal differences between the subsets of only .02, .007, and .06 respectively. This high agreement indicates that the indices are actually able to account for the trade-off between the absolute goodness of fit of a surmise relation and the number of pairs in the relation (or the absolute goodness of fit of a knowledge space and the number of knowledge states).

Overall, the empirical validation of the hypothesized structures (via the surmise relation and the knowledge space) indicate that both the theory- and data-driven surmise relations are valid representations of the investigated knowledge domain. Hence, there seems to be no obstacle to apply the methods to more than two tests (given a fairly large set of data). Results with a set of four small tests are given in Wesiak (2003).

The theoretical one-to-one correspondence between surmise relations and knowledge spaces together with the obtained results also indicate in principle that the validation methods via the surmise relation and knowledge space lead to the same results. However, because of noisy data we still suggest to apply both methods.

All in all, the results of this application of surmise relations between tests demonstrate the applicability of the used methods for generating and validating hypotheses on surmise relations between items and tests, at least in the applied domain. Further applications are needed. Furthermore probabilistic surmise relations and knowledge spaces for sets of tests have to be developed as well as the corresponding statistical methods for generating and validating hypotheses on probabilistic surmise relations across, between and within tests.

## Acknowledgements

The research reported in this article was supported by the Austrian Science Foundation (FWF) and by a visiting professorship to Dietrich Albert at Hiroshima University, Japan. We are grateful to the "Heerespsychologischer Dienst des Bundesministeriums für Landesverteidigung" (Psychological Service of the Austrian Ministry of Defense) for providing the tests and the data set. We also want to thank Theo Held and Klaus Korossy for providing a data set and the first version of the ITA program, Ivo Ponocny for providing the algorithm for the frequency simulations simf4, as well as Cord Hockemeyer and Alexander Wenzel who adapted the ITA and the simulation program. If there is interest in testing the programs used in this work, please contact one of the authors.

## 10.6 List of Abbreviations

| | |
|---|---|
| KST | Knowledge Space Theory |
| SRbT | Surmise Relation between Tests |
| ITA | Item Tree Analysis |
| $SRbI$ | Surmise Relation between Items |
| $SRwT$ | Surmise Relation within Tests |
| $SRxT$ | Surmise Relation across Tests |
| $KSbI$ | Knowledge Space between Items |
| $KSwT$ | Knowledge Space within Tests |
| $KSxT$ | Knowledge Space across Tests |
| $MT$ | Matrix Test |
| $AN$ | Analogy Test |
| $MMSD$ | Mean minimal symmetric distance |
| $\Delta$ | set of pairs $(q, q)$ |

# 11

# Skills, Competencies and Knowledge Structures

Jürgen Heller[1]

Ali Ünlü[2]

Dietrich Albert[3, 4]

## 11.1 Introduction

From its beginning knowledge space theory was developed from a purely behavioristic point of view. It focused on the solution behavior exhibited on the instances of a set of items constituting a knowledge domain. This kind of stimulus-response consideration lead to very successful applications. Knowledge space theory has most effectively been applied especially in educational contexts where there is a curriculum prescribing the content to be covered, allowing for a fairly obvious and more or less complete definition of the relevant knowledge domain. There are, however, good reasons not to limit knowledge space theory to the kind of operationalism that identifies the state of knowledge with the subset of items an individual is capable of solving. The framework offered by knowledge space theory is able to integrate psychological theory by bringing into the picture the underlying cognitive abilities (skills, competencies, ...) responsible for the observable behavior. This kind of development may be seen somewhat analogous to traditional mental testing (Falmagne and Doignon, 2011). In this context, psychometric models referring to latent variables are preferred to purely operationalistic approaches, like classical test theory (cf. Borsboom, 2006). The extended framework provides

[1]Department of Psychology, University of Tübingen, Germany
[2]TUM School of Education, Technische Universität München, Germany
[3]Department of Psychology, University of Graz, Austria
[4]Knowledge Management Institute, Graz University of Technology, Austria

J.-C. Falmagne et al. (eds.), *Knowledge Spaces: Applications in Education*,
DOI 10.1007/978-3-642-35329-1_11, © Springer-Verlag Berlin Heidelberg 2013

a surplus of explanatory power whenever the solution behavior on a large set of items can be predicted by a limited set of basic skills. Moreover, it allows for adding new items to a knowledge domain without explicitly considering their relation to all the other items in the knowledge domain. Issues like this will be addressed in Section 11.6 after introducing the key concepts of the skill-based extensions of knowledge space theory.

Falmagne et al. (1990) sketched a first approach that links the observed solution behavior to some underlying cognitive constructs by assigning to each item a subset of skills that are relevant for mastering it. There are several largely independent major contributions to this extended framework. They are due to Doignon (1994b), see also Falmagne and Doignon (2011)[5], and to Düntsch and Gediga (1995) and Gediga and Düntsch (2002). A similar and actually chronologically preceding approach was suggested by Korossy (1993), but was published only a few years later (Korossy, 1997, 1999). The subsequent sections review central notions and results of these extensions of knowledge space theory. As the terminology and notation used widely differs across the cited papers, a main goal of the present chapter is to present the theories within a unified framework. This paves the way for discussing the sometimes subtle differences between the approaches.

## 11.2 Relating Skills to Knowledge Structures

Let the nonempty set $\mathcal{Q}$ be a knowledge domain and let $\mathcal{S}$ be a nonempty set of abstract skills that are assumed to be relevant for solving the items in $\mathcal{Q}$. In general, both sets $\mathcal{Q}$ and $\mathcal{S}$ may be infinite. This allows for linking the approach to latent variable models, such as the psychometric models formulated in the tradition of latent trait theory, or latent class analysis (see Section 9.2.8 of this volume). For doing so one might want to identify the set $\mathcal{S}$ with the set of real numbers, which are conceived as representing some kind of ability levels. The results of Doignon (1994b) and Falmagne and Doignon (2011) are obtained within this general context allowing for infinite basic sets $\mathcal{Q}$ and $\mathcal{S}$. In practical applications, however, these sets will be finite. This case is treated by Düntsch and Gediga (1995) and Gediga and Düntsch (2002) as well as by Korossy (1997, 1999). It may lead to results that do not hold in general, so in the sequel we will have to explicitly mention the scope of the results reported.

The main idea formalized below is as follows: Identifying the skills that are sufficient for solving each of the considered items provides a complete characterization of the solution behavior. The set of items which can be solved within a given set of skills is uniquely defined. This subset of items constitutes a possible knowledge state, and their collection forms a knowledge structure.

---

[5]In the sequel we refer to the terminology of Falmagne and Doignon (2011) which slightly differs from that introduced by Doignon (1994b).

### SKILL MULTIMAPS AND SKILL FUNCTIONS

The definition below introduces a way to model different solution paths available for solving an item.

**11.2.1 Definition.** A *skill multimap* is a triple $(\mathcal{Q}, \mathcal{S}, \mu)$, where $\mu$ is a mapping from $\mathcal{Q}$ to $2^{2^{\mathcal{S}}}$ such that each $\mu(q)$, $q \in \mathcal{Q}$, is a nonempty collection of nonempty subsets of $\mathcal{S}$. The elements $C \in \mu(q)$ are called *competencies*. If, additionally, the competencies in each $\mu(q)$ are pairwise incomparable (with respect to set inclusion) then $(\mathcal{Q}, \mathcal{S}, \mu)$ is called a *skill function*. Whenever the basic sets $\mathcal{Q}$ and $\mathcal{S}$ are clear from the context, we simply use $\mu$ to refer to the respective skill multimap, or skill function.

A skill multimap, or a skill function may assign more than one competency to an item, representing the fact that there may be more than one set of cognitive operations for solving it. These alternatives can, for example, correspond to different solution paths that may be observed. The skills contained in any of the competencies of a skill multimap are assumed to be sufficient for solving the item. Introducing the notion of a skill function is motivated by the idea that the assigned competencies should even be minimally sufficient for solving the item. This means that any proper subset of skills is no longer sufficient for the solution. Minimality in this sense suggests the assumption that competencies assigned to an item are pairwise incomparable, a property which subsequently will be called *incomparability condition*. Notice that a skill function $\tilde{\mu}$ can be associated to each skill multimap $\mu$ whenever the set $\mathcal{S}$ is finite. This is achieved by simply discarding the competencies that are not minimal (with respect to set inclusion) in $\mu(q)$, $q \in \mathcal{Q}$. Let us call $\tilde{\mu}$ a *reduction of* $\mu$. A reduction of a skill multimap may not exist if $\mathcal{S}$ is infinite. Defining $\mu(q)$ to be the set of all open real intervals containing a certain real number $r_q$ provides an example where no reduction exists.

Consider the following example with $\mathcal{Q} = \{a, b, c, d\}$ and $\mathcal{S} = \{s, t, u\}$. Let the skill multimap $\mu$ be defined by

$$\begin{array}{ll} \mu(a) = \{\{s,t\},\{s,u\}\}, & \mu(b) = \{\{u\},\{s,u\}\}, \\ \mu(c) = \{\{s\},\{t\}\}, & \mu(d) = \{\{t\}\}. \end{array} \tag{11.1}$$

According to this definition, either of the pairs of skills $s$ and $t$, or $s$ and $u$ are sufficient for solving item $a$. Conversely, from observing a correct answer to item $a$ it cannot be decided whether the mastery is based on the competency $\{s, t\}$, or $\{s, u\}$. Because there is more than one way to solve the item, observing a correct answer does not automatically lead to identifying a unique set of skills. Notice that the skill multimap $\mu$ does not satisfy the incomparability condition as the two competencies in $\mu(b)$ are nested. The skill function $\tilde{\mu}$ defined by

$$\begin{array}{ll} \tilde{\mu}(a) = \{\{s,t\},\{s,u\}\}, & \tilde{\mu}(b) = \{\{u\}\}, \\ \tilde{\mu}(c) = \{\{s\},\{t\}\}, & \tilde{\mu}(d) = \{\{t\}\} \end{array} \tag{11.2}$$

is the reduction of $\mu$. Notice also that formally there is more than one way to associate a skill function to a skill multimap that does not satisfy the incomparability condition. Consider the skill function $\hat{\mu}$ defined by

$$\begin{aligned} \hat{\mu} &= \{\{s,t\},\{s,u\}\}, & \hat{\mu} &= \{\{s,u\}\}, \\ \hat{\mu} &= \{\{s\},\{t\}\}, & \hat{\mu} &= \{\{t\}\} \end{aligned} \tag{11.3}$$

as another example. Constructing the skill function in this way, however, is not in line with the interpretation of minimal sufficient competencies.

**11.2.2 Problem Functions.** Each skill multimap $(\mathcal{Q}, \mathcal{S}, \mu)$ induces a mapping $p\colon 2^{\mathcal{S}} \to 2^{\mathcal{Q}}$ defined by

$$p(T) = \{q \in \mathcal{Q} \mid \text{there is a } C \in \mu(q) \text{ such that } C \subseteq T\} \tag{11.4}$$

for all $T \subseteq \mathcal{S}$. We will call $p$ the *problem function induced by* the skill multimap $\mu$. It is easily seen that the problem function $p$ induced by any skill function $(\mathcal{Q}, \mathcal{S}, \mu)$ is monotonic with respect to set inclusion, and satisfies $p(\emptyset) = \emptyset$ and $p(\mathcal{S}) = \mathcal{Q}$ (see Lemma 2.1 of Düntsch and Gediga, 1995, for the finite case). This result motivates the following definition.

**11.2.3 Definition.** A *problem function* is a triple $(\mathcal{Q}, \mathcal{S}, p)$, where $p$ is a mapping from $2^{\mathcal{S}}$ to $2^{\mathcal{Q}}$ that is monotonic with respect to set inclusion, and satisfies $p(\emptyset) = \emptyset$ and $p(\mathcal{S}) = \mathcal{Q}$.

Consider the problem function $p$ induced by the above defined skill multimap $(\mathcal{Q}, \mathcal{S}, \mu)$ with $\mathcal{Q} = \{a, b, c, d\}$ and $\mathcal{S} = \{s, t, u\}$. Then $p$ is given by

$$\begin{aligned} p(\emptyset) &= \emptyset, & p(\{s,t\}) &= \{a,c,d\}, \\ p(\{s\}) &= \{c\}, & p(\{s,u\}) &= \{a,b,c\}, \\ p(\{t\}) &= \{c,d\}, & p(\{t,u\}) &= \{b,c,d\}, \\ p(\{u\}) &= \{b\}, & p(\mathcal{S}) &= \mathcal{Q}. \end{aligned}$$

Notice that $p$ is also the problem function induced by the reduction $\tilde{\mu}$ of the skill function $\mu$ defined in Eqs. (11.2) and (11.1). This is not incidental as it is immediate from the definitions that a skill multimap and its reduction induce the same problem function. This, however, is not true of the skill function $\hat{\mu}$ as specified by Equation 11.3, which was derived from $\mu$, too. In particular, for the problem function induced by $\hat{\mu}$ we have $p(\{u\}) = \emptyset$.

**11.2.4 Properties of Skill and Problem Functions.** Given a knowledge domain $\mathcal{Q}$ and a skill set $\mathcal{S}$, then assigning the induced problem function to a given skill function defines a mapping from the set of all skill functions $\mu\colon \mathcal{Q} \to 2^{2^{\mathcal{S}}}$ to the set of all problem functions $p\colon 2^{\mathcal{S}} \to 2^{\mathcal{Q}}$. Düntsch and Gediga (1995, Proposition 2.3) show that for finite $\mathcal{Q}$ and $\mathcal{S}$ this mapping actually forms a bijection. In this case the notions of a skill function and a problem function thus are equivalent. Notice that the proof of this result draws upon the finiteness of the set $\mathcal{S}$. This assumption guarantees that to any $q \in Q$ there can be assigned a collection of pairwise incomparable subsets of $\mathcal{S}$ by picking the minimal elements of $\{T \subseteq \mathcal{S} \mid q \in p(T)\}$. These minimal elements need not exist in the infinite case. So the proof of Düntsch and Gediga (1995) does not survive the generalization to infinite skill sets $\mathcal{S}$. In general, the relationship between the set of all skill multimaps and the set of all problem functions is many-to-one. Forming the reduction, whenever possible, singles out a representative of the equivalence class of all skill multimaps that induce a certain problem function.

Consider the following results that answer the question in how far properties of skill functions are mirrored by the induced problem functions, and vice versa. Let $\mu\colon \mathcal{Q} \to 2^{2^{\mathcal{S}}}$ be a skill function and $p$ its induced problem function. Then the following two statements are equivalent.

$$\text{For all } q \in \mathcal{Q} \text{ the competencies } C \in \mu(q) \text{ are singletons,} \tag{11.5}$$
$$p(T_1 \cup T_2) = p(T_1) \cup p(T_2) \text{ for all } T_1, T_2 \subseteq \mathcal{S}. \tag{11.6}$$

In general a problem function $p$ will not preserve union. The reason for this is that the combination of two subsets of skills may enable an individual to solve additional items which cannot be solved given either of the subsets only. This fact is captured by the inclusion $p(T_1) \cup p(T_2) \subseteq p(T_1 \cup T_2)$ holding for all $T_1, T_2 \subseteq \mathcal{S}$. Moreover, the following two statements are equivalent.

$$\text{For all } q \in \mathcal{Q} \text{ we have } \mu(q) = \{C\} \text{ for some } C \subseteq \mathcal{S}, \tag{11.7}$$
$$p(T_1 \cap T_2) = p(T_1) \cap p(T_2) \text{ for all } T_1, T_2 \subseteq \mathcal{S}. \tag{11.8}$$

Condition (11.7) requires that each $\mu(q)$ for $q \in \mathcal{Q}$ contains only a single competency. In this case the problem function $p$ not only satisfies the generally valid inclusion $p(T_1 \cap T_2) \subseteq p(T_1) \cap p(T_2)$ for all $T_1, T_2 \in \mathcal{S}$, but preserves intersections.

Gediga and Düntsch (2002, Theorem 3.1) as well as Heller and Repitsch (2008, Lemmas 5 and 6) provide proofs of these two equivalences assuming finite sets $\mathcal{Q}$ and $\mathcal{S}$. While both results hold for the infinite case with arbitrary unions and intersections, too, they do not readily generalize to skill multimaps. Both implications from (11.6) to (11.5) and from (11.8) to (11.7) draw upon the incomparability condition and thus do not hold for skill multimaps. The converse implications, however, remain valid. In fact, any skill multimap satisfying (11.5) or (11.7) is a skill function. For characterizing this case let us introduce the following notions.

**11.2.5 Definition.** Let $(\mathcal{Q}, \mathcal{S}, \mu)$ be a skill function. Then $\mu$ is said to be a *disjunctive skill function* whenever it satisfies (11.5), and is said to be a *conjunctive skill function* whenever it satisfies (11.7).

**11.2.6 Delineated Knowledge Structure.** For a given skill multimap $(\mathcal{Q}, \mathcal{S}, \mu)$ and its induced problem function $p$ the items in $p(T)$ are exactly those that can be solved within the subset $T$ of skills. Notice that the properties of the problem function imply that $\emptyset, \mathcal{Q} \in \mathcal{K}$. Thus, the range of the problem function $p$ forms a knowledge structure $(\mathcal{Q}, \mathcal{K})$ consisting of the knowledge states that are possible given the skill function $\mu$. The knowledge structure $(\mathcal{Q}, \mathcal{K})$ is said to be *delineated* by the skill multimap $\mu$ (Falmagne and Doignon, 2011).

In general, a knowledge structure delineated by a skill multimap is not necessarily closed under union or intersection. As an example consider the skill multimap $\mu$ defined in (11.1). The knowledge structure $\mathcal{K}$ on the domain $\mathcal{Q} = \{a, b, c, d\}$ delineated by $\mu$ is defined as the range of its induced problem function $p$, and is given by

$$\mathcal{K} = \{\emptyset, \{b\}, \{c\}, \{c, d\}, \{a, b, c\}, \{a, c, d\}, \{b, c, d\}, \mathcal{Q}\}.$$

It is easily seen that $\mathcal{K}$ is neither closed under union, nor closed under intersection.

It is natural to ask which kind of skill multimaps delineate knowledge spaces and closure spaces, which were introduced as knowledge structures that are closed under union and intersection, respectively. In the sequel a partial answer to this question is provided.

In order to treat these special cases Doignon (1994b) introduce a skill assignment $\tau$ as a mapping from $\mathcal{Q}$ to $2^{\mathcal{S}}$ that associates to each item $q$ a nonempty subset $\tau(q)$ of $\mathcal{S}$. The subsets $\tau(q)$ receive different interpretations, each of them coming with its own definition of how a knowledge structure is delineated. Within the so-called *disjunctive model* each single element of $\tau(q)$ is considered to be sufficient for solving the item $q$, but it need not be a necessary requirement for doing so. Within the so-called *conjunctive model* the elements of $\tau(q)$ are considered to form a set of skills which are necessary and sufficient for solving the item $q$. These models can easily be integrated into the present framework, resulting in skill functions characterized by Def. 11.2.5. A disjunctive model $\tau$ corresponds to a skill multimap $\mu$ where for all $q \in \mathcal{Q}$ each of the competencies $C \in \mu(q)$ is a singleton as required in condition (11.5) and thus forms a disjunctive skill function. A conjunctive model $\tau$ corresponds to a skill multimap $\mu$ where $\mu(q)$ consists of a single competency only for all $q \in \mathcal{Q}$ as specified in condition (11.7) and thus is a conjunctive skill function.

Drawing upon results by Doignon and Falmagne (1999), Gediga and Düntsch (2002), and Heller and Repitsch (2008), one arrives at the following conclusions. If $(\mathcal{Q}, \mathcal{S}, \mu)$ is a disjunctive skill function (or, equivalently, the induced problem function $p$ preserves unions) then it delineates a knowledge space. Whenever $\mu$ is a conjunctive skill function (or, equivalently, the induced problem function $p$ preserves intersections) then it delineates a closure space. Notice that in both cases the given conditions are sufficient but not necessary for the delineated knowledge structure being closed under union, or closed under intersection, respectively. See Düntsch and Gediga (1995, Proposition 2.4) and Heller and Repitsch (2008, p. 154) for counter-examples.

## 11.3 Relating Knowledge Structures to Skills

We now consider the situation that we are given a certain knowledge structure $(\mathcal{Q}, \mathcal{K})$. In this context some questions naturally arise: Is $\mathcal{K}$ delineated by some appropriately chosen skill multimap? If yes, what is the number of skills needed to do so? Are there particular results for $\mathcal{K}$ being a knowledge space, or a closure space?

Doignon (1994b, Proposition 6) shows that each arbitrary knowledge structure $\mathcal{K}$ is delineated by at least one skill multimap (see also Falmagne and Doignon, 2011). For the case of both $\mathcal{Q}$ and $\mathcal{S}$ being finite an analogue result was proven by (Düntsch and Gediga, 1995, Theorem 3.1). So, the answer to the first question is positive. But what about the number of skills needed? The proofs of the just mentioned results are constructive and proceed by choosing an abstract set of skills $\mathcal{S}$ of cardinality $|\mathcal{K}|$ or $|\mathcal{K}| - 2$, respectively. Most certainly, proceeding in this way will not provide a parsimonious description of a knowledge structure by means of underlying skills. Düntsch and Gediga (1995) study the question concerning a minimal number of skills for finite sets $\mathcal{Q}$ and $\mathcal{S}$. They argue that the number $\log_2 |\mathcal{K}|$ constitutes a lower bound for $|\mathcal{S}|$, which suffices in case $\mathcal{K}$ is a Boolean algebra. Moreover, they show that for a knowledge structure $\mathcal{K}$ with cardinality $|\mathcal{K}| \geq 8$ there is a skill function on a set $\mathcal{S}$ with $|\mathcal{S}| \leq \lceil \log_2(|\mathcal{K}| - 2) + \log_2(\log_2(|\mathcal{K}| - 2)) \rceil$ that delineates $\mathcal{K}$ (Düntsch and Gediga, 1995, Proposition 3.3). Here, by $\lceil x \rceil$ we denote the smallest integer greater than or equal to the real number $x$.

The results of this section reported so far are valid for arbitrary knowledge structures. For knowledge spaces a more specific structural characterization is available. We first generalize some notions that were introduced for skill assignments underlying a disjunctive model (cf. Doignon, 1994b).

**11.3.1 Definition.** The skill multimaps $(\mathcal{Q}, \mathcal{S}, \mu)$ and $(\mathcal{Q}, \mathcal{S}', \mu')$ (on the same knowledge domain $\mathcal{Q}$) are said to be *isomorphic* if and only if there exists a one-to-one mapping $f$ from $\mathcal{S}$ onto $\mathcal{S}'$ such that for all $q \in Q$

$$\mu'(q) = \{f(C) \mid C \in \mu(q)\}.$$

The following result is obvious from the definition.

**11.3.2 Proposition.** *Two isomorphic skill multimaps* $(\mathcal{Q}, \mathcal{S}, \mu)$ *and* $(\mathcal{Q}, \mathcal{S}', \mu')$ *delineate the same knowledge structure* $(\mathcal{Q}, \mathcal{K})$.

From what was said above on the many-to-one relationship between skill multimaps and problem functions, it is also obvious that two skill multimaps may delineate the same knowledge structure without being isomorphic. Consider the skill multimaps $\mu$ and its reduction $\tilde{\mu}$ defined by (11.1) and (11.2) as an example.

**11.3.3 Definition.** The skill multimap $(\mathcal{Q}', \mathcal{S}', \mu')$ *prolongs* (resp. *strictly prolongs*) the skill multimap $(\mathcal{Q}, \mathcal{S}, \mu)$ if the following conditions hold:

1. $\mathcal{Q}' = \mathcal{Q}$
2. $\mathcal{S}' \supseteq \mathcal{S}$
3. $\mu(q) = \{C' \cap \mathcal{S} | C' \in \mu'(q)\}$.

The skill multimap $(\mathcal{Q}, \mathcal{S}, \mu)$ is *minimal* if there is no skill multimap delineating the same knowledge structure while being strictly prolonged by $(\mathcal{Q}, \mathcal{S}, \mu)$.

The following results are due to Falmagne and Doignon (2011, Theorem 4.11). A knowledge space is delineated by some minimal disjunctive skill function if and only if it admits a base. The *base* of a knowledge space $\mathcal{K}$ is a minimal subfamily $\mathcal{B}$ of $\mathcal{K}$ (with respect to set inclusion) spanning $\mathcal{K}$, i.e., any knowledge state is the union of a subfamily of sets in $\mathcal{B}$. If a base exists, its cardinality equals that of the set of skills. Any disjunctive skill function $(\mathcal{Q}, \mathcal{S}, \mu)$ delineating a knowledge space $(\mathcal{Q}, \mathcal{K})$ having a base prolongs a minimal disjunctive skill function delineating the same space. For the case of the sets $\mathcal{Q}$ and $\mathcal{S}$ being finite this implies that any knowledge space is delineated by a minimal disjunctive skill function, where the number of skills equals the cardinality of the base of the knowledge space. The cardinality of the base thus provides a lower bound for the number of skills needed to delineate a knowledge space, and this value is attained by a minimal disjunctive skill function. Moreover, any two minimal disjunctive skill functions delineating the same knowledge space are isomorphic. Notice that this result does not generalize to minimal skill multimaps. For an example consider the skill multimaps $\mu$ and its reduction $\tilde{\mu}$ defined by (11.1) and (11.2), which both are minimal but not isomorphic in the sense of Def. 11.3.1.

## 11.4 Korossy's Competence-Performance Approach

Korossy (1993, 1997, 1999) suggested a skill-based extension of knowledge space theory which he calls competence-performance approach, following Chomsky (1965). This approach will now be recast within the suggested unified framework by using the above introduced notation and terminology. Korossy distinguishes two different levels. The performance level refers to the observable behavior and is characterized by a finite knowledge domain $\mathcal{Q}$ and a knowledge structure $\mathcal{K}$ on $\mathcal{Q}$. The competence level refers to theoretical entities capturing cognitive abilities, the presence or absence of which may explain the observable behavior. This level is characterized by a finite set $\mathcal{S}$ of skills (*elementary competencies* in Korossy's words) and a collection $\mathcal{C}$ of subsets of $\mathcal{S}$. In contrast to the theories described above the *competence structure* $\mathcal{C}$ puts constraints on the possible sets of skills that may occur. To avoid trivialities assume that a competence structure $\mathcal{C}$ is a collection of subsets of $\mathcal{S}$ containing $\emptyset$ and $\mathcal{S}$. The elements $C$ of $\mathcal{C}$ are conceived as *competence states*, very much in analogy to the knowledge states at the performance level. This formal correspondence also gives rise to an interpretation of the competence structure in terms of the possible learning paths from the naive state $\emptyset$ to the state $\mathcal{S}$ of full competency. Here, learning refers to the acquisition of the underlying skills rather than the mastery of items.

The competence and performance levels are interconnected by means of two mappings. On the one hand, each item is linked to a subset of competence states, each of which is sufficient for solving it. This defines a mapping $k$ from $\mathcal{Q}$ into $2^{\mathcal{C}}$ assigning to each item the nonempty collection of competence states within which the item can be solved. Korossy (1997) calls this mapping $k$ an *interpretation function*, and considers it to coincide with the notion of a skill multimap as introduced in Def. 11.2.1. Notice, however, that the ranges of the two mappings differ. While a skill multimap assigns to any item a nonempty collection of nonempty subsets of $\mathcal{S}$, the interpretation function $k$ associates to each item a nonempty collection of nonempty competence states from $\mathcal{C}$. Moreover, on the more informal side the competencies assigned to an item by a skill multimap can be interpreted as subsets of skills that are sufficient for solving an item. This includes to confine consideration to minimal subsets of skills that still are sufficient (as, for example, in a skill function), irrespective of the question whether these skills actually co-occur with others. The interpretation function on the contrary provides a complete list of competence states within which an item can be solved. This means that whenever $C \in k(q)$ for some $q \in \mathcal{Q}$ and $C \subseteq C'$ for some $C' \in \mathcal{C}$ then we ought to have $C' \in k(q)$. It will be shown below that this distinction is by no means marginal. On the other hand, a *representation function* $r$ is defined as a mapping from $\mathcal{C}$ to $2^{\mathcal{Q}}$ by

$$r(C) = \{q \in \mathcal{Q} \mid C \in k(q)\}. \tag{11.9}$$

This notion is closely related to the concept of a problem function as defined by Eq. (11.4), but differs from it in at least two respects. First, its domain is the competence structure $\mathcal{C}$ rather than the powerset of $\mathcal{S}$. Second, Equation (11.9) relies upon the fact that an interpretation function assigns to an item a collection of competence states that is complete in the above specified sense.

Recasting Korossy's interrelation of the competence and performance levels within the present framework is straightforward. Concerning the interpretation function $k$ we confine consideration to competence states in $k(q)$, $q \in \mathcal{Q}$, that are minimal with respect to set inclusion. In other words, we consider the skill function $\mu_k$ from $\mathcal{Q}$ into $2^{\mathcal{C}}$ defined as the reduction $\tilde{k}$. Consequently, we restate Eq. (11.9) in accordance with Eq. (11.4). Defining

$$p_k(C) = \{q \in \mathcal{Q} \mid \text{there is a } C' \in \mu_k(q) \text{ such that } C' \subseteq C\}.$$

results in a mapping $p_k$ from $\mathcal{C}$ to $2^{\mathcal{Q}}$ that coincides with $r$. It is easily seen that $p_k$ is monotonic with respect to set inclusion and satisfies $p_k(\emptyset) = \emptyset$ and $p_k(\mathcal{S}) = \mathcal{Q}$, and may thus be interpreted as a problem function with its domain restricted to $\mathcal{C}$. In the sequel we refer to $p_k$ as the problem function induced by $\mu_k$. Again, the range of $p_k$ is a knowledge structure $\mathcal{K}$ on $\mathcal{Q}$, which is called the knowledge structure delineated by $\mu_k$.

The reformulation of Korossy's approach through $\mu_k$ and $p_k$ makes evident that it is a generalization of the competence-based extensions of Doignon (1994b) and Düntsch and Gediga (1995). The arbitrary competence structure $\mathcal{C}$ takes over the role of the powerset $2^{\mathcal{S}}$ in the Definitions 11.2.1 and 11.2.3, where obviously the latter is included as a special case. This raises the question whether the results obtained above for $\mathcal{C} = 2^{\mathcal{S}}$ survive the generalization to arbitrary competence structures. In order to answer this question the subsequently presented results go well beyond those of Korossy (1993, 1997, 1999). They cover a special case that received particular attention, where the structure $(\mathcal{S}, \mathcal{C}, \mathcal{Q}, \mathcal{K}, k, r)$ is called a *union-stable diagnostic* whenever the following conditions hold. The competence structure $\mathcal{C}$ on $\mathcal{S}$ is closed under union, and thus forms a so-called *competence space*; the representation function $r$ preserves unions in the sense that $r(C_1 \cup C_2) = r(C_1) \cup r(C_2)$ for all $C_1, C_2 \in \mathcal{C}$; $\mathcal{K}$ forms a knowledge space on $\mathcal{Q}$. Moreover, we consider what might be called an *intersection-stable diagnostic*, in which the competence structure $\mathcal{C}$ on $\mathcal{S}$ is closed under intersection, the representation function $r$ preserves intersections such that $r(C_1 \cap C_2) = r(C_1) \cap r(C_2)$ for all $C_1, C_2 \in \mathcal{C}$, and $\mathcal{K}$ forms a closure space on $\mathcal{Q}$. As outlined above, in both cases we may reformulate the basic structure by $(\mathcal{S}, \mathcal{C}, \mathcal{Q}, \mathcal{K}, \mu_k, p_k)$.

The following lemmas generalize the equivalences of (11.5) to (11.6), and (11.7) to (11.8), respectively. Remember that a state $C$ in a competence space $\mathcal{C}$ is an *atom* if there is some skill $s \in \mathcal{S}$ for which $C$ is minimal (with respect to set inclusion) among those competence states in $\mathcal{C}$ containing $s$.

**11.4.1 Lemma.** *Let the competence structure $\mathcal{C}$ be closed under union. Let $\mu_k \colon Q \to 2^{\mathcal{C}}$ be a skill function and $p_k \colon \mathcal{C} \to 2^{\mathcal{Q}}$ its induced problem function. Then the following two statements are equivalent.*

1. *For all $q \in Q$ each of the competencies $C \in \mu_k(q)$ is an atom of $\mathcal{C}$;*
2. *$p_k(C_1 \cup C_2) = p_k(C_1) \cup p_k(C_2)$ for all $C_1, C_2 \in \mathcal{C}$.*

Proof. Since $\mathcal{S}$ is finite the competence structure $\mathcal{C}$ has a base, which is formed by the collection of all the atoms (Falmagne and Doignon, 2011, Theorem 1.22, Theorem 1.26).

The crucial inclusion in 2. is $p_k(C_1 \cup C_2) \subseteq p_k(C_1) \cup p_k(C_2)$ for all $C_1, C_2 \in \mathcal{C}$, as the converse inclusion follows from the monotonicity of $p_k$. So, assume 1. and $q \in p_k(C_1 \cup C_2)$. Then there is an atom $C$ of $\mathcal{C}$ with $C \in \mu_k(q)$ such that $C \subseteq C_1 \cup C_2$. Let $\mathcal{A}_1$ and $\mathcal{A}_2$ denote the collection of atoms of $\mathcal{C}$ that are included in $C_1$ and $C_2$. Then $C_1 = \bigcup \mathcal{A}_1$ and $C_2 = \bigcup \mathcal{A}_2$, and $C$ lies in $\mathcal{A}_1$ or $\mathcal{A}_2$. This implies that $C \subseteq C_1$ or $C \subseteq C_2$ providing $q \in p_k(C_1) \cup p_k(C_2)$. Conversely, assume 2. and $C$ is not an atom of $\mathcal{C}$ for some $C \in \mu_k(q)$ and $q \in Q$. Then we have $C = C_1 \cup C_2$ with nonempty competence states $C_1, C_2 \in \mathcal{C}$ distinct from $C$. Thus, we infer $q \in p_k(C_1 \cup C_2)$, and 2. yields $q \in p_k(C_1)$ or $q \in p_k(C_2)$. In the first case there is a $C' \in \mu_k(q)$ such that $C' \subseteq C_1 \subset C$, which contradicts the incomparability condition. The second case is analogous. □

Notice that the equivalence of (11.5) and (11.6) results for the special case $\mathcal{C} = 2^{\mathcal{S}}$, where the atoms of $\mathcal{C}$ coincide with the singleton subsets of $\mathcal{S}$.

**11.4.2 Lemma.** *Let the competence structure $\mathcal{C}$ be closed under intersection. Let $\mu_k \colon Q \to 2^{\mathcal{C}}$ be a skill function and $p_k \colon \mathcal{C} \to 2^{\mathcal{Q}}$ its induced problem function. Then the following two statements are equivalent.*

1. *For all $q \in Q$ we have $\mu_k(q) = \{C\}$ for some $C \in \mathcal{C}$;*
2. *$p(C_1 \cap C_2) = p(C_1) \cap p(C_2)$ for all $C_1, C_2 \in \mathcal{C}$.*

Proof. The critical inclusion in 2. is $p_k(C_1) \cap p_k(C_2) \subseteq p_k(C_1 \cap C_2)$ for all $C_1, C_2 \in \mathcal{C}$, as the converse inclusion follows from the monotonicity of $p_k$. So, assume 1. and $q \in p_k(C_1) \cap p_k(C_2)$. As there is only a single competency $C \in \mu_k(q)$ we have $C \subseteq C_1$ and $C \subseteq C_2$. This means that $C \subseteq C_1 \cap C_2$ providing $q \in p_k(C_1 \cap C_2)$. Conversely, assume 2. and that there are distinct $C_1, C_2 \in \mu_k(q)$ for some $q \in Q$. Then $q \in p_k(C_1) \cap p_k(C_2) = p_k(C_1 \cap C_2)$, and there is a $C \in \mu_k(q)$ such that $C \subseteq C_1 \cap C_2$, which contradicts the incomparability condition. □

The following result relates competence and performance structures.

**11.4.3 Proposition.** *Let $\mathcal{C}$ be a competence structure. Let $\mu_k \colon Q \to 2^{\mathcal{C}}$ be a skill function, $p_k \colon \mathcal{C} \to 2^{\mathcal{Q}}$ its induced problem function, and $\mathcal{K}$ the knowledge structure delineated by $\mu_k$. Then the following two statements hold.*

1. *If $\mathcal{C}$ is closed under union and $\mu_k$, $p_k$ satisfy the equivalent conditions of Lemma 11.4.1 then $\mathcal{K}$ is a knowledge space.*
2. *If $\mathcal{C}$ is closed under intersection and $\mu_k$, $p_k$ satisfy the equivalent conditions of Lemma 11.4.2 then $\mathcal{K}$ is a closure space.*

Proof. Let $K, L \in \mathcal{K}$. If $p_k$ satisfies condition 2. of Lemma 11.4.1 then

$$K \cup L = p_k(C_1) \cup p_k(C_2) = p_k(C_1 \cup C_2) \in \mathcal{K}$$

for some $C_1, C_2 \in \mathcal{C}$. Accordingly, if $p_k$ satisfies condition 2. of Lemma 11.4.2 then

$$K \cap L = p_k(C_1) \cap p_k(C_2) = p_k(C_1 \cap C_2) \in \mathcal{K}$$

for some $C_1, C_2 \in \mathcal{C}$. □

Again the given conditions are sufficient but not necessary for the knowledge structure $\mathcal{K}$ being closed under union, or closed under intersection, respectively. For this, again, refer to the counter-examples for the special case $\mathcal{C} = 2^{\mathcal{S}}$ in Düntsch and Gediga (1995) and Heller and Repitsch (2008).

## 11.5 Extending the Competence-Performance Approach

The notion of a competence structure as conceived in the competence-performance approach actually lumps together two different conceptions. On the one hand, the competence structure is considered the collection of possible competence states. On the other hand, it is interpreted as providing the subsets of skills sufficient for solving particular items. There is no reason that these two collections of subsets of skills should coincide. The competence structure $\mathcal{C}$ forms the domain of the representation function $r$, or, equivalently, the problem function $p_k$ restricted to $\mathcal{C}$. Here, it appears in the first of the two mentioned roles. Delineation of a knowledge structure is constrained to the possible competence states, an idea that clearly goes beyond the approaches detailed in the previous sections. In fact, the assumptions that not every subset of skills actually occurs, and that there are some particular learning paths along which the acquisition of the skills proceeds, will be highly plausible in many contexts. Confining consideration to the possible competence states then may have positive effects. It may lead to a reduction of the number of potential knowledge states compared to the knowledge structures delineated through a problem function defined on $2^{\mathcal{S}}$. In turn, this will help speeding up knowledge assessments, for example. Derivation of the competence states, however, should not be linked to a particular set of items. Nor should the characterization of the items be linked to a certain competence structure. It is by no means clear that a subset of skills sufficient for solving an item should be a possible competence state. Imagine a hard item that can be solved through a single skill which is acquired only after a number of other skills have been learned. In order to conceptually separate the competence states from the competencies related to solving the items of a particular knowledge domain the interpretation function may be redefined as a skill multimap $k$ from $\mathcal{Q}$ to $2^{2^{\mathcal{S}}}$ according to Definition 11.2.1. Then $\mu_k$, defined as the reduction of $k$, forms a skill function. The competencies in $\mu_k$ can be interpreted as subsets of skills that are minimally sufficient for solving the items. Consider the following example. Let the competence structure $\mathcal{C}$ on $\mathcal{S} = \{s, t, u\}$ be given by

$$\mathcal{C} = \{\emptyset, \{s\}, \{u\}, \{s, t\}, \{s, u\}, \mathcal{S}\}.$$

This means that skill $t$ is available only if skill $s$ is present too. On the knowledge domain $\mathcal{Q} = \{a, b, c\}$ then define the skill function $\mu_k$ from $\mathcal{Q}$ to $2^{2^{\mathcal{S}}}$ by

$$\mu_k(a) = \{\{s\}, \{u\}\}, \quad \mu_k(b) = \{\{t\}\}, \quad \mu_k(c) = \{\{u\}\}.$$

Notice that the competencies in $\mu_k$ constitute the subsets of skills minimally sufficient for solving the items, which may be distinct from the competence states. In particular, item $b$ may be solved drawing upon the single skill $t$, although the skill $t$ will always co-occur with skill $s$. We have $p_k(\mathcal{C}) = \{\emptyset, \{a\}, \{a, b\}, \{a, c\}, \mathcal{Q}\}$, while the knowledge structure delineated by $\mu_k$ (on the basis of $2^{\mathcal{S}}$) is

$$\mathcal{K} = \{\emptyset, \{a\}, \{b\}, \{a,b\}, \{a,c\}, \mathcal{Q}\}.$$

For the subsets of skills $\{t\}$ and $\{t, u\}$, which are not competence states, the delineated knowledge states are $\{b\}$ and $\mathcal{Q}$, respectively. The former is not a possible knowledge state whenever we confine consideration to the competence states in $\mathcal{C}$, and provides an example for the reduction of the size of the resulting knowledge structure. Reformulating the competence-performance approach within a unified theoretical framework as suggested above thus points to the essential conceptual distinction between competence states on the one hand, and subsets of skills sufficient for solving an item on the other hand. This adds a new perspective to skill- and competence-based extensions of knowledge space theory.

## 11.6 Conclusions

The previous sections have shown that knowledge space theory can be extended to incorporate psychological theories underlying the observed behavior. Including abstract skills and competencies into the consideration introduces a theoretical level, which goes well beyond the operationalist conception of a knowledge state. Several largely independent major contributions to this extension of knowledge space theory were reviewed, which are due to Doignon (1994b), see also Falmagne and Doignon (2011), as well as to Düntsch and Gediga (1995) and Gediga and Düntsch (2002), and finally Korossy (1993, 1997, 1999).

For the first time these approaches are systematically related to each other by recasting them within a unified framework, which allows for identifying their differences. While the first approach treats the infinite case, the other two are confined to finite knowledge domains and skill sets. Apart from this distinction, the competence-performance approach of Korossy (1993, 1997, 1999) may be conceived as a generalization of the other two approaches. It brings in the possibility to distinguish between subsets of skills that are plausible characterizations of an individuals competency (so-called competence states), and those that are not. Formally, this amounts to replace the powerset $2^{\mathcal{S}}$ of the set of skills $\mathcal{S}$ by a competence structure $\mathcal{C} \subseteq 2^{\mathcal{S}}$, with $\emptyset, \mathcal{S} \in \mathcal{C}$. Sections 11.4 and 11.5 not only reformulate Korossy's approach within the unified framework, but also provide new results and insights. Lemmas 11.4.1 and 11.4.2, and Proposition 11.4.3 generalize previous results (Gediga and Düntsch, 2002; Heller and Repitsch, 2008) based on considering $2^{\mathcal{S}}$ to arbitrary competence structures $\mathcal{C}$. Moreover, the developments also emphasize the conceptual distinction between competence states on the one hand, and subsets of skills sufficient for solving an item on the other hand.

The discussed approaches exemplify how to specify the relationship of underlying skills to the solution of the items. The framework, however, is highly flexible with respect to the specific form of this relationship. For example Suck extended the skill construction of Doignon (1994b) in a way not covered in this paper (cf. Suck, 2003, 2004, 2011). Moreover, the framework is not limited to the delineation of knowledge structures as defined in Section 11.2.6; other variants are conceivable. Falmagne and Doignon (2011), for example, suggest to define a knowledge state as a subset $K$ of $\mathcal{Q}$ consisting of all items $q$ whose associated competencies all meet a particular subset of $\mathcal{S}$ (depending on $K$).

There are a number of successful applications of the skill-based extension of knowledge space theory (see Chapter 12 in this volume).

Apart from the already mentioned advantages that the extended framework offers from an epistemological point of view, it also has important consequences on the pragmatic side. Consider a knowledge domain for which the set of items is not fixed once for all, where it may be reasonable to open up the possibility to include additional items. Even in educational contexts with prescriptive curricula, one may want to extend the knowledge domain, so that, for example, the teachers can bring in their own items. In 'classical' knowledge space theory this would require to relate the new item to all the old ones, in order to localize it within the knowledge structure. Having identified the relevant skills underlying the solution behavior, and having assigned them to the existing items (e.g. by means of a skill multimap), considerably simplifies integration of a new item that refers to the same set of skills (derived from the curriculum, for example). This task only requires to assign the relevant competencies to the new item. Once this is done, the relationship of the new item to the old ones is uniquely determined. This property is essential for any application of knowledge space theory to technology-enhanced learning within an open systems architecture, where the item repositories are not static but may be modified by adding or removing items. Heller and Repitsch (2008) further show how to apply the skill-based extension to distributed systems, in which the services as well as the resources (e.g., item repositories) they access may reside at different locations. The underlying theory of consistent aggregation of skill functions is covered in Section 12.2.

# 12

# Recent Developments in Competence-based Knowledge Space Theory

Jürgen Heller[1]
Thomas Augustin[2]
Cord Hockemeyer[3]
Luca Stefanutti[4]
Dietrich Albert[3, 5]

## 12.1 Introduction

A saying attributed to Kurt Lewin (1951) states 'There is nothing so practical as a good theory.' Accordingly, the theory and models outlined in Chapter 11 of this volume have many practical consequences and can be applied for instance in personalized competence assessment, individualized eTeaching and eLearning, and expert query. Furthermore, for practical reasons recent developments in Competence-based Knowledge Space Theory (CbKST) have to be taken into account.

In applied eTeaching and eLearning settings like ALEKS, a huge number of learning objects (LOs) are used, e.g. test items, explanations, exercises. The competence-based approach assigns skills and competencies to those learning objects for generating the structures used by an adaptive system for personalized assessment and teaching. These assignments are often established by different experts for overlapping subsets of LOs and skills. The resulting assignments may be distributed among several repositories. Furthermore, in applied settings modifying, removing and adding LOs and skill assignments are necessary.

[1]Department of Psychology, University of Tübingen, Germany
[2]Joanneum Research, Graz, Austria
[3]Department of Psychology, University of Graz, Austria
[4]Department of Psychology, University of Padua, Italy
[5]Knowledge Management Institute, Graz University of Technology, Austria

J.-C. Falmagne et al. (eds.), *Knowledge Spaces: Applications in Education*,
DOI 10.1007/978-3-642-35329-1_12, © Springer-Verlag Berlin Heidelberg 2013

Already the meshing of different structures of the same type, e.g. item sets, is a non-trivial problem (see e.g. Falmagne and Doignon, 1998). Combining structures of different types (e.g. items and skills) is even more demanding and described in Section 12.2 on distributed skill functions and the meshing of knowledge structures.

The notion of careless errors and lucky guesses and their probabilities has been introduced in Falmagne and Doignon (1988a,b) in order to create a more realistic and applicable model for knowledge-based performances. Also in case of competence-based performance differences between the predicted and the observed performance needs to be interpreted in a probabilistic way. A student may give the wrong answer although he/she has all the necessary skills for answering correctly—and the other way round. Furthermore, Doignon and Falmagne (1985) already pointed out in their fundamental paper on knowledge spaces that not only one but several solution strategies have to be taken into account for realistic applications. Therefore, in case of the skill approach, not only exactly one but one of several sets of skills can result in a correct answer. For creating a probabilistic skill and competence model also a prerequisite structure on the set of skills—introduced by Korossy (1993, 1997) and described in Section 11.4—has to be taken into account. This is realized in a probabilistic version of skill mappings which is described in Section 12.3.

Traditional adaptive eLearning systems (Intelligent Tutoring Systems, ITS) guide the individual learner in an "intelligent" way. More recently also self-regulated learning (SRL) is required for technology-enhanced learning systems. Thus in the future the whole spectrum from strict guidance to total freedom in learning and navigating will be available for the learner. Total freedom in terms of navigating causes, however, serious problems for assessing the individual learning progress. In order to solve these problems, three basic ideas are used: (a) Combining the link-structure of hypertexts with the structure of knowledge spaces Hockemeyer et al. (1998), (b) interpreting the navigation behavior and studying the visited webpages (learning objects) in terms of learning, and (c) assigning to the movement from webpage to another the moving from one cognitive state to another. A stochastic model linking navigation behavior to the learner's cognitive state is outlined in Section 12.4 below.

In Game-based-Learning (GBL) and Complex Problem Solving (CPS) the individual acts more or less freely and self-regulated. However, instead of moving from one distinct, static webpage to another one, in GBL and CPS the currently presented webpage is dynamically changing in accordance with the individual's activities. The individual is acting by moving virtual 3D objects in a so called problem space. Interpreting the individual's actions in terms of missing or present competencies and skills allows for (a) assessing the learning progress and (b) presenting personalized feedback. These feedback events may be of different character and supporting different aims (teaching, motivating, stimulating, assessing etc.). In a GBL environment, assessments and interventions have to be computed in real time in order not to interrupt the flow experience of the gamer/learner. Furthermore, the intervening events have to fit into the interactively generated game story. Thus the modeling is rather demanding, described in section 12.5 on Micro-Adaptivity: Non-invasive skill assessment in educational games.

The above mentioned theoretical developments have been stimulated by practical requirements. Thus, the question arises which of these recent theoretical developments have already been applied in practical settings. The answer can be found in the last section (12.6) of this chapter on applications in learning and competence management.

## 12.2 Distributed Skill Functions and the Meshing of Knowledge Structures

A major challenge to competence-based extensions of knowledge space theory comes from the need to integrate distributed information on (partially) overlapping domains and skill sets, which arises in a number of scenarios. Consider a situation in which various experts independently identify and assign relevant skills to the items in a domain. In realistic applications, where the number of items tends to be very large, the experts may not be able to cover the whole set, but only subsets with sufficient overlap. In any case, most certainly the expert assignments will not completely coincide. The question then is whether the assignments are at least consistent, so that the local information provided by the experts can be aggregated into a global skill assignment. Similar problems emerge when implementing the competence-based extension of knowledge space theory in technology-enhanced learning within a distributed and open systems architecture. In these systems, the services as well as the resources (e.g. repositories of items) they access may reside at different locations. Moreover, the local item repositories need not be static but may be modified by adding or removing items. Again, the question is whether the skills assigned to the items (through appropriate metadata tags) can be integrated in a consistent way. The situation is formalized by a distributed skill function, which generalizes the notion of a distributed skill map introduced by Stefanutti et al. (2005). The subsequently outlined results are due to Heller and Repitsch (2008), where further details and the proofs omitted below are provided.

### Basic concepts

We begin with reminding the reader of a few concepts. The following notion was introduced by Doignon and Falmagne (1999, Definition 1.17).

**12.2.1 Definition.** Let $(\mathcal{Q}, \mathcal{K})$ be a knowledge structure, and let $A$ be some nonempty subset of $\mathcal{Q}$ and $\mathcal{H} = \{H \in 2^A \mid H = A \cap K \text{ for some } K \in \mathcal{K}\}$. Then $(A, \mathcal{H})$ is called a *substructure* of the *parent structure* $(\mathcal{Q}, \mathcal{K})$. Sometimes the substructure $\mathcal{H}$ is referred to as the *trace* of $\mathcal{K}$ on $A$, and is denoted by $\mathcal{K}|_A$.

The notion of a substructure of a knowledge structure will play a prominent role in the sequel. The central concept of this paper, however, is that of a skill function as introduced in Definition 11.2.1. A skill function is a triple $(\mathcal{Q}, \mathcal{S}, \mu)$, where $\mathcal{S}$ is a set of skills and $\mu$ is a mapping from $\mathcal{Q}$ to $2^{2^{\mathcal{S}}}$ such that each $\mu(q)$, $q \in \mathcal{Q}$, is a nonempty collection of nonempty pairwise incomparable subsets of $\mathcal{S}$ (with respect to set inclusion). The elements $C \in \mu(q)$ are called competencies. The skills contained in each competency are assumed to be minimally sufficient for solving the item. This minimality motivates the assumption that competencies are pairwise incomparable, a property which was called incomparability condition. Notice that each skill function $(\mathcal{Q}, \mathcal{S}, \mu)$ induces a mapping $p\colon 2^{\mathcal{S}} \to 2^{\mathcal{Q}}$ defined by

$$p(T) = \{q \in \mathcal{Q} \mid \text{there is a } C \in \mu(q) \text{ such that } C \subseteq T\}$$

for all $T \subseteq \mathcal{S}$, which is a problem function according to Definition 11.2.3 . The items in $p(T)$ are exactly those that can be solved within the subset $T$ of skills. Thus, the range of the problem function $p$ forms a knowledge structure $(\mathcal{Q}, \mathcal{K})$ consisting of the knowledge states which are possible given the skill function $\mu$. The knowledge structure $(\mathcal{Q}, \mathcal{K})$ is said to be *delineated* by the skill function $\mu$.

**12.2.2 Distributed skill functions.** Let us now turn to the situation where a collection of skill functions $(\mathcal{Q}_i, \mathcal{S}_i, \mu_i)$ is given, with $i$ an element of a finite or infinite index set $I$. Assume the most general case, in which knowledge domains $\mathcal{Q}_i$ as well as skill sets $\mathcal{S}_i$ may show (partial) overlap. Notice that the set $\mathcal{S}_i$ may also be taken as the union of all competencies $C \in \mu_i(q)$, $q \in \mathcal{Q}_i$. These sets of actually assigned skills may differ over components, even if the skills are drawn from a common set $\mathcal{S}$.

The following definition provides a straightforward construction for integrating the different skill functions into a single skill assignment defined on the union of the domains and the skill sets, respectively.

**12.2.3 Definition.** Given a collection of skill functions $(\mathcal{Q}_i, \mathcal{S}_i, \mu_i)$, $i \in I$, their *merge* $(\mathcal{Q}, \mathcal{S}, \mu)$ is defined by

1. $\mathcal{Q} = \bigcup_{i \in I} \mathcal{Q}_i$;
2. $\mathcal{S} = \bigcup_{i \in I} \mathcal{S}_i$;
3. For all $q$ in $\mathcal{Q}$, $\mu(q) = \bigcup_{i \in I} \mu_i^*(q)$, with $\mu_i^*(q) = \mu_i(q)$ if $q$ is an element of $\mathcal{Q}_i$, and $\mu_i^*(q) = \emptyset$ otherwise.

By this definition $\mu$ is a mapping from $\mathcal{Q}$ to $2^{2^{\mathcal{S}}}$ such that each $\mu(q)$, $q \in \mathcal{Q}$, is a nonempty collection of nonempty subsets of $\mathcal{S}$. For all $C \in \mu(q)$ there exists some $j$ with $q \in \mathcal{Q}_j$ and $C \in \mu_j(q)$. In other words, there is at least one skill function $(\mathcal{Q}_j, \mathcal{S}_j, \mu_j)$ according to which the competency $C$ is sufficient for solving $q$. Notice that the competencies in $\mu(q)$ need not satisfy the incomparability condition. So, the merge of skill functions is not necessarily a skill function. The term introduced for the subsequently defined notion was coined by Stefanutti et al. (2005).

**12.2.4 Definition.** Let $(\mathcal{Q}_i, \mathcal{S}_i, \mu_i)$, $i \in I$, be skill functions. If their merge $(\mathcal{Q}, \mathcal{S}, \mu)$ is a skill function then it is called a *distributed skill function.*

In connection with this notion a number of questions naturally arise:

1. Which conditions ensure that the global information represents a consistent aggregation of the local information?
2. To which extent can we retrieve the local information on the components from the global information?

The problems raised by these questions have three facets. A distributed skill function $\mu$ induces a problem function $p\colon 2^{\mathcal{S}} \to 2^{\mathcal{Q}}$ in the same way as the component skill functions $\mu_i$ induce the problem functions $p_i\colon 2^{\mathcal{S}_i} \to 2^{\mathcal{Q}_i}$. Moreover, $\mu$ delineates a knowledge structure $\mathcal{K}$ on the knowledge domain $\mathcal{Q}$ that we obtain together with the component knowledge structure $\mathcal{K}_i$ on $\mathcal{Q}_i$ delineated by the $\mu_i$. Now, answering the above questions requires to clarify how the local entities $\mu_i$, $p_i$ and $\mathcal{K}_i$ are related to their global counterparts $\mu$, $p$ and $\mathcal{K}$. The following example demonstrates that these problems are far from being trivial.

**12.2.5 Example.** We consider two skill functions $(\mathcal{Q}_1, \mathcal{S}_1, \mu_1)$ and $(\mathcal{Q}_2, \mathcal{S}_2, \mu_2)$ which are defined on $\mathcal{Q}_1 = \{a, b, c, d\}$, $\mathcal{S}_1 = \{x, y, z\}$ and $\mathcal{Q}_2 = \{a, b, e, f\}$, $\mathcal{S}_2 = \{w, x, y\}$ by

$$\begin{array}{ll}
\mu_1(a) = \{\{x, y\}, \{x, z\}\} & \mu_2(a) = \{\{w, y\}, \{x, y\}\} \\
\mu_1(b) = \{\{x\}, \{y\}, \{z\}\} & \mu_2(b) = \{\{w\}, \{x\}\} \\
\mu_1(c) = \{\{x\}, \{y\}\} & \mu_2(e) = \{\{x\}, \{w, y\}\} \\
\mu_1(d) = \{\{y, z\}\} & \mu_2(f) = \{\{y\}, \{w, x\}
\end{array}$$

They delineate the knowledge structures (see Fig. 12.1)

$$\mathcal{K}_1 = \{\emptyset, \{b\}, \{b, c\}, \{a, b, c\}, \{b, c, d\}, \mathcal{Q}_1\},$$
$$\mathcal{K}_2 = \{\emptyset, \{b\}, \{f\}, \{b, e\}, \{b, e, f\}, \mathcal{Q}_2\}.$$

Applying the above defined construction provides the distributed skill function with $\mathcal{Q} = \{a, b, c, d, e, f\}$, $\mathcal{S} = \{x, y, w, z\}$ and

$$\begin{array}{ll}
\mu(a) = \{\{x, y\}, \{x, z\}, \{w, y\}\}, & \mu(b) = \{\{w\}, \{x\}, \{y\}, \{z\}\}, \\
\mu(c) = \{\{x\}, \{y\}\}, & \mu(d) = \{\{y, z\}\}, \\
\mu(e) = \{\{x\}, \{w, y\}\}, & \mu(f) = \{\{y\}, \{w, x\}\}.
\end{array}$$

The distributed skill function $\mu$ delineates the knowledge structure

$$\mathcal{K} = \{\emptyset, \{b\}, \{b, c, e\}, \{b, c, f\}, \{a, b, c, e\}, \{b, c, d, f\}, \{b, c, e, f\}, \{a, b, c, e, f\}, \mathcal{Q}\},$$

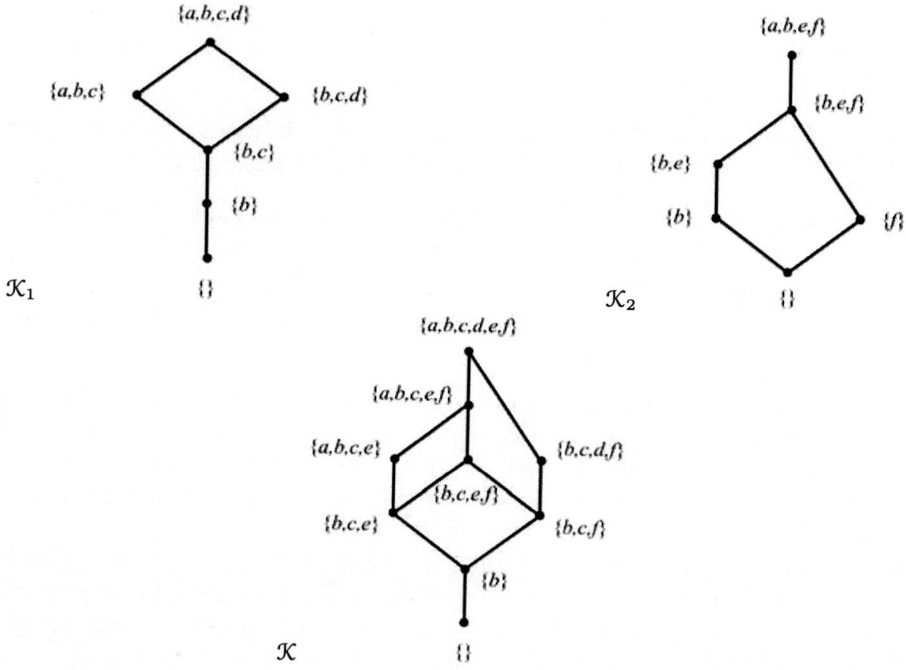


**Figure 12.1.** Graphical representation of the knowledge structures $\mathcal{K}_1$, $\mathcal{K}_2$, and $\mathcal{K}$ of Example 12.2.5.

which is also illustrated in Fig. 12.1. Now, forming the trace of $\mathcal{K}$ on $\mathcal{Q}_1$ and $\mathcal{Q}_2$ yields the substructures

$$\mathcal{K}|_{\mathcal{Q}_1} = \{\emptyset, \{b\}, \{b, c, \}, \{a, b, c\}, \{b, c, d\}, \mathcal{Q}_1\},$$
$$\mathcal{K}|_{\mathcal{Q}_2} = \{\emptyset, \{b\}, \{b, e\}, \{b, f\}, \{a, b, e\}, \{b, e, f\}, \mathcal{Q}_2\}.$$

This means that we have $\mathcal{K}_1 = \mathcal{K}|_{\mathcal{Q}_1}$, but neither $\mathcal{K}_2 \subseteq \mathcal{K}|_{\mathcal{Q}_2}$ nor $\mathcal{K}|_{\mathcal{Q}_2} \subseteq \mathcal{K}_2$, and thus $\mathcal{K}_2 \neq \mathcal{K}|_{\mathcal{Q}_2}$.

Example 12.2.5 shows that the local knowledge structures do not necessarily coincide with the restricted global knowledge structures. In order to get a grip on the impact of this fact, consider the assessment of an individual's knowledge, which may proceed at two different levels: At a local level based on one of the local domains $\mathcal{Q}_1$ or $\mathcal{Q}_2$ and their associated knowledge structures, and at a global level through the knowledge structure delineated by the distributed skill function on $\mathcal{Q} = \mathcal{Q}_1 \cup \mathcal{Q}_2$. The following questions arise in this context.

1. Can an assessment at the global level build upon the information collected within a local assessment?

2. Is it possible to localize a given global assessment to any of the local domains?

The following exemplifies that the requested smooth transition between local and global knowledge assessment is by no means trivial. In order to assess their knowledge individuals are presented with a sequence of items. Furthermore, for each item $q$, it is recorded whether it is solved or not and let these two cases be denoted by $q$ and $\overline{q}$, respectively. We also use the notation $\mathcal{K}_q = \{K \in \mathcal{K} \mid q \in K\}$ and $\mathcal{K}_{\overline{q}} = \{K \in \mathcal{K} \mid q \notin K\}$ for a knowledge structure $(\mathcal{Q}, \mathcal{K})$. Referring to Example 12.2.5, an assessment may proceed at the local level (i.e. based on the knowledge structures $\mathcal{K}_1$ or $\mathcal{K}_2$) or at the global level (i.e. with respect to $\mathcal{K}$).

Assume that assessing an individual on $\mathcal{Q}_1$ provides the sequence $b\overline{d}a$ of responses. There are exactly two knowledge states in $\mathcal{K}$ that are consistent with this information. The collection of states satisfying these constraints is

$$\mathcal{K}_b \cap \mathcal{K}_{\overline{d}} \cap \mathcal{K}_a = \{\{a, b, c, e\}, \{a, b, c, e, f\}\}.$$

The state of the individual in the knowledge structure $\mathcal{K}$ on $\mathcal{Q}$ may be identified by continuing the assessment, i.e. by presenting items from $\mathcal{Q} \setminus \mathcal{Q}_1$. Assessing the performance on item $f \in \mathcal{Q}_2$, for example, yields either $b\overline{d}af$ corresponding to state $\{a, b, c, e, f\}$, or $b\overline{d}a\overline{f}$ corresponding to state $\{a, b, c, e\}$, depending on whether the item is solved or not. Conversely, truncating the sequence that results from an assessment on $\mathcal{Q}$ to the elements of $\mathcal{Q}_1$ gives a sequence that designates a state in $\mathcal{K}_1$. Applying this truncation to the sequences $b\overline{d}a\overline{f}$ and $b\overline{d}af$, for example, gives back $b\overline{d}a$ in both cases.

On the contrary, assume that assessing an individual on the domain $\mathcal{Q}_2$ resulted in the sequence $\overline{a}f\overline{b}$, which corresponds to the state $\{f\}$ in $\mathcal{K}_2$. Then this local assessment cannot be extended to a global assessment on $\mathcal{Q}$. There is no knowledge state in $\mathcal{K}$ that is consistent with this information. The collection of states $\mathcal{K}_{\overline{a}} \cap \mathcal{K}_f \cap \mathcal{K}_{\overline{b}}$ satisfying the constraints in $\overline{a}f\overline{b}$ is empty. Conversely, consider the sequence $ae\overline{f}$ resulting from an assessment on $\mathcal{Q}$ which corresponds to the state $\{a, b, c, e\}$ of $\mathcal{K}$. Actually, this sequence may in principle result in an assessment on $\mathcal{Q}_2$, but there is no state in $\mathcal{K}_2$ that corresponds to this sequence. The key to explaining these differences in the relation between local and global assessments is the fact that we have $\mathcal{K}_1 = \mathcal{K}|_{\mathcal{Q}_1}$, but $\mathcal{K}_2 \not\subseteq \mathcal{K}|_{\mathcal{Q}_2}$ as well as $\mathcal{K}|_{\mathcal{Q}_2} \not\subseteq \mathcal{K}_2$.

**12.2.6 Meshing knowledge structures.** Falmagne and Doignon (1998)—see also Doignon and Falmagne (1999) and Falmagne and Doignon (2011)—introduced the notion of a so-called mesh for describing the integration of two knowledge structures on the union of their domains. A knowledge structure $(\mathcal{Q}, \mathcal{K})$ is called a *mesh* of the knowledge structures $(\mathcal{Q}_1, \mathcal{K}_1)$ and $(\mathcal{Q}_2, \mathcal{K}_2)$ if $\mathcal{Q} = \mathcal{Q}_1 \cup \mathcal{Q}_2$, and $\mathcal{K}_1$ and $\mathcal{K}_2$ are the traces of $\mathcal{K}$ on $\mathcal{Q}_1$ and $\mathcal{Q}_2$, respectively. Thus, a mesh allows for recovering its component knowledge structures. It is easily seen that two knowledge structures can have one or more meshes, or no mesh at all. A necessary and sufficient condition for the existence of a mesh of two knowledge structures has been identified (see Doignon and Falmagne, 1999, Theorem 5.14). This *compatibility condition* requires that the knowledge structure $(\mathcal{Q}_1, \mathcal{K}_1)$ and $(\mathcal{Q}_2, \mathcal{K}_2)$ have the same trace on $\mathcal{Q}_1 \cap \mathcal{Q}_2$. For explicitly constructing a mesh of two compatible knowledge structures, Falmagne and Doignon (1998) provide the concept of a maximal mesh. We generalize the notions of a mesh and a maximal mesh to a finite or even infinite number of knowledge structures. As already introduced above, let $I$ denote a finite or infinite index set.

**12.2.7 Definition.** The knowledge structure $(\mathcal{Q}, \mathcal{K})$ is called a *mesh* of the knowledge structures $(\mathcal{Q}_i, \mathcal{K}_i), i \in I$, if

1. $\mathcal{Q} = \bigcup_{i \in I} \mathcal{Q}_i$;
2. $\mathcal{K}_i = \mathcal{K}|_{\mathcal{Q}_i}$ for all $i \in I$.

The construction of a maximal mesh may be generalized accordingly. Whenever there exists a mesh of the knowledge structures $(\mathcal{Q}_i, \mathcal{K}_i), i \in I$, we obtain the *maximal mesh* by defining

$$\mathcal{K}^\star = \{K \in 2^{\bigcup_{i \in I} \mathcal{Q}_i} \mid K \cap \mathcal{Q}_i \in \mathcal{K}_i, \text{ for all } i \in I\}.$$

**12.2.8 Definition.** A collection $(\mathcal{Q}_i, \mathcal{K}_i)$, $i \in I$, of knowledge structures is said to be *pairwise compatible* if any two of the knowledge structures have the same trace on the intersection of their domains, i.e. $\mathcal{K}_j|_{\mathcal{Q}_j \cap \mathcal{Q}_k} = \mathcal{K}_k|_{\mathcal{Q}_j \cap \mathcal{Q}_k}$ for all $j, k \in I$.

A collection of knowledge states $K_i$, $i \in I$, with $K_i \in \mathcal{K}_i$ is said to be *pairwise compatible* if for all $j, k \in I$ we have $K_j \cap \mathcal{Q}_k = K_k \cap \mathcal{Q}_j$.

Pairwise compatibility still remains to be a necessary condition for the existence of a mesh of the knowledge structures $(\mathcal{Q}_i, \mathcal{K}_i), i \in I$. In case of $|I| > 2$, however, pairwise compatibility is no longer sufficient for the existence of a mesh (see Heller and Repitsch, 2008). Proposition 12.2.9 provides a necessary and sufficient condition for the meshability of any arbitrary collection of knowledge structures.

**12.2.9 Proposition.** *For any collection of knowledge structures* $(\mathcal{Q}_i, \mathcal{K}_i), i \in I$, *the following statements are equivalent.*

1. *The knowledge structures are meshable.*
2. *For all* $j \in I$ *and all* $K_j \in \mathcal{K}_j$ *there exists a collection* $K'_i$, $i \in I$, *of pairwise compatible knowledge states* $K'_i \in \mathcal{K}_i$ *such that* $K'_j = K_j$.

For a formal proof see Heller and Repitsch (2008, Proposition 1). Proposition 12.2.9 shows that meshability of a collection of knowledge structures is equivalent to requiring that each of their states can be "covered" by a collection of pairwise compatible knowledge states in the sense of Definition 12.2.8. See Heller and Repitsch (2008) for various alternative sets of conditions for the meshability of knowledge structures.

**12.2.10 Skills and problem functions.**
The general assumption in this section is that we are given a collection $(\mathcal{Q}_i, \mathcal{S}_i, \mu_i)$, $i \in I$, of skill functions, which give rise to a distributed skill function $(\mathcal{Q}, \mathcal{S}, \mu)$. The problem functions that correspond to the $\mu_i$ and $\mu$ are denoted by $p_i$ and $p$, respectively, while the delineated knowledge structures are denoted by $\mathcal{K}_i$ and $\mathcal{K}$, respectively. For notational convenience reference to the basic sets $\mathcal{Q}_i$, $\mathcal{S}_i$ as well as $\mathcal{Q}$, $\mathcal{S}$ is omitted whenever possible. The following Lemma is immediate from Definitions 12.2.1 and 12.2.3.

**12.2.11 Lemma.** *For all $i \in I$ the following inclusions hold.*

1. $\mu_i(q) \subseteq \mu(q) \cap 2^{\mathcal{S}_i} \subseteq \mu(q)$ *for all* $q \in \mathcal{Q}_i$;
2. $p_i(T \cap \mathcal{S}_i) \subseteq p(T \cap \mathcal{S}_i) \cap \mathcal{Q}_i \subseteq p(T) \cap \mathcal{Q}_i$ *for all subsets* $T \subseteq \mathcal{S}$.

*Moreover, we have*

3. $p(T) = \cup_{i \in I} p_i(T \cap \mathcal{S}_i)$ *for all subsets* $T \subseteq \mathcal{S}$.

Point 3, of this lemma shows that the problem function $p$ induced by the distributed skill function $\mu$ can be computed directly from the component problem functions $p_i$ by taking their union. An interesting special case results if the sequences of inclusions in conditions 1. and 2. of this lemma are restricted to equality.

**12.2.12 Proposition.** *For all $i \in I$ the following statements are equivalent.*

1. $\mu_i(q) = \mu(q)$ *for all* $q \in \mathcal{Q}_i$;
2. $p_i(T \cap \mathcal{S}_i) = p(T) \cap \mathcal{Q}_i$ *for all subsets* $T \subseteq \mathcal{S}$.

*Moreover, both statements imply* $\mathcal{K}_i = \mathcal{K}|_{\mathcal{Q}_i}$.

For a formal proof see Heller and Repitsch (2008, Proposition 7). Notice that the proof of the implication from statement 2. to statement 1. draws upon incomparability, and thus cannot be generalized to skill multimaps (see Definition 11.2.1). Statement 1. of Proposition 12.2.12 formalizes the most stringent notion of consistency that a collection of skill functions may satisfy: Whenever the domains of any skill functions overlap then they all assign the same competencies to the common items. Proposition 12.2.12 asserts that in this case the knowledge structure $\mathcal{K}$ delineated by the distributed skill function $\mu$ is a mesh of the knowledge structures $\mathcal{K}_i$ delineated by the component skill functions $\mu_i$. The converse of this implication is not true, as is shown by Example 12.2.5. This gives room for the following definition characterizing consistency of skill functions through meshability of their induced knowledge structures.

**12.2.13 Definition.** A collection of skill functions $(\mathcal{Q}_i, \mathcal{S}_i, \mu_i)$, $i \in I$, is called *consistent* whenever the collection of their delineated knowledge structures $\mathcal{K}_i$, $i \in I$, is meshable.

As the knowledge structure delineated by a skill function is nothing but the range of its associated problem function, we can arrive at a characterization of consistent collections of skill functions by the subsequent corollary to Proposition 12.2.9. We prepare this result by introducing the following notion.

**12.2.14 Definition.** Let $(\mathcal{Q}_i, \mathcal{S}_i, \mu_i)$, $i \in I$, be a collection of skill functions. A collection of skill sets $T_i$, $i \in I$, with $T_i \subseteq \mathcal{S}_i$ is called *pairwise compatible* if the following equivalent conditions hold.

1. For all $j, k \in I$ we have $p_j(T_j) \cap \mathcal{Q}_k = p_k(T_k) \cap \mathcal{Q}_j$;
2. For all $j, k \in I$ and all $q \in \mathcal{Q}_j \cap \mathcal{Q}_k$ there is a $C_j \in \mu_j(q)$ with $C_j \subseteq T_j$ if and only if there is a $C_k \in \mu_k(q)$ with $C_k \subseteq T_k$.

Definition 12.2.14 recasts the notion of a collection of pairwise compatible knowledge states (Definition 12.2.8) in terms of skill sets. Notice that the second condition results from simply plugging in the definition of the problem function into the first one.

**12.2.15 Corollary.** *A collection of skill functions $(\mathcal{Q}_i, \mathcal{S}_i, \mu_i), i \in I$, is consistent if and only if for all $j \in I$ and any skill set $T_j \subseteq S_j$ there exists a collection $T_i'$, $i \in I$, of pairwise compatible skill sets $T_i' \subseteq \mathcal{S}_i$ such that $p_j(T_j') = p_j(T_j)$.*

Corollary 12.2.15 shows that consistency of a collection of skill functions is equivalent to requiring that each "local" skill set can be "covered" by a collection of pairwise compatible skill sets in the sense of Definition 12.2.14. Heller and Repitsch (2008) also treat a variety of special cases, including situations where knowledge domains or skill sets are pairwise disjoint, and where skill and problem functions delineate (quasi ordinal) knowledge spaces. Here, we only provide the following result.

**12.2.16 Corollary.** *Assume a collection of skill functions $(\mathcal{Q}_i, \mathcal{S}_i, \mu_i)$, $i \in I$, and the corresponding distributed skill function $(\mathcal{Q}, \mathcal{S}, \mu)$.*

1. *If the knowledge domains $\mathcal{Q}_i$ are pairwise disjoint then the knowledge structure $\mathcal{K}$ delineated by $\mu$ is a mesh of the component knowledge structures $\mathcal{K}_i$, i.e. $\mathcal{K}_i = \mathcal{K}|_{\mathcal{Q}_i}$ for all $i \in I$.*
2. *If the skill domains $\mathcal{S}_i$ are pairwise disjoint then $\mathcal{K}_i \subseteq \mathcal{K}|_{\mathcal{Q}_i}$ for all $i \in I$.*

**12.2.17 Conclusions.** The previous sections introduced the notion of a distributed skill function for treating the consistent integration of a collection of skill functions on (partially) overlapping knowledge domains. These skill functions may be conceived as representing assignments of skills to items that come from different sources (e.g. different domain experts). In the present approach their consistency is captured by the meshability of the delineated knowledge structures (see Definition 12.2.13). This draws upon a characterization of the meshing of arbitrary collections of knowledge structures, which is developed in Section 12.2.6 and extends and generalizes previous results on the binary case (Doignon and Falmagne, 1999; Falmagne and Doignon, 1998).

Proposition 12.2.9 provides necessary and sufficient conditions for meshability in this situation, which do not hold in general for knowledge domains with nonempty overlap. Consistency in distributed skill functions is linked to the meshability of the delineated knowledge structures by Proposition 12.2.12. Corollary 12.2.15 then identifies properties of a collection of skill functions that are necessary and sufficient for their consistency.

The presented results also suggest mechanisms for reconciling skill functions in order to ensure consistency. Proposition 12.2.12 offers a simple recipe for such an updating mechanism. It consists of making the local skill function coincide with the (global) distributed skill function by assigning to each item all the competencies that are associated to it by any of the skill functions. However, proceeding in this way results in a notion of consistency that is too restrictive. Corollary 12.2.15 shows that an appropriate updating mechanism has to check whether each local skill set can be "covered" by a collection of pairwise compatible skill sets (see Definition 12.2.14). It remains to efficiently implement the steps that need to be taken in order to render a collection of skill functions consistent, which is highly relevant for applications. Reconciling the local information in the indicated way, for instance, is useful in both of the scenarios that were sketched at the beginning of this section. In the context of integrating expert queries on overlapping subsets of items, the updating mechanism suggests how to resolve inconsistencies in the resulting skill assignments. A more cautious application is confined to identifying the inconsistent skill assignments, which the experts then may reconsider. The quasi-automatical nature of the outlined updating may be exploited when integrating distributed resources in technology-enhanced learning, especially in open systems, where the local repositories of assessment items may not be static. In such a setting the consistency of the local skill assignments (implemented through metadata tags) needs to be established at runtime. Complications may arise in applications like that whenever identical skills appear under different names or labels. Resolving this issue requires to define an equivalence relation on the set of all skills prior to applying the above developed framework.

## 12.3 Probabilistic Competence Approach

**12.3.1 Introduction.** In KST the relationship between problems and skills is studied through the concept of a *skill map*, (Albert et al., 1994; Doignon, 1994b; Doignon and Falmagne, 1999; Düntsch and Gediga, 1995; Falmagne et al., 1990; Korossy, 1997, 1999; Lukas and Albert, 1993, see also Chapter 11). Behind this concept there is the essential idea of identifying and describing the cognitive components and procedures—the *skills*—that are needed to solve a given problem or to carry out a specific observable task. The interest of the researcher thus switches from a behavioral, observable level to a latent, cognitive one.

The relationship between these two levels has been thoroughly studied both inside and outside KST, giving rise to a variety of deterministic and probabilistic models for cognitive diagnosis. Perhaps, the most prominent developments outside KST come from the area of Item Response Theory (see e.g. de la Torre and Douglas, 2004; DiBello and Stout, 2007; Embretson, 1995; Junker and Sijtsma, 2001; Maris, 1999; Tatsuoka, 2002). The Bayesian network approach has also to be mentioned (Almond et al., 2007; Mislevy, 1996). For instance, Tatsuoka's $Q$-matrix, which can be found at the basis of many conjunctive non-compensatory models in IRT, is equivalent to the concept of a conjunctive skill map, developed inside KST. The latter is defined to be a triple $(X, S, f)$, where $X$ is a finite nonempty set of problems, $S$ a finite nonempty set of skills, and $f : X \to 2^S \setminus \{\emptyset\}$ maps each problem $i \in X$ to a subset $f(i)$ of skills. The interpretation of $f(i)$ is that a student correctly solves problem $i$ only if he or she possesses all skills in $f(i)$. If a student has exactly the skills in $T \subseteq S$, then the collection $K$ of all problems in $X$ that this student is capable of solving (i.e. his or her knowledge state) is derived from $T$ by

$$K = F(T) = \{i \in X : f(i) \subseteq T\}, \tag{12.1}$$

and $K$ is called the knowledge state delineated by $T$ via the conjunctive skill map $(X, S, f)$. The collection $\mathbb{K} = \{F(T) : T \subseteq S\}$ is the knowledge structure delineated by the skill map $(X, S, f)$. It is easily shown that forms a closure system on $X$, i.e., a collection of subsets of $X$ which is closed under intersection (Doignon and Falmagne, 1999).

Tatsuoka's $Q$-matrix, on the other hand, is a binary $m \times n$ matrix, where $m$ is the number of problems in $X$ and $n$ the number of skills in $S$. Given any row (problem) $i \in X$ and any column (skill) $j \in S$, the entry $Q_{ij}$ of the matrix contains a 1 if skill $j$ is needed for solving item $i$. Provided that there is at least a 1 in each row of the matrix, the conjunctive skill map $f$ corresponding to $Q$ is easily derived by setting

$$f(i) = \{j \in S : Q_{ij} = 1\}. \tag{12.2}$$

**12.3.2 Probabilistic approach to skill maps.** A typical probabilistic conjunctive model in IRT, which makes use of the $Q$-matrix, is the so-called DINA model (Deterministic Input Noisy AND-gate) model (Junker and Sijtsma, 2001). The DINA assumes the existence of a deterministic internal response of a student to a problem. Such internal response acts as an "AND" gate which returns 1 only if all skills for solving the particular problem are possessed by that student. For $T \subseteq S$ and $j \in S$ let $w_j(T) \in \{0, 1\}$ be an indicator function such that $w_j(T) = 1$ if $j \in T$. If a student, who has exactly the skills in $T$, deals with problem $i$, then the deterministic, internal response of this student to problem $i$ is defined to be

$$\xi_i(T) = \prod_{j \in S} w_j(T)^{Q_{ij}}. \tag{12.3}$$

It is easily seen that $\xi_i(T)$ is either 0 or 1. In particular $\xi_i(T) = 0$ if and only if there is some skill $j \in S$ such that $w_j(T) = 0$ and $Q_{ij} = 1$, meaning that the student does not have a skill which is required by problem $i$. Stated another way $\xi_i(T) = 1$ only if problem $i$ can be solved from the skill set $T$. Therefore, if $f$ is defined as in (12.2) then the relationship between $f$ and $\xi$ is as follows:

$$\xi_i(T) = 1 \text{ iff } f(i) \subseteq T,$$

and $F(T)$ can also be written as $F(T) = \{i \in X : \xi_i(T) = 1\}$. Thus it is clear that the types of knowledge structures captured by the DINA model are just the closure systems on $X$.

The DINA model postulates a probabilistic relationship between the internal response $\xi_i(T)$ and the observed (either correct or wrong) response $y_i \in \{0, 1\}$ of the student. For $i \in X$, let $Y_i$ be a discrete random variable whose realizations are the observed responses $y_i \in \{0, 1\}$ of a student. Two conditional probabilities are defined for each of the problems in $X$:

- a *careless error* parameter $\alpha_i = P(Y_i = 0|\xi_i(T) = 1)$ (also called the *slip* parameter of the model), which specifies the conditional probability that the response of a student to problem $i$ is wrong given that he or she has all skills that are necessary and sufficient for solving $i$;
- a *lucky guess* parameter $\beta_i = P(Y_i = 1|\xi_i(T) = 0)$, which specifies the conditional probability that the response of a student to problem $i$ is correct, given that he or she does not have all the skills needed for solving $i$.

A basic assumption of the DINA model is the so-called *local independence* among the responses provided by a student, given a fixed skill state of the student. Under this assumption, if a student is characterized by the skill set $T \subseteq S$, then the conditional probability that this student produces the response pattern $\mathbf{y} = (y_1, y_2, \ldots, y_n)$ is given by

$$\begin{aligned} &P(\mathbf{Y} = \mathbf{y}|\mathbf{T} = T) \\ &= \prod_{i \in X} \left[\alpha_i^{1-y_i}(1-\alpha_i)^{y_i}\right]^{\xi_i(T)} \left[\beta_i^{y_i}(1-\beta_i)^{1-y_i}\right]^{1-\xi_i(T)}, \end{aligned} \tag{12.4}$$

where $\mathbf{Y}$ is a random variable whose realizations are $n$-dimensional binary vectors, and $\mathbf{T}$ a random variable whose realizations are subsets of $S$. Then, indicating with $P(\mathbf{T} = T)$ the probability of sampling a student whose skill set is $T$ (assuming thus a probability distribution on the power-set of $S$), the marginal probability of observing the response pattern $\mathbf{y}$ is given by the unrestricted latent class model

$$P(\mathbf{Y} = \mathbf{y}) = \sum_{T \subseteq S} P(\mathbf{Y} = \mathbf{y}|\mathbf{T} = T)P(\mathbf{T} = T). \tag{12.5}$$

where the latent classes of the model are just the subsets of $S$.

The DINA model has thus three types of parameters: a careless error $\alpha_i$ and a lucky guess $\beta_i$ for each of the problems in $X$, plus a probability $\pi(T) = P(\mathbf{T} = T)$ for each of the skill sets $T \subseteq S$. Given a suitable empirical data set, the parameters of the DINA model can be estimated by maximum likelihood procedures, e.g. the Expectation-Maximization algorithm (Dempster et al., 1977).

Another conjunctive model based on the $Q$-matrix, which however will be not discussed here, is the so-called NIDA (Noisy Input Deterministic AND-gate) model (Junker and Sijtsma, 2001; Maris, 1999). Both the DINA and the NIDA models can be seen as probabilistic models for conjunctive skill maps. As such they can only be applied to a restricted class of knowledge structures, namely the closure systems on a set $X$ of problems. The main limitation of this class of knowledge structures is that they do not allow to model multiple solution strategies. Each of the problems $i \in X$ can be solved by exactly one strategy represented by the skill set $f(i)$. Actually, when fitting a conjunctive model to the data this limitation could be crucial: the existence of solution strategies that are not taken into account by the model might lead to a misfit of the model itself, or the lucky guess and careless error probabilities might become too large to be credible as noise parameters (Tatsuoka, 2002).

**12.3.3 Modeling multiple solution strategies.** Actually, conjunctive skill maps are special cases of the more general concept of a skill multimap (Doignon and Falmagne, 1999; Düntsch and Gediga, 1995, see also Definition 11.2.1). The latter is a triple $(X, S, g)$, where $g$ assigns a nonempty collection $g(i)$ of nonempty subsets of $S$ to each of the problems $i \in X$. The collection $g(i)$ is usually interpreted as the set of alternative strategies for solving problem $i$, in the sense that each $G \in g(i)$ (also called a *competency*) is a set of skills that are necessary and sufficient for solving $i$.

It has been stated before that conjunctive skill maps can only delineate knowledge structures that are closure systems on the set $X$ of items. Doignon and Falmagne (1999) show that every knowledge structure on the set $X$ is, in fact, delineated by some skill multimap. Düntsch and Gediga (1995) provide an alternative representation of a skill multimap through what they call a *problem function*, which is a mapping $h : 2^S \to 2^X$ which satisfies the following three properties:

1. $h(\emptyset) = \emptyset$;
2. $h(S) = X$;
3. $A \subseteq B \Rightarrow h(A) \subseteq h(B)$ for all $A, B \subseteq S$.

Düntsch and Gediga show that, for a fixed skill set $S$ and a fixed problem set $X$, there exists a bijection between the collection of skill multimaps and the collection of problem functions. In particular, if the problem function $h$ corresponds to the skill multimap $g$, then the collection $\{h(T) : T \subseteq S\}$ is the knowledge structure delineated by $g$ where $h(T)$ is just the collection of all problems in $X$ that can be solved by the skill set $T$.

Given this result, it is not difficult to extend the DINA model to any skill multimap and thus, to any knowledge structure on the set $X$ of problems: Let $z_i(T) \in \{0,1\}$ be an indicator such that $z_i(T) = 1$ if and only if $i \in h(T)$. Then Equation (12.4) can simply be modified in the following way:

$$P(\mathbf{Y} = \mathbf{y} | \mathbf{T} = T) = \prod_{i \in X} \left[\alpha_i^{1-y_i}(1-\alpha_i)^{y_i}\right]^{z_i(T)} \left[\beta_i^{y_i}(1-\beta_i)^{1-y_i}\right]^{1-z_i(T)}. \tag{12.6}$$

In the sequel this will be referred to as the *saturated model* for probabilistic skill multimaps.

A straightforward and, actually, rather strong restriction of this model is the assumption that the skills in $S$ are stochastically independent:

$$P(\mathbf{T} = T) = \prod_{s \in T} \pi_s \prod_{t \in S \setminus T} (1 - \pi_s), \tag{12.7}$$

where $\pi_s$ is the probability of sampling a student who has skill $s \in S$.

**12.3.4 Modeling dependence among skills.** Models examined so far are based on the implicit assumption that the skill set of an individual is any subset $T \subseteq S$ of skills. In certain circumstances, however, some of these subsets can be excluded for theoretical reasons. If, for instance, two skills $s, t \in S$ are such that $s$ is a prerequisite for $t$ (possessing skill $t$ implies possessing skill $s$) then every subset $T \subseteq S$ containing $t$ and not containing $s$ should be excluded from the collection of the possible skill sets.

In order to explain possible dependencies among the skills, (Korossy, 1997, see also Section 11.4) postulates a *competence structure*, which is a collection $\mathcal{S} \subseteq 2^S$ of subsets of $S$, containing at least the empty set and $S$. The latent class model (12.5) can be easily adapted to this situation by assuming a probability distribution on the subsets in $\mathcal{S}$. Clearly in this case the independence model (12.7) can no longer be applied.

In the sequel, a restriction of the saturated model is examined, which is discussed in Stefanutti (2006). It naturally applies to a special class of competence structures, called *graded competence structures*, but it can be easily extended to any arbitrary competence structure.

The notion of a *graded competence structure* (not to be confused with *well-graded competence structure*[6]) is defined in the following way: For a competence structure $\mathcal{S}$ on $S$, let $\mathbb{C} = \{K_1, K_2, \ldots, K_p\}$ be any maximal

[6]The class of the well-graded competence structures is a subfamily of the graded competence structures.

chain[7] of skill sets (henceforth called *competence states*) in $\mathcal{S}$. The chain $\mathbb{C}$ is a *gradation* if for any $i \in \{1, 2, \dots, p-1\}$ there exists a skill $s \in S$ such that $K_{i+1} = K_i \cup \{s\}$. A competence structure $\mathcal{S}$ is *graded* if any arbitrary maximal chain in $\mathcal{S}$ is a gradation. In the following example, $\mathcal{S}'$ is a graded competence structure, while $\mathcal{S}''$ is not:

$$\mathcal{S}' = \{\emptyset, \{a\}, \{b\}, \{a,c\}, \{b,d\}, \{a,c,d\}, \{b,c,d\}\{a,b,c,d\}\},$$

$$\mathcal{S}'' = \{\emptyset, \{a\}, \{b\}, \{a,b\}, \{b,c,d\}, \{a,b,c,d\}\}.$$

The competence structure $\mathcal{S}'$ contains exactly two maximal chains and both of them are gradations. The competence structure $\mathcal{S}''$ contains three maximal chains, but the chain

$$\{\emptyset, \{b\}, \{b,c,d\}, \{a,b,c,d\}\}$$

is not a gradation, since $\{b\}$ and $\{b,c,d\}$ differ by more than one element.

The restriction of the saturated model to graded competence structures is based on the following assumption: (henceforth, we refer to the probability $P(\mathbf{T} = K)$ with the shortcut $P(K)$)

(A1) Let $\mathcal{S}$ be a graded competence structure on the finite set $S$ of skills, and $s \in S$ be a skill. Then for every pair $K, K \cup \{s\} \in \mathcal{S}$ the odds $P(K \cup \{s\})/P(K)$ only depends on skill $s$. That is, there exists a positive parameter $\psi_s$ such that

$$\frac{P(K \cup \{s\})}{P(K)} = \psi_s$$

for all pairs $K, K \cup \{s\} \in \mathcal{S}$.

Under Assumption **(A1)**, given that a student possesses all skills in $K$, the probability that this subject possesses also $s$ only depends on $\psi_s$. It is important to note that this is a special case of the constant ratio rule: Adding the same skill to both states $K$ and $K'$, the ratio between the corresponding probabilities does not change.

The introduction of Assumption **(A1)** yields the following result concerning the probabilities of the states.

**12.3.5 Theorem.** *Let $\mathcal{S}$ be a competence structure on the set $S$ of skills and let $K \in \mathcal{S}$ be a competence state. If $\mathcal{S}$ is graded then Assumption* **(A1)** *implies that the probability of $K$ takes on the form*

$$P(K) = \frac{\prod_{s \in K} \psi_s}{\sum_{K' \in \mathcal{S}} \prod_{s \in K'} \psi_s}.$$

[7] A chain in a partially ordered set $(P, \leqslant)$ is any subset $C = \{c_1, c_2, \dots, c_n\} \subseteq P$ such that $(C, \leqslant)$ is a linear order, i.e. $c_1 \leqslant c_2 \leqslant \dots \leqslant c_n$. The chain is maximal if for all chains $C' \subseteq P$, $C \subseteq C'$ implies $C = C'$.

A formal proof of Theorem 12.3.5 can be found in Stefanutti (2006). A suitable log-linear version of this model is obtained by introducing the new parameters $\eta_a = \ln \psi_a$. In this way the following equivalent model is obtained:

$$P(K) = \frac{\exp\left(\sum_{s \in K} \eta_s\right)}{\sum_{K' \in \mathcal{S}} \exp\left(\sum_{s \in K'} \eta_s\right)}. \tag{12.8}$$

This model is characterized by a parameter $\alpha_i$ and a parameter $\beta_i$ for each item $i$, and a parameter $\eta_s$ for each of the skills in $S$. The model parameters can be estimated by means of a maximum likelihood procedure, e.g. an application of the Expectation-Maximization algorithm.

**12.3.6 Conclusions.** In many areas of psychology, like e.g. the educational, cognitive, and organizational ones, the researchers or professionals are aimed at recovering the specific and unobservable cognitive operations and procedures that a human being carries out in performing a given observable task. Such cognitive operations—often referred to as *skills*—are regarded as discrete units that can be either present or absent in an individual.

A number of different approaches, including KST and IRT, are built upon the very essential idea that solving a problem requires that a certain set of skills is available to an individual. Therefore a basic question of almost all these approaches is how problems are related to skills.

In this chapter the question of assigning skills to problems is tackled from a probabilistic perspective. A brief overview is given in Section 12.3.1, which points out some of the connections between IRT and KST, that can be found in the area of probabilistic skill modeling. For example, an important probabilistic model on the IRT side is the DINA (discussed in Section 12.3.2), a conjunctive model whose application scope is limited to problems with a single solution strategy. The deterministic part of this model is formally equivalent to the concept of a conjunctive skill map developed inside KST.

Section 12.3.3 deals with an extension of the DINA to skill multimaps and, thus, to problems having more than one solution strategy. The question of modeling possible dependencies among the skills is discussed in Section 12.3.4. The resulting probabilistic model is capable of capturing problems with multiple solution strategies and it accounts for specific dependencies among the skills.

A final remark concerns dependence. A competence structure, as much as a knowledge structure, is a model where the specific dependencies among the skills are expressed in a deterministic (algebraic) fashion (Düntsch and Gediga, 1995; Korossy, 1993, 1997). There is no randomness implied by this model. On the other hand it would be rather difficult to test the empirical validity of a competence structure without resorting to some suitable probabilistic framework. For this reason probabilistic models come about.

Typically a researcher would like to construct the deterministic version of the model and to validate its probabilistic counterpart. However it must be recognized that such two versions, although closely connected, are not exactly the same thing. It is thus desirable to have clearly and easily interpretable relationships between them. One such relationship concerns dependence. In particular the requirement seems important that the deterministic model and its counterpart are in agreement with respect to how skills depend on each other. This question has been addressed in Stefanutti (2006). The answer is that there is no correspondence between probabilistic and algebraic dependence in general. There are however special cases where such correspondence holds. The type of models presented in Section 12.3.4 is one of these special cases.

## 12.4 Linking Navigation Behavior to the Cognitive State in eLearning

While the navigation behavior of users of an eLearning environment can be directly observed, for modeling this behavior it seems reasonable to refer to some kind of non-observable (latent) states of the learner. Cognitive states, like the learner's state of knowledge, will influence the concrete actions performed by the learner. Conversely, the actual trail of learning objects (LOs) visited will affect the cognitive state of a learner. In the sequel we outline a formal framework developed by Heller et al. (2007) that models both directions of this mutual dependence, and offers methods and tools that are needed for applying the model.

A hypertext can be modeled as a network (graph), whose nodes are the pages of the hypertext and whose edges are the hyperlinks between pages (Halasz and Schwartz, 1994). When the hypertext in question is a *web learning environment*, where each page is a learning object (LO) then the network graph is a model of the learning environment, and the trails through it are LO trails. Navigation through the environment is then represented by paths through the graph. These paths—hypertext trails (and so our trails of LOs)—were stochastically modeled as Markov chains (Kemeny and Snell, 1960) on the network, where the probability of moving from one node (i.e. one page of the hypertext) to another is determined by which page the user is currently visiting. In order to deduce a Markov model of learners' observable behavior, web data mining techniques can be used to deduce navigation sessions from the server log files (Borges and Levene, 2000), including the statistics of how many times learners choose particular links when confronted with a choice (i.e. from the frequencies of page transitions as recorded in the log file).

On the other hand, the consideration of learning processes in a knowledge structure is based on the interpretation of the upwards directed line segments in the Hasse diagram as the possible learning paths. Falmagne (1993, 1994) suggested Markov models of the learning processes that proceed along these learning paths. The simplest such model identifies the states of the Markov chain with the knowledge states of the knowledge structure. As an initial condition we may assume that the probability mass is concentrated on the naïve state $\emptyset$. Moreover, a transitions from state $K$ to state $K'$ is only possible if $K'$ is a neighboring state that covers $K$. Various such models are reviewed in Doignon and Falmagne (1999).

**12.4.1 Theoretical framework.** The following suggests a formal framework capable of capturing the mutual dependence between observable trails of LOs and cognitive trails in the associated knowledge structure by means of a stochastic model.

Two sets form the basis of the stochastic modeling in the subsequently developed theory. On the one hand, we consider the set $\mathcal{L}$ of LOs (e.g. webpages) that constitute the learning environment. On the other hand, we refer to a knowledge structure $\mathcal{K}$ on a knowledge domain $\mathcal{Q}$. The knowledge domain $\mathcal{Q}$ consists of problems that test the skills and competencies taught by the LOs in $\mathcal{L}$. The two sets $\mathcal{L}$ and $\mathcal{K}$ thus are intimately related. This relationship is mediated by the underlying skills and competencies (see Chapter 11 this volume). It can be characterized by a map $g$ that associates to each problem in $\mathcal{Q}$ a collection of subsets of LOs. Each of these subsets assigned to a problem is interpreted as a minimal subset of LOs that provide the content sufficient for solving it. More than one such subset may be assigned to a problem to reflect the fact that there may be different ways to solve it, which are taught in different subsets of LOs. Formally, a *prerequisite map* on the knowledge domain $\mathcal{Q}$ is defined as a map $g\colon \mathcal{Q} \to 2^{2^{\mathcal{L}}}$ that associates to each problem in $\mathcal{Q}$ a nonempty collection of nonempty subsets of $\mathcal{L}$. Notice that this notion is formally equivalent to a skill multimap (see Definition 11.2.1), but assigns subsets of LOs instead of skills to problems.

Consider the following simple example. The set of LOs for the learning environment illustrated in Figure 12.2 is given by

$$\mathcal{L} = \{E, O_1, O_2, O_3, O_4, O_5\}.$$

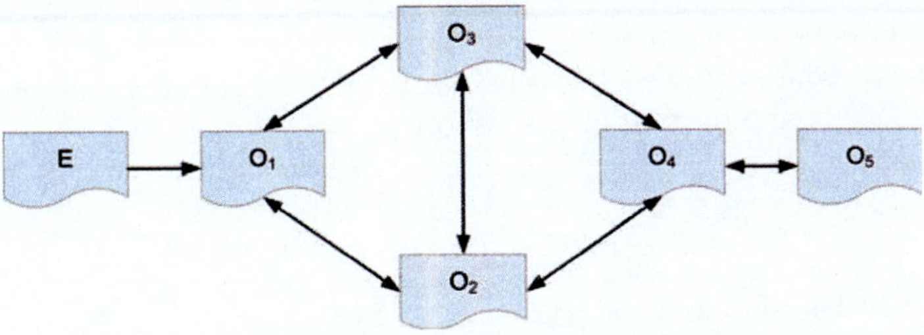


**Figure 12.2.** Example of a learning environment.

The symbol $E$ denotes an entry point to the learning environment. In contrast to $E$, the LOs $O_1$ to $O_5$ teach some content, and may be accessed in accordance with the depicted link structure.

Suppose that the problems in the knowledge domain $\mathcal{Q} = \{a, b, c, d\}$ test the contents taught by the LOs in $\mathcal{L}$. In particular, let the prerequisite map on $\mathcal{Q}$ be given by

$$\begin{aligned} g(a) &= \{\{O_1, O_2\}\}, \\ g(b) &= \{\{O_1, O_3\}\}, \\ g(c) &= \{\{O_1, O_2, O_3, O_4\}, \{O_1, O_2, O_5\}\}, \\ g(d) &= \{\{O_1, O_2, O_3, O_5\}\}. \end{aligned} \tag{12.9}$$

According to this assignment, the problem $a$ can be solved using the content taught by the two LOs $O_1, O_2$. Problem $c$ can be solved in two different ways. One of the solutions requires the content of the LOs $O_1, O_2, O_3, O_4$, and the other one the content of $O_1, O_2, O_5$.

The prerequisite map $g$ is actually mediated by the skills and competencies associated to both problems and LOs. Assume that there is a set $\mathcal{S}$ of skills that are taught in the learning environment, and tested in the knowledge assessment. The assignment of these skills to LOs and problems in the knowledge domain $\mathcal{Q}$ can be formalized by two mappings. First, consider a mapping $\tau\colon \mathcal{L} \to 2^{\mathcal{S}}$ associating to each LO the set of skills that it teaches. We will assume that $\tau(\mathcal{L}) = S$. Furthermore, in the example presented above we have $\tau(E) = \emptyset$, and $\tau(L) \neq \emptyset$ for all $L \in \mathcal{L}$ with $L \neq E$. Second, consider a mapping $\mu$ from $\mathcal{Q}$ to $2^{2^{\mathcal{S}}}$ associating to each problem $q$ in $\mathcal{Q}$ a nonempty collection of nonempty subsets of $\mathcal{S}$, which is a skill multimap according to Definition 11.2.1. Now, given a problem $q \in \mathcal{Q}$ for each competency $C \in \mu(q)$ consider those subsets $M$ of LOs such that

$$\bigcup_{L \in M} \tau(L) \supseteq C,$$

and $M$ is minimal with respect to set-inclusion. Finally, this defines a mapping assigning to each problem a collection of subsets $M$ of LOs providing the content sufficient for solving the problem, which is the prerequisite map $g$ introduced above.

For the present example let $\mathcal{S} = \{s, t, u, v, w\}$ form the set of relevant skills, and define the mapping $\tau$ from $\mathcal{L}$ to $2^{\mathcal{S}}$ by

$$\begin{aligned} &\tau(E) = \emptyset, &&\tau(O_1) = \{s\}, &&\tau(O_2) = \{t\}, \\ &\tau(O_3) = \{u\}, &&\tau(O_4) = \{v\}, &&\tau(O_5) = \{w\}. \end{aligned}$$

Moreover, let the skill multimap $\mu$ be given by

$$\begin{aligned} \mu(a) &= \{\{s, t\}\}, \\ \mu(b) &= \{\{s, u\}\}, \\ \mu(c) &= \{\{s, t, u, v\}, \{s, t, w\}\}, \\ \mu(d) &= \{\{s, t, u, w\}\}. \end{aligned}$$

Then following the route outlined above the prerequisite map defined by (12.9) results.

The skill multimap $\mu$ delineates a knowledge structure on $\mathcal{Q}$. For the example this gives the knowledge structure

$$\mathcal{K} = \{\emptyset, \{a\}, \{b\}, \{a, b\}, \{a, c\}, \{a, b, c\}, \{a, b, d\}, \mathcal{Q}\},$$

which is illustrated in Figure 12.3. The diagram of the knowledge structure shows the possible learning paths that a user of the above displayed learning environment can take. We subsequently outline a stochastic model capable of predicting the overt navigation behavior and the underlying learning process captured by the sequence of knowledge states, as well as their mutual dependencies.

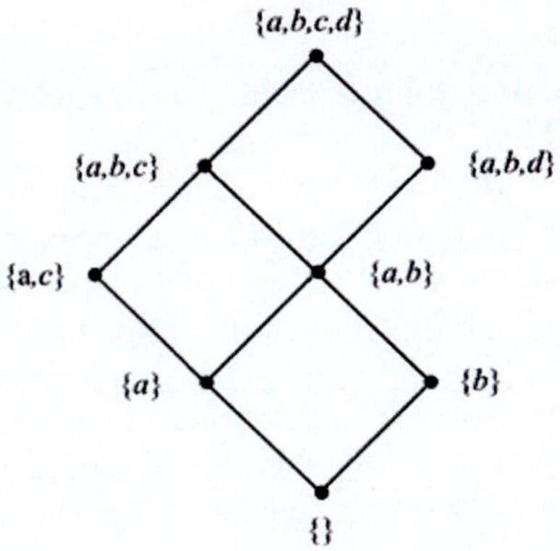


**Figure 12.3.** Example of a knowledge structure on the knowledge domain $\mathcal{Q} = \{a, b, c, d\}$.

**12.4.2 A Markovian model for learning.** Within a Markov chain model we identify the set of Markov states (M-states) with the Cartesian product $\mathcal{L} \times \mathcal{K}$. This choice reflects the fact that only knowledge of both the current LO and the current knowledge state leads to a proper characterization of the process at any point in time. The Markov property that we assume implies that the future of the process is completely determined by the current M-state. In particular, it is not important how the process got to this state, or, in other words, all information about the past is embodied in the current state.

We consider discrete time, which will be indicated by a subscript $t = 0, 1, 2, \ldots$. This index is incremented whenever the learner selects a link to another learning object from those available at a time. For all points $t$ in time let $L_t$ and $K_t$ denote random variables that take their values in $\mathcal{L}$ and $\mathcal{K}$, respectively. The Markov chain then consists of a sequence of pairs

$$(L_0, K_0), (L_1, K_1), (L_2, K_2), (L_3, K_3), \ldots,$$

to which we will subsequently refer to as a *trail*. To identify this sequence, information from different sources has to be integrated. The trail of visited LOs can be obtained by log data analysis. Assessing the knowledge states of $\mathcal{K}$ requires to test the learner on the items in the knowledge domain $\mathcal{Q}$. We will provide details on the requested knowledge assessment later.

The Markov model is defined by specifying an initial probability distribution on the set of M-states $\mathcal{L} \times \mathcal{K}$ (i.e. by specifying $\mathbb{P}(L_0, K_0)$), and by giving the conditional probabilities $\mathbb{P}(L_t, K_t \mid L_{t-1}, K_{t-1})$ of a transition from state $(L_{t-1}, K_{t-1})$ at time $t-1$ to state $(L_t, K_t)$ at time $t$, for all $t = 1, 2, \ldots$. By drawing upon the Markov property this allows for computing the probability of any trail $(L_0, K_0), \ldots, (L_n, K_n)$ by

$$\mathbb{P}((L_0, K_0), \ldots, (L_n, K_n)) = \mathbb{P}(L_0, K_0) \cdot \prod_{t=1}^{n} \mathbb{P}(L_t, K_t \mid L_{t-1}, K_{t-1}). \quad (12.10)$$

Specifying an initial probability distribution on $\mathcal{L} \times \mathcal{K}$ will pose no problems in most of the cases. The LO $L_0$ is taken to be nothing else than the entry point of the learning environment, the common starting point from which all learners depart (denoted by $E$ in Figure 12.2). Consequently, LO $L_1$ represents the first LO providing content that is actually inspected by the learner. This interpretation is of purely technical nature, and is in line with considering $K_0$ as the knowledge state before being exposed to the content. This knowledge state is either the naïve state $\emptyset$ for all learners (e.g. whenever the material is completely new to them), or is assumed to be any other state in $\mathcal{K}$, which may differ over learners. In the first case, we have the initial condition

$$\mathbb{P}(L_0, K_0) = \begin{cases} 1 & \text{if } L_0 = E, K_0 = \emptyset; \\ 0 & \text{otherwise,} \end{cases}$$

where $E$ denotes the entry point of the learning environment. In the second case only the probabilities $\mathbb{P}(L_0 = E, K_0)$ can be non-zero, and their actual values may be estimated from data of an assessment that precedes access to the learning environment (pre-assessment).

Defining the conditional probabilities $\mathbb{P}(L_t, K_t \mid L_{t-1}, K_{t-1})$ for all $t = 1, 2, \ldots$ requires to take into account their interpretation in the present context. In fact, the transition $(L_{t-1}, K_{t-1}) \to (L_t, K_t)$ may be interpreted as:

> A person visiting LO $L_{t-1}$ and having knowledge state $K_{t-1}$ at time $t-1$ selects LO $L_t$ and, as a consequence of this, moves into knowledge state $K_t$ at time $t$.

This interpretation suggests to decompose the transition from state $(L_{t-1}, K_{t-1})$ at time $t-1$ to state $(L_t, K_t)$ at time $t$ into two sub-processes, or stages

1. the selection of the next LO,

2. the learning process induced by the selected LO.

These two stages are reflected in the formula

$$\mathbb{P}(L_t, K_t \mid L_{t-1}, K_{t-1}) = \mathbb{P}(L_t \mid L_{t-1}, K_{t-1}) \cdot \mathbb{P}(K_t \mid L_t, K_{t-1}), \qquad (12.11)$$

whenever we subscribe to the the straightforward and quite plausible assumption that only the current LO will affect the transition between knowledge states. The conditional probability $\mathbb{P}(L_t \mid L_{t-1}, K_{t-1})$ in (12.11) refers to the first of the above listed stages, and describes the impact of the knowledge state $K_{t-1}$ on choosing the link from LO $L_{t-1}$ to $L_t$. The conditional probability $\mathbb{P}(K_t \mid L_t, K_{t-1})$ in (12.11) models the second stage, and captures the impact of the LO $L_t$ on the transition of knowledge states from $K_{t-1}$ to $K_t$ (i.e. the learning process).

Various aspects of the learning environment put restrictions on the values of the conditional probabilities $\mathbb{P}(L_t \mid L_{t-1}, K_{t-1})$, and $\mathbb{P}(K_t \mid L_t, K_{t-1})$, among them the link structure on the set of LOs, and the relationship between the knowledge states considered. Due to the link structure on $\mathcal{L}$ on the one hand, and the knowledge structure $\mathcal{K}$ on the other hand, some of the conditional probabilities will have zero value ('structural zeros'). These parameters need not be estimated in the applications. We illustrate the resulting reduction of the number of free parameters for the learning environment depicted in Figure 12.2 with $|\mathcal{L}| = 6$, and the knowledge structure $\mathcal{K}$ of Figure 12.3 with $|\mathcal{K}| = 8$.

The probability $\mathbb{P}(L_t \mid L_{t-1}, K_{t-1})$ can be non-zero only if there is a direct link from LO $L_{t-1}$ to $L_t$. There are 13 direct links between the 6 LOs. This means that instead the $|\mathcal{L}|^2 \cdot |\mathcal{K}| = 6^2 \cdot 8 = 288$ potential parameters, only $13 \cdot 8 = 104$ conditional probabilities have to estimated from the data. Similarly, for $\mathbb{P}(K_t \mid L_t, K_{t-1})$ to be non-zero we have to have $K_{t-1} \subseteq K_t$ (see condition C3 below), and the difference between the knowledge states $K_{t-1}$ and $K_t$ has to be related to the content taught in LO $L_t$ (see condition C2 below). If we consider the relation of set-inclusion then, instead of $|\mathcal{K}|^2 \cdot |\mathcal{L}| = 8^2 \cdot 6 = 384$ potential parameters only $31 \cdot 6 = 186$ non-zero conditional probabilities remain (31 is the number of pairs $K, K' \in \mathcal{K}$ for which $K \subseteq K'$ holds). If the learners can only move from a knowledge state into one of its immediate successors (with respect to $\subseteq$), then there are no more than $10 \cdot 6 = 60$ non-zero conditional probabilities.

**12.4.3 Inferring cognitive trails.** The basic problem is as follows. Given a scenario in which we have a pre-assessment before the LOs are inspected, and a post-assessment after finishing the interaction with the learning environment, how to infer the sequence of knowledge states in the trail of M-states $(L_t, K_t)$? For any learner we observe the trail $L_0, \ldots, L_n$ of visited LOs from $\mathcal{L}$, and we have available a pre-assessment $K_0$ and a post-assessment $K_n$. Given a prerequisite map $g$ on $\mathcal{Q}$ capturing the relation between the LOs and assessment items, we want to identify those trails of knowledge states that can be inferred by the observed trail $L_0, \ldots, L_n$ of LOs. The conditions C1, C2, and C3 listed below provide a minimal set of compatibility requirements to characterize this relationship. After stating these conditions, we provide a few remarks that clarify their motivation as well as their impact.

C1. Solving an item cannot be learned before visiting a set of relevant LOs, which are sufficient for its solution. Formally, this can be expressed as follows:

$$q \in K_t \setminus K_0 \text{ implies that there is a subset } M \in g(q) \text{ such that } M \subseteq \{L_0, \ldots, L_t\};$$

C2. Learning to solve an item can only occur when visiting a relevant LO:

$$q \in K_t \setminus K_{t-1} \text{ implies that there is a subset } M \in g(q) \text{ such that } L_t \in M;$$

C3. There is no forgetting, i.e. the trail of knowledge states $K_0, \ldots, K_n$ is increasing (with respect to set-inclusion):

$$K_0 \subseteq \ldots \subseteq K_n.$$

**12.4.4 Definition.** A trail $(L_0, K_0), \ldots, (L_n, K_n)$ on $\mathcal{L} \times \mathcal{K}$ is called *consistent* whenever the trail of knowledge states $K_0, \ldots, K_n$ is compatible to the trail $L_0, \ldots, L_n$ of LOs, i.e. whenever the compatibility conditions C1-C3 are satisfied. Let $\mathcal{T}$ denote the set of consistent trails on $\mathcal{L} \times \mathcal{K}$.

Condition C1 assumes that learning to solve an item the learner was not capable of solving in the pre-assessment cannot occur before visiting a subset of LOs that provide content sufficient for solving it. This reasonable assumption is related to the *scope* of the given trail of LOs, i.e. what in principle can be learned from it. An item $q \in \mathcal{Q}$ lies within this scope (which means that we may have $q \in K_n \setminus K_0$) if and only if there exists $M \in g(q)$ such that $M \subseteq \{L_0, \ldots, L_n\}$. A learner following this trail of LOs, however, does not necessarily learn to solve all the items within its scope.

Condition C2 means that learning to solve an item can only occur if the currently visited LO is relevant for its solution. This means that the solution is learned as soon as the last portion of the relevant and sufficient information is considered, and excludes effects based on unrelated material mediating learning.

Condition C3 implies $K_0 \subseteq K_n$, which, in principle, may be contradicted by data. In empirical applications, however, we may avoid this problem by simply identifying $K_0$ with $K_0 \cap K_n$. Proceeding in this way implicitly interprets the correct response to the items in $K_0 \setminus K_n$ as lucky guesses.

Notice that the compatibility conditions C1-C3 are mutually independent, and, in general, there will be more than one trail of knowledge states satisfying these conditions. The non-uniqueness is illustrated by the following trails.

$$(E,\emptyset),(O_1,\emptyset),(O_2,\{a\}),(O_3,\{a,b\}),(O_4,\{a,b,c\}),\\ (O_5,\{a,b,c,d\}) \tag{12.12}$$

$$(E,\emptyset),(O_1,\emptyset),(O_2,\{a\}),(O_3,\{a,b\}),(O_4,\{a,b\}),\\ (O_5,\{a,b,c,d\}) \tag{12.13}$$

$$(E,\emptyset),(O_1,\emptyset),(O_2,\{a\}),(O_3,\{a,b\}),(O_2,\{a,b\}),\\ (O_4,\{a,b,c\}),(O_5,\{a,b,c\}) \tag{12.14}$$

$$(E,\emptyset),(O_1,\emptyset),(O_2,\emptyset),(O_3,\{b\}),(O_2,\{a,b\}),\\ (O_4,\{a,b,c\}),(O_5,\{a,b,c\}) \tag{12.15}$$

Both pairs of consistent trails (12.12), (12.13) and (12.14), (12.15) are based on identical trails of LOs as well as coinciding pre- and post-assessment. However, the solution to item $c$ is learned at different points in time in the trails (12.12) and (12.13), and trails (12.14) and (12.15) differ with respect to learning to solve item $a$. A more detailed interpretation of the differences between the trails is given below. In particular, the examples demonstrate that the compatibility conditions C1-C3 will not suffice to reconstruct a single trail of knowledge states from the given data. Although this does not pose a fundamental problem to deriving model predictions and empirical testing of the stochastic model, it seems interesting to point out special cases that allow for uniquely identifying a single consistent trail. In these cases, parameter estimation techniques based on log data analysis can be used. The remainder of this section provides a discussion of two different sets of assumptions either of which is sufficient to guarantee a unique consistent trail.

First, we assume that learning can occur as soon as the relevant content has been exposed. Given the observable trail $L_0, \ldots, L_n$ of LOs from $\mathcal{L}$, a prerequisite map $g$ on $\mathcal{Q}$, and $K_0, K_n \subseteq \mathcal{Q}$. Consider a trail $K_0, \ldots, K_n$ of knowledge states in the domain $\mathcal{Q}$, which is defined in the following way: For all $q \in K_n \setminus K_0$ and $t \in \{1, \ldots, n-1\}$ we have

$$q \in K_t \setminus K_0 \text{ if and only if there is a subset } M \in g(q)\\ \text{such that } M \subseteq \{L_0, \ldots, L_t\}.$$

This condition is called *Strict Learning Assumption* (SLA). It is easily seen SLA defines a uniquely determined and consistent trail (see Prop. 12.4.7 in Section **??**).

From the above listed trails only (12.12) and (12.14) are in accordance with SLA. The essential assumption is that learning occurs as soon as the relevant material is available. This may be too optimistic. In particular, SLA lacks plausibility if LOs are visited more than once. Consider the trail (12.14), where the LO $O_2$ is revisited, although its contents have already been learned. In fact, revisiting an LO may be interpreted as an indication of the fact that the material has not been learned during previous visits. This is taken into account in the subsequently outlined Weak Learning Assumption.

For all $q \in K_n \setminus K_0$ and $t \in \{1, \ldots, n-1\}$ we have

$$q \in K_t \setminus K_0 \text{if and only if there is a subset } M \in g(q)$$
$$\text{such that } (M \subseteq \{L_0, \ldots, L_t\} \text{ and } M \cap \{L_{t+1}, \ldots, L_n\} = \emptyset).$$

This condition is called *Weak Learning Assumption* (WLA). As in case of SLA, WLA defines a unique consistent trail (see Prop. 12.4.8 in Section **??**). Still, learning occurs whenever the relevant contents has been exposed, only deferred to the last occurrence of content visited multiple times.

From the above listed trails only (12.12) and (12.15) satisfy WLA. Notice that for (12.12) both SLA and WLA hold, since there are no multiple visits, while for (12.15) SLA does not hold (compare to (12.14)). Moreover, trail (12.13) satisfies neither SLA, nor WLA. In principle, we can formulate an even more relaxed assumption where learning occurs at the latest point in time for which a consistent trail results. Stating this hypothesis as a general rule, however, would mean adopting an overly pessimistic point of view.

**12.4.5 How to apply the model.** In the sequel we briefly discuss some aspects of the application of the model, covering parameter estimation, model validation, and assessing the effectiveness of trails of LOs.

Estimation of the initial probability distribution $\mathbb{P}(L_0, K_0)$ is straightforward. The LO $L_0$ can be read off from the log data, and the knowledge state $K_0$ is obtained through the pre-assessment. These parameters may thus be estimated by the respective observed relative frequencies. Due to the complexity of the model it will not be possible to obtain analytic solutions for the estimates of the conditional probabilities. Thus, in the general case, we will have to resort to numerical optimization in a parameter space that tends to be very large in practical applications. Parameter estimation, however, is greatly simplified whenever the underlying assumptions warrant a unique identification of the consistent trail $(L_0, K_0), \ldots, (L_n, K_n)$ given the trail $L_0, \ldots, L_n$ of visited LOs from $\mathcal{L}$, a pre-assessment $K_0$, and a post-assessment $K_n$. In this case, which holds under SLA and WLA, the conditional probabilities can be estimated via log data analysis by the corresponding relative frequencies.

Within the discussed scenario, for each learner we observe the trail $L_0, \ldots, L_n$ of visited LOs from $\mathcal{L}$, and we have available a pre-assessment $K_0$ and a post-assessment $K_n$. In order to test the Markov chain model, Heller et al. (2007) derive a prediction of the marginal distribution of the post-assessment given the observed trails of visited LOs and the pre-assessment. This prediction may be contrasted with the marginal distribution that is estimated directly from the log data by the observed relative frequencies within a cross-validation design. Methods based on the Kullback–Leibler divergence (sometimes also called relative entropy) may be used to evaluate the resulting discrepancy (Kullback and Leibler, 1951; Kullback, 1959).

The proposed model allows for judging the effectiveness of certain trails of LOs. This information may provide guidelines for an optimization of the learning environment, e.g. by reshaping the link structure in the learning environment by eliminating links that belong to ineffective trails. In general, the effectiveness of a trail of LOs has to evaluate the actual performance of the learners relative to what in principle can be learned from the particular trail. The notion of the scope of a trail of LOs was introduced above to capture the latter. The scope $s(L_0, \ldots, L_n) \in \mathcal{K}$ of the trail $L_0, \ldots, L_n$ is the set of items $q \in \mathcal{Q}$ for which there exists $M \in g(q)$ such that $M \subseteq \{L_0, \ldots, L_n\}$. As learners following the trail of LOs do not necessarily learn to solve all the items within its scope, we may contrast it with the actual solution behavior represented by $K_n \setminus K_0$. A whole variety of numerical indices may be devised to capture the resulting discrepancy. As a first example we propose to consider the probability $\mathbb{P}(s(L_0, \ldots, L_n) \mid L_0, \ldots, L_n)$ of solving all items within the scope given a certain trail, which means $K_n \setminus K_0 = s(L_0, \ldots, L_n)$. A trail of LOs clearly is effective, if this probability is close to 1, and ineffective if it is close to 0. Other aspects of the observed trail, like its length $n$, may also be taken into account. Given the same effectiveness it seems natural to consider a shorter trail superior to a longer one.

**12.4.6 Conclusions.** The presented Markov chain model interlinks the observable trails of learning objects (LOs) with the associated latent trails of cognitive states in an underlying knowledge structure. Its main contribution is the combination of Knowledge Space Theory with web data mining techniques to produce a new model of the relationship between the cognitive processes of users of online environments and their observable navigation behavior. Application of the model is exemplified by Heller et al. (2007) in two case studies. The model is a definite improvement of the current state of both data mining (which does not consider cognitive states), and learner assessment (which does not usually consider learner navigation). However, there is still potential to further improve predictive power of the model, e.g. by taking into account the effect of the time spent on a particular LO (visiting time). Within an extended framework the dependence of the probability of a transition between knowledge states on the current LO's visiting time may be modeled by a learning curve, the possible forms of which are well understood in psychology.

PROOFS

**12.4.7 Proposition.** *Given an observable trail $L_0, \ldots, L_n$ of LOs from $\mathcal{L}$, a prerequisite map $l$ on $Q$, and $K_0, K_n \subseteq Q$.*

*Let $K_0, \ldots, K_n$ be a trail of knowledge states in the domain $Q$, which is defined in the following way. For all $q \in K_n \setminus K_0$ and $t \in \{1, \ldots, n-1\}$ we have $q \in K_t \setminus K_0$ if and only if there exists $M \in g(q)$ such that $M \subseteq \{L_0, \ldots, L_t\}$.*

*Then the trail $K_0, \ldots, K_n$ satisfies the compatibility conditions C1-C3.*

PROOF. C1 is immediate from the definition.

For proving C2 let $q \in K_t \setminus K_{t-1}$ for some $t \in \{1, \ldots, n\}$, which means that $q \in K_t$ and $q \notin K_{t-1}$. So there exists $M \in g(q)$ such that $M \subseteq \{L_0, \ldots, L_t\}$, and $M \not\subseteq \{L_0, \ldots, L_{t-1}\}$, because the latter holds for all subsets in $g(q)$. This, however, implies that $L_t \in M$, which proves C2.

In order to prove C3 assume that $q \in K_{t-1} \setminus K_0$ for some $t \in \{1, \ldots, n\}$. This means that there exists $M \in g(q)$ such that $M \subseteq \{L_0, \ldots, L_{t-1}\}$. Obviously, any such $M$ also satisfies $M \subseteq \{L_0, \ldots, L_{t-1}, L_t\}$, and thus $q \in K_t \setminus K_0$. □

**12.4.8 Proposition.** *Given an observable trail $L_0, \ldots, L_n$ of LOs from $\mathfrak{L}$, a prerequisite map $l$ on $Q$, and $K_0, K_n \subseteq Q$.*

*Let $K_0, \ldots, K_n$ be a trail of knowledge states in the domain $Q$, which is defined in the following way. For all $q \in K_n \setminus K_0$ and $t \in \{1, \ldots, n-1\}$ we have $q \in K_t \setminus K_0$ if and only if there exists $M \in g(q)$ such that $M \subseteq \{L_0, \ldots, L_t\}$ and $M \cap \{L_{t+1}, \ldots, L_n\} = \emptyset$.*

*Then $K_0, \ldots, K_n$ satisfies the compatibility conditions C1-C3.*

PROOF. C1 is immediate from the definition.

For proving C2 let $q \in K_t \setminus K_{t-1}$ for some $t \in \{1, \ldots, n\}$, which means that $q \in K_t$ and $q \notin K_{t-1}$. So, on the one hand there exists $M \in g(q)$ such that $M \subseteq \{L_0, \ldots, L_t\}$ and $M \cap \{L_{t+1}, \ldots, L_n\} = \emptyset$. On the other hand, for all $M \in g(q)$ we have $M \not\subseteq \{L_0, \ldots, L_{t-1}\}$, or $M \cap \{L_t, \ldots, L_n\} \neq \emptyset$. We have to treat two cases. First, let $M \in g(q)$ be such that $M \subseteq \{L_0, \ldots, L_t\}$, $M \cap \{L_{t+1}, \ldots, L_n\} = \emptyset$, and $M \not\subseteq \{L_0, \ldots, L_{t-1}\}$. As in Proposition 12.4.7 $M \subseteq \{L_0, \ldots, L_t\}$ and $M \not\subseteq \{L_0, \ldots, L_{t-1}\}$ implies $L_t \in M$, which together with $M \cap \{L_{t+1}, \ldots, L_n\} = \emptyset$ provides $L_t \notin \{L_{t+1}, \ldots, L_n\}$. Second, let $M \in g(q)$ be such that $M \subseteq \{L_0, \ldots, L_t\}$, $M \cap \{L_{t+1}, \ldots, L_n\} = \emptyset$, and $M \cap \{L_t, \ldots, L_n\} \neq \emptyset$. The last two assumptions then imply $L_t \in M$ and $L_t \notin \{L_{t+1}, \ldots, L_n\}$, which concludes the proof of C2.

In order to prove C3 assume that $q \in K_{t-1} \setminus K_0$ for some $t \in \{1, \ldots, n\}$. This means that there exists $M \in g(q)$ such that $M \subseteq \{L_0, \ldots, L_{t-1}\}$ and $M \cap \{L_t, \ldots, L_n\} = \emptyset$. Obviously, any such $M$ also satisfies $M \subseteq \{L_0, \ldots, L_{t-1}, L_t\}$ and $M \cap \{L_{t+1}, \ldots, L_n\} = \emptyset$. Thus, we have $q \in K_t \setminus K_0$. □

## 12.5 Micro-Adaptivity: Non-invasive Skill Assessment in Educational Games

Research on adaptive and intelligent tutoring basically focused on adaptive presentation and adaptive navigation support de Bra (2008). In classical approaches to adaptivity and personalization, there is no need to adapt to the learners' behavior during their consumption of the current learning object. This changes in game and simulation based learning environments where the learning objects are rather few but have a long consumption time. Adaptivity should occur here *within* and not only *between* learning objects. This distinction has led to coining the term of *micro-adaptivity* to be differentiated from the classical *macro-adaptivity*.

The data basis for adaptation is most often querying the learner, asking for preferences, or providing typical test items for assessing the user's knowledge and learning progress. This strategy is not feasible in an immersive learning game, however. In contrast to conventional adaptive tutoring and knowledge testing, the adaptive knowledge assessment within such games is massively restricted by the gameplay, the game's narrative, and the game's progress. Typical methods of knowledge assessment would suddenly and seriously destroy immersion and, consequently, also the gaming and learning process. What is required is an assessment procedure that is strictly embedded in the game's story and covered by game play activities. The core idea is to avoid any queries or interruptions but to monitor and interpret the learner's behavior in gaming situations. Subsequently, psycho-pedagogical interventions (e.g., providing the learner with appropriate guidance, feedback, cheer, or hints) can be triggered on the basis of the system's conclusions.

The present approach has been developed in two projects focusing on game-based learning, ELEKTRA and 80Days (see Section 12.6 below) and builds on earlier research of Stefanutti and Albert (2003) and Stefanutti and Koppen (2003). Its main idea is to connect the concept of *problem spaces* developed by Simon (1978) to CbKST. This is done by identifying competencies for observable, problem state changing actions.

The following Sections 12.5.1 to 12.5.3 formalize this general idea, while Section 12.5.5 provides a series of simulations aiming at the precision and efficiency of the postulated assessment procedure.

**12.5.1 Basic definitions.** Imagine a certain gaming situation with an educational intention. Let us, for instance, imagine a slope device, balls of different materials, a magnet and a fan. The learner's task might be to adjust the magnet and the fan in such a way that the balls fall in a given bin (cf. Figure 12.6, right blue hole). In each situation, the learner can perform an action (e.g., selecting a ball, turning on the fan, or increasing the magnetic force). Each action, in turn, is interpreted regarding its correctness or appropriateness for accomplishing the task (e.g., altering the trajectories of the balls).

To describe the learner's solution behavior in a formal way, let $\mathcal{S}$ be the set of all possible gaming situations. Then a *problem situation* or *task* is defined as a tuple $(i, \mathcal{T}) \in \mathcal{S} \times 2^{\mathcal{S}}$, where $i \in \mathcal{S}$ is the initial state a user is confronted with and $\mathcal{T} \subset \mathcal{S}$ is the set of solution states. If one of the solution states in $\mathcal{T}$ is reached by the learner, then the task is completed successfully. Finally, the set of different problem situations (tasks) is denoted by $\mathcal{P}$.

To describe the learner's actions in a formal way, let $\mathcal{A}$ be a nonempty set of actions with $q \in \mathcal{A}$. The element $q$ refers to that action that quits the current task. Furthermore, let $R \subset \mathcal{S} \times \mathcal{A}$ denote a "compatibility relation" in the following sense: $(s, a) \in R$ if and only if action $a$ is performable in the gaming situation $s$. To describe the effect of an action $a \in \mathcal{A}$, let $f : R \to \mathcal{S}$ be a "transition function" with the following interpretation in mind: If a user performs action $a$ in the gaming situation $s$, then the gaming situation $f(s,a)$ results.

In the following, let us consider a problem situation $(i, \mathcal{T}) \in \mathcal{P}$, and a natural number $n \in \mathbb{N}$. Then a finite sequence of the form

$$X_n(i, \mathcal{T}) = \langle (s_1, a_1), (s_2, a_2), \ldots, (s_n, a_n) \rangle$$

is called a *solution process* for problem $(i, \mathcal{T})$ if the following conditions are satisfied: (1) $s_1 = i$; (2) $(s_t, a_t) \in R$ for all $t = 1, \ldots, n$; (3) $f(s_t, a_t) = s_{t+1}$ for all $t = 1, \ldots, n-1$; (4) $s_t \notin \mathcal{T}$, for all $1 \leq t \leq n$; (5) $a_t \neq q$, for all $1 \leq t \leq n-1$.

The index $n$ is referred to as the *length* of the solution process. If $f(s_n, a_n) \in \mathcal{T}$, then the task is completed successfully. If, on the other hand, $a_n = q$, then the user quits the task before completion.

**12.5.2 The updating rule.** In order to interpret the learner's behavior in terms of his or her knowledge, it is necessary to link the observed actions to the underlying skills (see Chapter 11). To this end, let us assume a set $E$ of *skills* (in the following also referred to as *elementary competencies*), which are required to proceed successfully through the game, and a *prerequisite relation* on $E$, which reflects the dependencies between the skills in $E$. Note that this is in accordance with the assumptions of Knowledge Space Theory (e.g., Doignon & Falmagne, 1985, 1999; Falmagne, Koppen, Villano, Doignon, & Johannesen, 1990), and Competence-based Knowledge Space Theory (e.g., Albert & Lukas, 1999; Doignon, 1994; Düntsch & Gediga, 1995; Heller, Steiner, Hockemeyer, & Albert, 2006; Korossy, 1997, 1999). Furthermore, by using Birkhoff's Theorem (Doignon & Falmagne, 1999, Theorem, 1.49), we can construct a uniquely determined competence structure $(E, \mathcal{C})$, with $\emptyset \in \mathcal{C}$ and $E \in \mathcal{C}$, and $\mathcal{C}$ a family of subsets of $E$ closed under union. Following the assumptions of Competence based Knowledge Space Theory, we assume that at a certain point in time, any person is in a specific, yet not directly observable, competence state $C \in \mathcal{C}$.

To specify the learner's competence state, we assume that for any problem solution process $X_n(i, \mathcal{T})$, there exists a conditional probability distribution

$$L(\bullet | X_n(i, \mathcal{T})) : \mathcal{C} \to [0, 1],$$

with the following interpretation in mind: $L(C | X_n(i, \mathcal{T}))$ is the conditional probability that a person is in the competence state $C \in \mathcal{C}$ given that the solution process $X_n(i, \mathcal{T})$ has been observed. Furthermore, let

$$L(\bullet | X_0(i, \mathcal{T})) \equiv L(\bullet | (i, \mathcal{T})) : \mathcal{C} \to [0, 1]$$

denote the initial distribution at the beginning of the task. At the beginning of the game (that is, prior to the first problem situation), the initial distribution might be estimated from an entry test or other relevant information, like school type, age etc.; or it might be (arbitrarily) chosen as a uniform distribution:

$$L(C|(i,\mathcal{T})) = \frac{1}{|\mathcal{C}|}, \quad \forall C \in \mathcal{C}.$$

Alternatively, let us assume that the user has already solved some of the problem situations in $\mathcal{P}$. If the final problem solution process is denoted as $X_n(i,\mathcal{T})$, then the conditional probability distribution $L(\bullet|X_n(i,\mathcal{T}))$ might be used as the initial distribution for the upcoming problem situation.

In the following, let us consider a problem situation $(i,\mathcal{T})$ and a problem solution process $X_{n+1}(i,\mathcal{T})$:

$$X_{n+1}(i,\mathcal{T}) = \langle (s_1,a_1),(s_2,a_2),\ldots,(s_n,a_n),(s_{n+1},a_{n+1})\rangle.$$

If also the conditional probability distribution $L(\bullet|X_n(i,\mathcal{T}))$ is given, then we are confronted with the problem of computing the probability distribution $L(\bullet|X_{n+1}(i,\mathcal{T}))$ from the given probability distribution $L(\bullet|X_n(i,\mathcal{T}))$ and the fact that the learner performed action $a_{n+1}$ in situation $s_{n+1}$. To this end, the multiplicative updating rule by Falmagne and Doignon (1988) is adapted to our needs:

1. If action $a_{n+1}$ in gaming situation $s_{n+1}$ provides evidence in favor of the elementary competency $c$, then increase the probability of all competence states containing $c$, and decrease the probability of all competence states not containing $c$.
2. If action $a_{n+1}$ in gaming situation $s_{n+1}$ provides evidence against $c$, then decrease the probability of all competence states containing $c$, and increase the probability of all competence states not containing $c$.

To formalize this general idea, let $E(i,\mathcal{T}) \subset E$ denote those elementary competencies in $E$ which are necessary to master the problem situation $(i,\mathcal{T})$. Furthermore, let us assume two "skill assignment" functions $\sigma^{(i,\mathcal{T})}: R \to 2^{E(i,\mathcal{T})}$ and $\rho^{(i,\mathcal{T})}: R \to 2^{E(i,\mathcal{T})}$ with the following interpretations: If action $a$ is performed in the gaming situation $s$, then we can surmise that the user has all the elementary competencies in $\sigma^{(i,\mathcal{T})}(s,a)$ which is equal to $\sigma(s,a)$, but none of the competencies in $\rho^{(i,\mathcal{T})}(s,a) = \rho(s,a)$.

Furthermore, to compute $L(\bullet|X_{n+1}(i,\mathcal{T}))$ from the given probability distribution $L(\bullet|X_n(i,\mathcal{T}))$, let us fix two parameters $\zeta_0$ and $\zeta_1$ with $\zeta_0 > 1$ and $\zeta_1 > 1$. Then, for a competence state $C \in \mathcal{C}$, let

$$L(C|X_{n+1}(i,\mathcal{T})) = \frac{\zeta^{(i,\mathcal{T})}(C)L(C|X_n(i,\mathcal{T}))}{\sum_{C' \in \mathcal{C}} \zeta^{(i,\mathcal{T})}(C')L(C'|X_n(i,\mathcal{T}))}, \tag{12.16}$$

with the parameter function $\zeta^{(i,\mathcal{T})}(C) \equiv \zeta(C)$ defined as

$$\zeta^{(i,\mathcal{T})}(C) = \prod_{c \in C \cap \sigma(s_{n+1},a_{n+1})} \zeta_0 \prod_{c \in (E(i,\mathcal{T}) \setminus C) \cap \rho(s_{n+1},a_{n+1})} \zeta_1. \tag{12.17}$$

Note that, by definition, the parameter $\zeta^{(i,\mathcal{T})}(C)$ is set to 1, if $C \cap \sigma(s_{n+1},a_{n+1}) = \emptyset$ and $(E(i,\mathcal{T}) \setminus C) \cap \rho(s_{n+1},a_{n+1}) = \emptyset$.

**12.5.3 Partitioning competence structures.** In real applications we are confronted with the problem that, in general, the number of competence states is huge, and that, consequently, the probability updates (in the sense of Formula (12.16)) cannot be realized at run time. If, as an example, we have a set $E$ of 100 elementary competencies, then the resulting competence structure $\mathcal{C}$ can have up to

$$|2^E| = 2^{100} = 1.267651 \times 10^{30}$$

different competence states, which is quite large.

Therefore, in order to simplify the probabilistic updating process, it is necessary to restrict the underlying competence structure $\mathcal{C}$ to a smaller set of elementary competencies. To this end, let $E(i, \mathcal{T})$ denote those elementary competencies in $E$ which are related to the problem situation $(i, \mathcal{T})$ (cf. Section 12.5.2). Furthermore, let $\mathcal{C}(i, \mathcal{T})$ denote the $(i, \mathcal{T})$-*restricted competence structure*:

$$\mathcal{C}(i, \mathcal{T}) = \{C \cap E(i, \mathcal{T}) : C \in \mathcal{C}\}.$$

For restricting the conditional probability distribution $L(\bullet|X_n(i, \mathcal{T}))$ to the $(i, \mathcal{T})$-restricted competence structure $\mathcal{C}(i, \mathcal{T})$, let

$$[C] = \{C' \in \mathcal{C} : C' \cap E(i, \mathcal{T}) = C \cap E(i, \mathcal{T})\}, \quad C \in \mathcal{C}.$$

Augustin et al. (2011) show that the following definition provides a probability distribution on $\mathcal{C}(i, \mathcal{T})$:

$$L^{(r)}(C \cap E(i, \mathcal{T})|X_n(i, \mathcal{T})) = \sum_{C' \in [C]} L(C'|X_n(i, \mathcal{T})), \quad C \in \mathcal{C}. \tag{12.18}$$

Furthermore, by applying Formula (12.16) to the restricted probability distribution $L^{(r)}(\bullet|X_n(i, \mathcal{T}))$, we obtain that for all $C \in \mathcal{C}(i, \mathcal{T})$

$$L^{(r)}(C|X_{n+1}(i, \mathcal{T})) = \frac{\zeta^{(i,\mathcal{T})}(C) L^{(r)}(C|X_n(i, \mathcal{T}))}{\sum_{C' \in \mathcal{C}(i,\mathcal{T})} \zeta^{(i,\mathcal{T})}(C') L^{(r)}(C'|X_n(i, \mathcal{T}))}, \tag{12.19}$$

with the parameter $\zeta^{(i,\mathcal{T})}(C) \equiv \zeta(C)$ defined according to Formula (12.17). In real applications, with millions of different competence states, it is of vital importance that an update of the (unrestricted) probability distribution $L$ can be accomplished by updating the restricted distribution $L^{(r)}$. The following theorem specifies the relationship between the (unrestricted) distribution $L$ and its restricted counterpart $L^{(r)}$:

**12.5.4 Theorem.** *For every competence state $C \in \mathcal{C}$, and every natural number $n \in \mathbb{N}_0$, we have*

$$L(C|X_n(i, \mathcal{T})) = \frac{L^{(r)}(C \cap E(i, \mathcal{T})|X_n(i, \mathcal{T}))}{L^{(r)}(C \cap E(i, \mathcal{T})|(i, \mathcal{T}))} L(C|(i, \mathcal{T})). \tag{12.20}$$

For a mathematical proof of Theorem 12.5.4, see Augustin et al. (2011). It is important to note that the initial distributions $L(\bullet|(i,\mathcal{T}))$ and $L^{(r)}(\bullet|(i,\mathcal{T}))$ are given by definition (cf. Section 12.5.2).

Furthermore, we have to note that, in general, the restricted competence structure $\mathcal{C}(i,\mathcal{T})$ is much smaller than the original structure $\mathcal{C}$. Note that, by definition, the corresponding skill set $E(i,\mathcal{T})$ contains only those elementary competencies in $E$ which are related to the problem situation $(i,\mathcal{T})$. If, for instance, $|E(i,\mathcal{T})| = 15$, then the maximum number of competence states is relatively small (as compared to the overall competence structure $(E,\mathcal{C})$ with $|E| = 100$):

$$|\mathcal{C}(i,\mathcal{T})| \leq |2^{E(i,\mathcal{T})}| = 2^{15} = 32768.$$

Therefore, in order to reduce the computational load of the updating process, the following strategy is advisable:

1. Within a given problem situation $(i,\mathcal{T})$, the *restricted* probability distributions on $\mathcal{C}(i,\mathcal{T})$ are updated according to Formula (12.19).
2. The *unrestricted* probability distribution on $\mathcal{C}$ is updated only after task completion. This is done according to Theorem 12.5.4.

**12.5.5 Simulation results.** In order to study the precision and efficiency of the multiplicative updating rule, Augustin et al. (2011) computed a series of 18000 different simulations. The simulations differed in the size of the underlying competence structure, the probabilities for a lucky guess (i.e., a correct action by a learner who does not have the underlying skills) and a careless error (i.e., an incorrect action by a learner who has the underlying skills), and the selection of positive and negative evidences: We considered three different competence structures $(E,\mathcal{C})$ of the following types: A small one $\mathcal{C}_1$ with 121943 competence states, a medium one $\mathcal{C}_2$ with 747283 states, and a large one $\mathcal{C}_3$ with 1310950 states. Every competence structure contained a total number of 77 different skills: $|E| = 77$. Furthermore, we set the probability of a lucky guess or a careless error to 0.05, 0.10, and 0.15, respectively. Finally, we used two different mechanisms to select the set of positive and negative evidences: Either, every single update (computed according to Formula (12.16)) is based on exactly one evidence—positive or negative—which is selected by chance, or the probability updates are based on a variable number $N$ of evidences, with $1 \leq N \leq 4$. Formally, this means that for every $(s,a) \in R$, either $|\sigma(s,a) \cup \rho(s,a)| = 1$ (Selection Type 1), or $|\sigma(s,a) \cup \rho(s,a)|$ is selected randomly from $\{1,2,3,4\}$ (Selection Type 2). Since for every combination of competence structure (small, medium, and large), error probability (0.05, 0.10, and 0.15), and selection type (1 and 2), 1000 separate simulations were computed, we performed a total number of 18000 different simulations.

Each simulation proceeded as follows: For the initial distribution $L(\bullet|X_0(i,\mathcal{T}))$ we assumed a uniform distribution:

$$L(C|X_0(i,\mathcal{T})) = \frac{1}{|\mathcal{C}|}, \quad \forall C \in \mathcal{C}.$$

Furthermore, we independently selected two updating parameters $\zeta_0, \zeta_1 \in [\frac{3}{2}, 4]$, which were held constant throughout a simulation (cf. Formula (12.17)). To simulate the problem solution behavior of a user, a competence state $K$ was selected randomly from the given competence structure $\mathcal{C}$. Then a set $E_1 \subset E$ of evidences was selected by chance, and the updating process proceeded as follows: If $e \in E_1 \cap K$, then

$$e \in \begin{cases} \sigma(s_1, a_1), & \text{with probability } 1-p, \\ \rho(s_1, a_1), & \text{with probability } p, \end{cases} \tag{12.21}$$

where $p$ corresponds to the given error probability (ranging from 0.05 to 0.15). Similarly, if $e \in E_1 \cap (E \setminus K)$, then

$$e \in \begin{cases} \sigma(s_1, a_1), & \text{with probability } p, \\ \rho(s_1, a_1), & \text{with probability } 1-p. \end{cases} \tag{12.22}$$

Then, in order to update the initial distribution $L(\bullet|X_0(i,\mathcal{T}))$, we applied the multiplicative updating rule (12.16), resulting in the conditional probability distribution $L(\bullet|X_1(i,\mathcal{T}))$. To update $L(\bullet|X_1(i,\mathcal{T}))$, we proceeded in a similar way: First of all, a set $E_2 \subset E$ of evidences was selected by chance. Subsequently, $\sigma(s_2, a_2)$ and $\rho(s_2, a_2)$ were specified according to Formulas (12.21) and (12.22). Finally, we computed the update of the given probability distribution $L(\bullet|X_1(i,\mathcal{T}))$ according to Formula (12.16). The updating process continued until one of the conditional probabilities $L(C|X_n(i,\mathcal{T}))$ exceeded the critical value 0.8.

In the following let us assume that $L(\hat{K}|X_n(i,\mathcal{T})) > 0.8$. Then, in order to evaluate the precision and efficiency of the multiplicative updating rule, we considered the symmetric distance

$$d(K, \hat{K}) = |K \Delta \hat{K}| = |(K \setminus \hat{K}) \cup (\hat{K} \setminus K)|,$$

and the total number of probability updates. Table 12.1 summarizes the results separately for every combination of competence structure (small, medium, and large), error probability (0.05, 0.10, and 0.15), and selection type (1 and 2).

First of all, we note that there is a clear dependency of the selection type on the efficiency of the assessment procedure (cf. Columns 5 to 7): For Selection Type 1, we observed an average number of 744 updates, as compared to an average number of 269 updates for Selection Type 2. This is in accordance with the fact that in case of Selection Type 1, a single evidence was used per update, as compared to an average of 2.5 evidences per update for Selection Type 2. Similarly, the number of probability updates slightly increases with the size of the underlying competence structure, and the value of the error probability.

Secondly, the averaged symmetric distances in Column 2 (ranging from 0 to 0.07) and the maximum distances in Column 4 (with a maximum value of 5) indicate an extraordinary high precision of the assessment procedure. Finally,

**Table 12.1.** Simulation results, separately for every combination of competence structure (small, medium, and large), error probability (0.05, 0.10, and 0.15), and selection type (1 and 2) (cf. Column 1). Columns 2 to 4 ("Precision") refer to the symmetric distance $|K \Delta \hat{K}|$, and Columns 5 to 7 ("Efficiency") to the total number of probability updates, respectively. The table entries are the corresponding mean values ($\mu$), standard deviations ($\sigma$), and maxima (max). Each table entry is based on a total number of 1000 separate simulations.

| | Precision | | | Efficiency | | |
|---|---|---|---|---|---|---|
| Simulation type | $\mu$ | $\sigma$ | max | $\mu$ | $\sigma$ | max |
| (s, 0.05, 1) | 0.003 | 0.055 | 1 | 569 | 191.700 | 1746 |
| (s, 0.05, 2) | 0 | 0 | 0 | 206 | 74.564 | 547 |
| (s, 0.10, 1) | 0.022 | 0.153 | 2 | 705 | 279.863 | 2553 |
| (s, 0.10, 2) | 0.020 | 0.153 | 2 | 245 | 104.607 | 885 |
| (s, 0.15, 1) | 0.060 | 0.269 | 3 | 868 | 425.933 | 4989 |
| (s, 0.15, 2) | 0.055 | 0.260 | 4 | 309 | 180.163 | 2534 |
| (m, 0.05, 1) | 0 | 0 | 0 | 609 | 202.064 | 1983 |
| (m, 0.05, 2) | 0.003 | 0.0547 | 1 | 213 | 75.042 | 617 |
| (m, 0.10, 1) | 0.031 | 0.179 | 2 | 745 | 293.049 | 2620 |
| (m, 0.10, 2) | 0.023 | 0.162 | 2 | 255 | 96.988 | 749 |
| (m, 0.15, 1) | 0.058 | 0.246 | 2 | 908 | 428.637 | 4082 |
| (m, 0.15, 2) | 0.066 | 0.289 | 3 | 319 | 165.838 | 1700 |
| (l, 0.05, 1) | 0.003 | 0.054 | 1 | 614 | 216.273 | 2298 |
| (l, 0.05, 2) | 0.007 | 0.083 | 1 | 230 | 81.983 | 619 |
| (l, 0.10, 1) | 0.019 | 0.143 | 2 | 749 | 316.346 | 2305 |
| (l, 0.10, 2) | 0.014 | 0.117 | 1 | 282 | 124.680 | 1131 |
| (l, 0.15, 1) | 0.070 | 0.314 | 5 | 933 | 489.031 | 5007 |
| (l, 0.15, 2) | 0.062 | 0.268 | 3 | 358 | 189.011 | 1350 |

it is quite remarkable that neither the size of the underlying competence structure, nor the error probability or the selection type have a considerable influence on the precision of the assessment.

To sum up, the results in Table 12.1 provide strong evidence that a reasonable number of probability updates is sufficient to diagnose the user's competence state with a high degree of accuracy. Since in state-of-the-art adventure games, the number of observable actions to solve a given task is usually as high as several hundreds, we are quite optimistic that the proposed assessment procedure could be successfully applied to most of today's digital educational games (cf. also Augustin et al., 2011).

**12.5.6 Conclusions.** The micro-adaptive model combines the based knowledge space theory CbKST with the concept of problem spaces. The main idea is by identifying competencies for observable, problem state changing actions. These actions are interpreted in terms of existing or missing skills and competencies. The remarkable progress is in modeling the solution process itself and the behavioral actions instead of interpreting only its final result, like as in the macro-adaptive approach. Assessing the existing and missing skills and competencies is the basis for adaptive psycho-pedagogical interventions in terms of specific feedback to actions. The application of the micro-adaptive model in problem solving and GBL settings requires not only specific but also immediate feedback, because the flow of thinking and gaming should not be interrupted. As a consequence, the assessment of a competence state has to be very fast as well as precise. In face of the huge competence structures, fast and precise assessment is highly challenging. That is the main reason for focusing on strategies for reducing computational load. The proposed algorithms have been tested in realistic simulations. The results provide strong evidence that the procedure is sufficient to diagnose the competence state of the learner/problem solver/gamer in a fast and accurate way. On the other hand, for real applications in e.g. highly interactive educational games further improvement of the assessment procedure maybe helpful. One possibility seems to be to take into account the expected effect of specific feedback and intervention presented by the system. The feedback of the system is twofold. (a) As a consequence of the learner's/gamer's behavior the problem situation is changing and signals success or failure. (b) Based on the assessment result and the psycho-pedagogical model, specific interventions are presented to the person. Taking both of these types of specific feedback into account, the resulting effects on competence states and behavioral actions maybe predicted and used for even more effective and efficient assessment.

## 12.6 Applications in Learning and Competence Management

The main focus of CbKST applications has been on the technology enhanced learning (TEL) from the very beginning on. In this section, we describe several such applications on the basis of the research projects in which they have been developed. This is done clustering the specific applications into three groups: (i) general personalized TEL applications, (ii) non-invasive TEL applications for game- and simulation-based learning, and (iii) applications in the context of competence management and workplace learning. The increasingly seen importance of competence models for TEL is also reflected by various recent standardization movements, e.g. by the IEEE Learning Technology Standards Committee (IEEE LTSC) or by the International Organization for Standardization (ISO). Besides the applications around learning and knowledge, the applicability of CbKST has been investigated also for quite different fields, e.g.

developmental psychology (see Albert et al., 2008) or philosophy for children (see Pilgerstorfer et al., 2006).

### 12.6.1 Personalized learning.

APeLS: Reusable adaptive learning objects

The APeLS system (Advanced personalized Learning System; see Hockemeyer et al., 2003) is a prototypical system developed within the EASEL[8] project. The core objective of EASEL was to develop mechanisms for the reusability of adaptive learning objects and learning services. In APeLS, the distinction between performances and competencies (or skills) was transferred to an explicit distinction between learning objects and competencies where the competencies were further divided into competencies required for understanding a learning object and competencies taught by the respective learning object. If such assignments of taught and required competencies are given for all learning objects within some course, a competence space and a learning object structure can be derived even if the competence assignments are incomplete, i.e. if they specify not all required competencies but only the more complex ones (Hockemeyer et al., 2003; Hockemeyer, 2003).

For developing APeLS, an existing course on Newtonian mechanics was taken and analyzed with respect to the competencies required and taught, respectively, by each of the existing learning objects. The APeLS system itself starts with the assumption of a novice in the field. The learner gets offered a rather small list of available learning objects. After consuming some of these objects, the system assumes the acquisition of the respective competencies[9]. At any time, the learner can actively refresh the list of available learning objects which will be increased according to the competencies which have been additionally acquired in the meantime. Figure 12.4 shows two screenshots of learners with different levels of expertise. In the top of the figure you see a rather novice learner's view of some learning object; the bottom of the figure shows the screen for a rather expert learner re-entering the system.

[8]Project funded by the European Commission within the FP5/IST Programme.

[9]This simplifying assumption is a tribute to the prototypical character of the system.

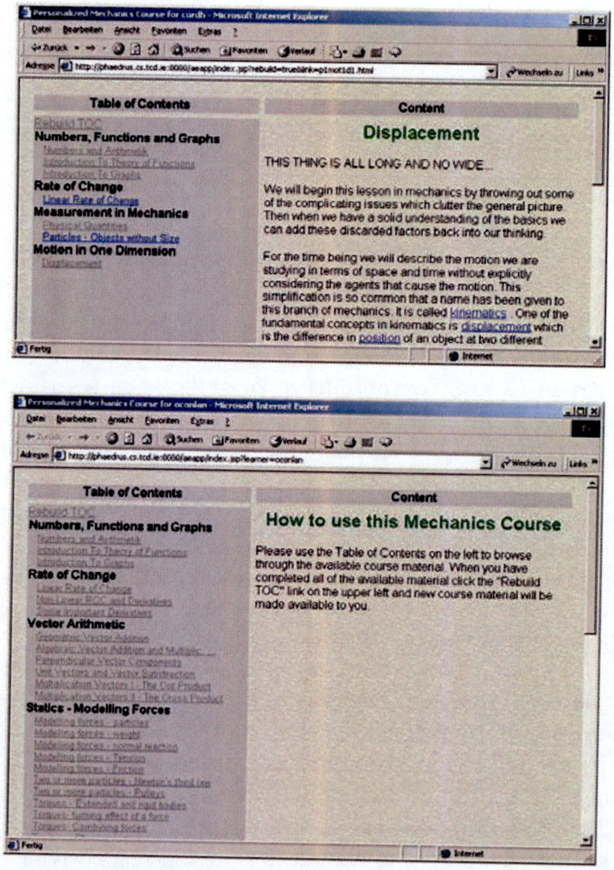


**Figure 12.4.** Two APeLS screenshots for learners at different competence levels.

### iClass: Self-regulated personalized learning

The iClass[10] project aimed at combining the pedagogical approach of self-regulated learning (i.e. the learner selects his/her learning path freely and autonomously) with the idea of personalization based on learners' competencies and learning material requirements. While these two concepts may appear contradictory at first sight, it became quickly obvious that they fit together very well merging into the concept of self-regulated personalized learning. Self-regulated learning requires informed learners in order to be successful, i.e. learners who can base their decision whether or not to consume a certain learning object on the meta-knowledge whether or not they have the prerequisites necessary for understanding this learning object.

[10]This project was funded by the European Commission within the FP6/IST programme.

In this context, a number of tools have been developed which allow authoring skill maps, assessing learners' competence states, and planning individual learning paths. Figure 12.5 shows a screenshot of the planning tool.

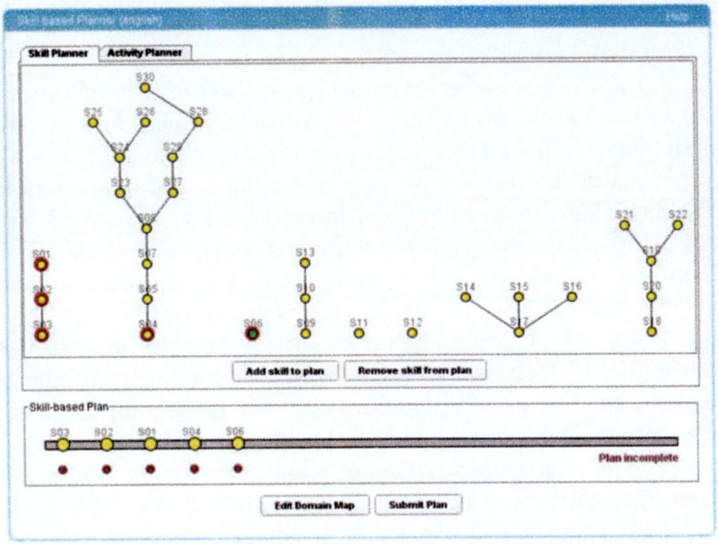


**Figure 12.5.** Screenshot of the iClass skill planning tool.

**12.6.2 Personalization in game- and simulation-based learning.** Games and simulations have been used to enhance learning success already for a long time. Connecting them with personalized TEL, however, brings some new challenges—and interesting results—which will be discussed in the subsequent sections.

### ELEKTRA: Non-invasive assessment in games through micro-adaptivity

Digital educational games (DEGs) are one of the hottest topics in current educational research and development (Kickmeier-Rust et al., 2007). The basic idea of digital educational games is to utilize the rich possibilities of state-of-the-art computer games for realizing entirely new and compelling learning experiences. The vision that at least a small portion of time spent on playing computer games can be used for learning is fascinating and desirable and, consequently, there is a rapidly increasing body of research and development in this field. The core strength of game-based learning is that such games—in a very natural way—are capable of making learning and knowledge appealing and important to the learner. Moreover, learning games serve the educational needs of the "Nintendo generation" and the "digital natives", who grew up with "twitch speed" computer games, MTV, action movies, and the Internet (Prensky, 2001). Authors like Mark Prensky argue that this context has emphasized certain cognitive aspects and de-emphasized others, thus, the demands on education have changed (Prensky, 2001).

One of the main advantages of digital educational games, namely their immersive motivational potential, is closely related to one of their main disadvantages, that is, their lack of adaptation to the learner in a psycho-pedagogical sense. Note that an appropriate adaptation to preferences and interests, motivational and emotional states, the observed learning progress, and above all, to the learner's abilities is of vital importance for being educationally effective and for retaining the user's motivation to play and to learn: If the player is bored by a too easy game play, or the challenges are too difficult to be accomplished, the player will quit playing the game very soon.

For obvious reasons, such an adaptation is not trivial: It requires a subtle interaction between the learner's behavior within the game and the challenges through the game. Unfortunately, such an interaction is very fragile and, due to the coexistence of learning- and gaming-aspects, more complex than personalization in conventional educational settings (Kickmeier-Rust et al., 2008a).

The concept of micro-adaptivity, i.e. adaptivity within a larger learning object or learning situation (see Section 12.5 above), was developed within the ELEKTRA[11] project. It is the response to the two challenges mentioned, i.e. (i) the fact that in learning games, we find a rather small set of complex learning objects (i.e. the gaming situations) such that adaptivity purely in the selection and order of these situations is insufficient and (ii) the specific need for a non-invasive assessment, i.e. assessing the learner's competence state should be done in a way not disturbing the flow of the game.

Using the mechanisms described in Section 12.5 above, all actions of the learner are interpreted with respect to the underlying competencies, and thus the learner's competence state is determined without explicitly asking questions. This competence assessment is then the basis for adaptive interventions which might, e.g., point the learner to missing or non-activated competencies.

To illustrate our approach, we refer to a concrete example from the ELEKTRA demonstrator game (cf. Fig. 12.6): The aim is to save Lisa and her uncle Leo, a researcher, who have been kidnapped by the evil Black Galileans. During this journey, the learner needs to acquire specific concepts from a 8th grade physics course. Learning occurs in different ways, ranging from hearing or reading to freely experimenting. One source of information is the ghost of Galileo Galilei shown in Fig. 12.6(a).

To learn about the straight propagation of light, for instance, the learner experiments with a torch and blinds on a table in the basement of uncle Leo's

[11] ELEKTRA was funded by the European Commission within the FP6/IST Programme (see http://www.elektra-project.org/).

(a) (b)

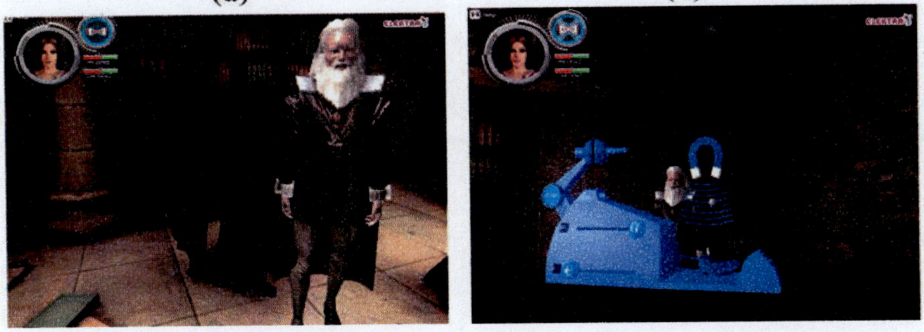

**Figure 12.6.** Screenshots from the ELEKTRA demonstrator game.

villa, or with a device that allows balls of different materials rolling down a slope as in Fig. 12.6(b). By performing these experiments, the user should understand that light propagates straight, as opposed to the curved trajectories of other objects. This, in turn, is important for the game play, because to continue in the game, the learner has to unlock a door by exactly hitting a small light sensor with a laser beam. The experimenting is accompanied and observed by Galileo who, if necessary, can also provide feedback or guidance. The goal of using the slope device is to make the various balls (of wood, plastic, or solid iron) fall in a given bin. As shown in Figure 12.6, the learner can adjust a magnet and a fan to alter the trajectories of the balls. On the contrary, a laser beam cannot be influenced by such external forces.

The important point is that by continuously interpreting the learner's actions in terms of his or her knowledge, the system gathers information about the learner's learning progress. For example, when the learner continuously tries to increase the magnetic force in order to affect the trajectory of a plastic ball, the system eventually concludes that the learner lacks the knowledge that plastic is not affected by magnetic force. Since, usually, a single observation cannot provide reliable evidence, we are relying on a probabilistic approach. This means that with each observation we are updating the probabilities that certain knowledge is available.

First empirical results show a positive effect of micro-adaptivity on the learning results (Kickmeier-Rust et al., 2008b). These investigations of applying CbKST for personalizing game-based learning have been continued in two other projects, namely 80Days and Target[12]. Their objectives include the development of faster assessment procedures (see also Section 12.5.3) as well as the connection of CbKST to pedagogical and to storytelling models.

[12]The 80Days and Target projects were funded by the European Commission within the FP6/IST and FP7/ICE Program, respectively (see http://www.eightydays.eu/ and http://www.reachyourtarget.org/)

MEDCAP: USING A HAPTIC SIMULATOR FOR COMPETENCE ASSESSMENT

The practical training of young doctors still follows the idea of master and disciple—a medical trainee is regarded as competent for performing certain tasks when his/her supervisor sees him/her fit. The MedCAP project[13] aims at changing this. Basis of the project was an existing haptic simulator for performing spinal anaesthesia. Figure 12.7 shows on the left side a schematic illustration of spinal anaesthesia and, on the right side, a person working with the haptic simulator device.

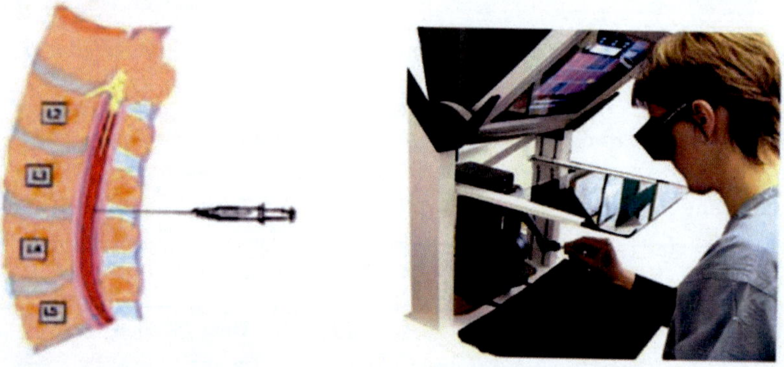

**Figure 12.7.** Schematic illustration of spinal anaesthesia and the haptic simulation device.

A challenge for MedCAP was to create a way for gathering and merging assessment information from various sources: In the training and assessment process, the young doctor is first confronted with theoretical questions through an elearning system. In a second step, s/he is directed to the simulator and asked to perform the spinal anaesthesia there. The simulator provides information on the trainee's performance back to the eLearning system. These first two steps are connected together by following the line of a case scenario. In a third, final step or phase, the trainee performs the operation on real patients. Here, s/he is supervised by an experienced doctor who can intervene whenever necessary and who is asked to answer a questionnaire on the trainee's performance and competencies afterwards. All these information are fed into the assessment procedure (Hockemeyer et al., 2009; Zhang et al., 2008).

Due to the variety of regarded competencies (e.g. medical knowledge, motor skills, and communication abilities), the competence spaces in MedCAP grew very large. As a consequence, the partitioning procedure suggested in Section 12.5.3 was applied with the different case scenarios serving as learning situation.

---

[13]MedCAP was funded by the European Commission within the LLP/LdV Programme.

MedCAP was only a first step into this new direction. Learning with haptic simulation devices has many similarities with game-based learning: we have also complex learning situations where the learner's (in this case physical) actions are interpreted with respect to competencies in order to draw conclusions on his/her competence state (Albert et al., 2007). In a micro-adaptive system, e.g., complications could be issued or omitted during the simulations based on the trainees performance in the first stages of the needle insertion.

**12.6.3 Competence management and business-related application.** Competence and knowledge management is an issue related to and—in a certain sense—going beyond pure learning (Hockemeyer et al., 2003). A major question from a psychological point of view is how to identify competencies in employees' daily work and to relate these to competencies previously acquired. Ley and Albert (2003; see also Ley et al. 2007), e.g., have analyzed documents produced by employees in the course of their work. They identified a priori competencies within the respective field of work and then asked experts to specify for the produced documents which competencies had been necessary for producing them. Based on this competence assignment to documents, prerequisite structures between the documents as well as between the competencies could be derived, and information on the employees' individual competencies could be gained.

A completely different, rather business-related area was approached by Stefanutti and Albert (2002). They applied CbKST for the analysis and assessment of processes in business environments. One aim of their analysis was structuring organizational actions where certain actions within a process require that other actions have been finished. Thus, they demonstrated, that CbKST can be applied to other areas structured by prerequisite relations.

## 12.7 General Conclusions and Discussion

Compared with the basic models in competence-based knowledge spaces (see Chapter 11 in this volume) remarkable progress has been made as described above. The theoretical progress was primarily stimulated by application requirements. For some of the theoretical results the applicability has been demonstrated already in section 12.6, for other results (Sections 12.2,12.3.2) the potential consequences are obvious. Furthermore, the scope of applications of the developed models might be even larger than expected. For instance, the stochastic model presented in section 12.4 that links navigation behavior to the learner's cognitive state can be applied in self-regulated learning settings as well as in informal learning environments (like workplaces and museums). This interpretation and application will become more and more important, because in the future not only formal but also informal learning has to be supported by adaptive systems and personalized certification.

These considerations refer to the current general tendency towards open systems and open environments. Traditional Intelligent Tutoring Systems have been monolithic adaptive systems with strict 'intelligent' guidance of the learner. The systems of the future are distributed, modular and allow strict guidance as well as total freedom in terms of navigation and interaction. The advantages of skill and competence orientation (see Chapter 11 this volume) are valid even for future developments on the basis of the theoretical results presented above. However, for real life applications in open learning environments with large numbers of learning objects, skills, competencies, learners, behavioral data, etc. even some more theoretical work, method improvements, applied research, and technical developments seem necessary.

Regarding future developments the general concepts of skills and competencies have to be clarified. For instance what are the advantages for separating and assigning skills instead of taking the basic elements of the knowledge and learning space suggested by Doignon (1994b). How can the huge amount of work for assigning skills to learning objects and learner's actions be reduced, e.g. by semantic technologies? Does it make sense to separate skills for declarative/conceptual knowledge from procedural/action knowledge? How to classify appropriately the skills (like taught, tested, ..) and how to model misconceptions (e.g. Lukas, 1997). What is the next step of the development after evaluating the task performance and interpreting the behavioral actions? Is the definition of the learner's state by a set of skills too trivial? Can the definition of states for learning and skills be improved by adding rich data like psychophysiological and eye-tracking data? Following the micro-adaptive approach, how can the system and the learning be modeled, if the learning environment changes even without any action of the learner and gamer, which is the case in many realistic applications? These and other topics call for further research, in order to create technology-enhanced, personalized, open teaching and learning systems of the future.

**Acknowledgements.** Part of the research reported here was financially supported through several projects from the Marie Curie and IST/ICT Programmes of the European Commission.

# 13

# Learning Sequences: An Efficient Data Structure for Learning Spaces

David Eppstein[1]

*Learning spaces* form the basis of a combinatorial theory of the possible states of knowledge of a human learner that has been successfully deployed in computerized assessment and learning systems such as ALEKS (Falmagne and Doignon, 2011). Until recently, however, both the computational efficiency of these systems and their ability to accurately assess the knowledge of their users have been hampered by a mismatch between theory and practice: they used a simplified version of learning space theory based on partially ordered sets and quasi-ordinal spaces, leading both to computational inefficiencies and to inaccurate assessments. In this chapter we present more recent developments in algorithms and data structures that have allowed learning systems to use the full theory of learning spaces. Our methods are based on *learning sequences*, sequences of steps through which a student, starting with no knowledge, could learn all the concepts in the space. We show how to define learning spaces by their learning sequences and how to use learning sequences to efficiently perform the steps of an assessment algorithm.

## 13.1 Overview

Rather than being based on numerical test scores and letter grades, systems such as ALEKS are based on a combinatorial description of the concepts known to the student. In this formulation, the current state of knowledge of a student is represented as the set of facts or concepts mastered by the student, drawn from a *learning space* that describes the sets that form valid knowledge states. The assessment task is to determine which concepts the student has mastered, by asking test questions and using the answers to narrow down the set of valid knowledge states that are consistent with the student's answers. Ideally, each test answer would be consistent with only half of the remaining knowledge states; if this could be achieved, the number of questions in an assessment would be the base-two logarithm of the number of valid knowledge states,

[1]Dept. of Computer Science, University of California, Irvine. Supported in part by NSF grants 0830403 and 1217322 and by the Office of Naval Research under grant N00014-08-1-1015.

J.-C. Falmagne et al. (eds.), *Knowledge Spaces: Applications in Education*,
DOI 10.1007/978-3-642-35329-1_13, 

a number that could be much smaller than the number of distinct concepts in the model. In this way, the learning space model allows the system to make inferences about whether the student has mastered concepts that have not been directly tested. In addition to speeding student interactions, this model also allows the system to determine sets of concepts which it judges the student ready to learn and to present to the student a selection of lessons based on those concepts, rather than forcing all students to proceed through the curriculum in a rigid linear ordering of lessons.

A student's answer to a test question may not be definitive in determining whether he or she understands the corresponding concepts. In order to take this uncertainty into account, these systems use a Bayesian methodology in which the system computes likelihoods of each valid knowledge state from the students' answers, aggregates these values to infer likelihoods that the student can answer each not-yet-asked question, and uses these likelihoods to select the most informative question to ask next. Once the student's knowledge is assessed, the system generates a *fringe* of concepts that it judges the student ready to learn, by determining which valid knowledge states differ from the assessed state by the addition of a single concept, and presents the students with a selection of online lessons based on that fringe.

Early versions of ALEKS represented its learning spaces via partially ordered sets describing prerequisite relations between concepts. However, this sort of representation cannot represent every learning space, but only a significantly restricted class of spaces known as *quasi-ordinal spaces* characterized by closure under both set unions and set intersections (cf. Falmagne and Doignon, 2011, Section 3.8). Closure under set unions is justified pedagogically, both at a local scale of learnability (learning one concept does not preclude the possibility of learning a different one) and more globally (if student $x$ knows one set of concepts, and student $y$ knows another, then it is reasonable to assume that there may be a third student who combines the knowledge of these two students). However, closure under set intersections seems much harder to justify pedagogically: it implies that every concept has a single set of prerequisites, all of which must be learned prior to the concept. On the contrary, in practice it may be that some concepts may be learned via multiple pathways that cannot be condensed into a single set of prerequisites. The restriction to quasi-ordinal spaces made it difficult to modify these spaces by adding or removing states, and also led to modeling inaccuracies in which the modeled learning space contained large numbers of extraneous states. These extra states, in turn, led to slower calculations within the knowledge assessment procedure, an unnecessarily large number of questions per student assessment, and inaccurate calculations of the fringe of the student's state.

In this chapter we outline algorithms for a more flexible representation, used in more recent versions of ALEKS, that allow a fully general definitions of learning spaces while preserving the scalability and efficient implementability of the earlier quasi-ordinal representation. We accomplish these goals by using a representation based of *learning sequences*: orderings of the learning space's concepts into linear sequences in which those concepts could be learned. A learning space may have an enormous number of possible learning sequences, but it is possible to correctly and accurately represent any learning space using only a subset of these sequences. We show how to generate efficiently all states of a learning space defined from a set of learning sequences, allowing for assessment procedures that are similar in their outlines and their efficiency to the ones already used for partial order based learning spaces. We will detail this learning sequence based representation, and the efficient algorithms based on it, later in this chapter, after first describing in more detail the previously used partial order based representation.

In Chapter 14, we will show how to find a concise set of learning spaces to represent any given learning space, and we will use the learning sequence representation as the basis for methods for adapting and decomposing learning spaces.

## 13.2 Learning Spaces and Quasi-Ordinal Spaces

Following Falmagne and Doignon (2011) and Chapter 8, we axiomatize the possible states of knowledge of a student learner as sets in a finite family of finite sets: each set may represent the concepts that a particular student has mastered. A set is said to be a *valid state of knowledge* if it belongs to the family. These set families may satisfy some or all of the following axioms:

**Accessibility.** A family $\mathcal{F}$ of sets is *accessible* if every nonempty set $S \in \mathcal{F}$ has an element $x \in S$ such that $S \setminus \{x\} \in \mathcal{F}$.

**Union Closure.** A family $\mathcal{F}$ of sets is *closed under unions* if, for every two sets $S$ and $T$ in $\mathcal{F}$, $S \cup T \in \mathcal{F}$.

**Intersection Closure.** A family $\mathcal{F}$ of sets is *closed under intersections* if, for every two sets $S$ and $T$ in $\mathcal{F}$, $S \cap T \in \mathcal{F}$.

In the context of learning space theory, accessibility is also known as downgradability, and is named Learning Smoothness in Section 1.2.4; it formalizes the notion that every state of knowledge can be reached by learning one concept at a time. For accessible families of sets, union closure is equivalent to the Learning Consistency property of Section 1.2.4, and formalizes the notion that the knowledge of two individuals may be pooled to form a state of knowledge that is also valid.

**13.2.1 Definition.** A *learning space* is a finite family of finite sets that is accessible and closed under unions. A *quasi-ordinal space* is a learning space that is closed under intersections.

Mathematically, a learning space is synonymous with an *antimatroid*[2], and a quasi-ordinal space is synonymous with a *lattice of sets*, or, equivalently, a

[2]See Theorem 8.1.3 on page 134 for a formal statement of the equivalence, and Korte et al. (1991) for a general reference on antimatroids.

*distributive lattice*. By Birkhoff's representation theorem (Birkhoff, 1937) every lattice of sets may be represented as the family of lower sets of a *partial order*, and as we describe in Section 13.4 this partial order representation of a quasi-ordinal space leads to efficient algorithms for knowledge assessment in these spaces. The intent of this chapter is to describe more general algorithms for knowledge assessment in arbitrary learning spaces, based on a representation of these spaces in terms of learning sequences.

## 13.3 From State Generation to Assessment

While a student is being assessed, he or she will have answered some of the assessment questions correctly and some incorrectly. We desire to infer from these results likelihoods that the student understands each of the concepts in the learning space, even those concepts not explicitly tested in the assessment.

The inference method of Falmagne and Doignon (1988a) and Falmagne et al. (1990) (cf. also Section 8.7) applies not only to learning spaces, but more generally to any system of sets, and can be interpreted using Bayesian probability methods. We begin with a prior probability distribution on the valid states of the learning space. In the most simple case, we can assume an uninformative uniform prior in which each state is equally likely, but the method allows us to incorporate prior probabilities based on any easily computable function of the state, such as the number of concepts in the set represented by the state. The prior probilities may also incorporate knowledge about the student's age or grades, or results from earlier assessments.

**13.3.1 A probabilistic model of student performance.** If a given learning space formed a completely accurate model of the knowledge it represents, and if students' performance on test questions were completely deterministic, then a student could be predicted to give a correct answer to a test question if and only if the concept tested by that question belonged to the student's state of knowledge. But because these models are not a perfect representation of the student, we instead use a probabilistic model of student performance, in which we define for each question $q$ and for each possible state $s$ a term $t_{q,s}$ measuring the probability that the student in state $s$ would give a correct answer to question $q$. If two states $s$ and $s'$ both contain the concept tested by question $q$ or both do not contain that concept, then their terms should be the same: $t_{q,s} = t_{q,s'}$. If the concept belongs to state $s$ then $t_{q,s}$ should be high, indicating that a student who has mastered this concept will be likely to answer the question correctly; however, the probability $1 - t_{q,s}$ of incorrect answers (careless mistakes) may be nonzero and in practice may be greater than 0.25. Similarly, if the concept does not belong to state $s$, then $t_{q,s}$ should be close to zero, indicating that a student who has not mastered the concept will most likely answer it incorrectly; however, again there may be a nonzero but small probability $t_{q,s}$ of a correct answer (a lucky guess).

As long as the test questions are designed so that lucky guesses are rare, the necessity for accurate assessment in the presence of careless mistakes will be of much greater significance to the design of the assessment procedure, but this also means that it is necessary to ascribe different rates to these two types of events. Additionally, we assume that the student's answers to different assessment questions are independent of each other, as systematic patterns of careless mistakes or lucky guesses would indicate a deficiency in the underlying learning space used to model the student's behavior. According to this assumption, for a student known to be in state $s$, the probability of seeing the given set of answers would be the product $\prod_q t_{q,s}$

**13.3.2 Posterior probabilities from test responses.** With these assumptions, we may use Bayes' rule to invert the conditional dependence from states to answers, and calculate posterior probabilities of each state, given the student's answers to the test questions. Let $p_s$ be a number associated to each state $s$, proportional to its prior probability; for the uninformative prior, we may take $p_s = 1$. We define the *likelihood* of state $s$ to be the value $p_s \prod_q t_{q,s}$. (These are not probabilities, because they have not been normalized to have sum one.) According to Bayes' rule, the posterior probability of being in state $s$ is proportional to this likelihood; that is, there exists a scalar $Z$ such that the posterior probability of being in state $s$ is $\frac{1}{Z} p_s \prod_q t_{q,s}$. Once we have calculated the likelihoods of each state, we may calculate the posterior probability of each state by dividing its likelihood by this scalar, which equals the sum of the likelihoods of all states.

From these posterior probabilities of states we may also calculate the posterior probability that the assessed student has mastered any individual concept: it is the sum of the probabilities of all of the states that contain that concept. The *most informative concept* is the one whose posterior probability of being mastered by the student, calculated in this way, is as close as possible to 50%. The assessment procedure of Falmagne and Doignon (1988a) repeatedly finds the most informative concept, tests the student on it, and uses the test answer to update the posterior probabilities of the states and concepts in the learning space. Eventually, all concepts will have probabilities bounded well away from 50%, at which point the evaluation procedure terminates.

**13.3.3 Single-pass calculation of posterior probabilities.** It would be straightforward, but inefficient, to calculate the posterior probability separately for each concept by listing all possible states, checking whether each state contains the concept, and (when it does) calculating the state's posterior probability and adding it to the total for the concept. However, by combining the independence assumption given above with additional assumptions about the order in which states are generated (assumptions that are valid both for quasi-ordinal spaces and for our learning sequence based methods) it is possible to perform a more efficient algorithm that calculates the posterior probabilities for all concepts using only a single pass through all of the states of the learning space. This algorithm also saves time by computing the likelihoods of each state incrementally from previous likelihoods rather than recomputing the likelihood from scratch for each state.

The key assumptions used by this more efficient assessment procedure are that the states of knowledge are organized into a tree, with the empty set as its root, and with each tree edge connecting two states that differ by a single concept belonging to the child state but not the parent state, and that the state generation procedure follows a traversal of this tree. (The accessibility property of learning spaces is equivalent to the statement that such a tree exists.) With these assumptions, it is straightforward to maintain the product $\prod_q t_{q,s}$ for the current state $s$ of the tree traversal: the value of this product for each child state $S \cup \{x\}$ may be calculated as the product of the corresponding value for the parent state $S$ with the terms for concept $x$. Multiplying by the prior probability $p_s$ gives the likelihood for each state. We can also maintain throughout the traversal subtotals of the likelihoods of all states within each subtree: to do so, when returning from a child state $S \cup \{x\}$ to a parent $S$, we add the child's total to the parent's total, and at the same time add the child's total likelihood to the total likelihood associated with concept $x$. In this way, the total likelihood for each concept is calculated in constant time per state: two multiplications for each step down the tree, and two additions for each step back up the tree. To turn these computed total likelihoods for each concept into posterior probabilities that the student understands the concept, we must divide them by the total likelihood for all states, but this is simply the total likelihood stored at the root of the state tree. Thus, the total time to calculate the posterior probabilities of all of the concepts is proportional only to the number of states to be generated, plus the amount of time that it takes to generate them.

## 13.4 Learning Spaces from Partial Orders

For quasi-ordinal spaces, a data representation based on partial orders can be used to develop highly efficient state generation algorithms.

**13.4.1 Definition.** A (strict) *partial order* is a relation $<$ among a set of objects, satisfying *irreflexivity* ($x \not< x$) and *transitivity* ($x < y$ and $y < z$ implies $x < z$). The *Hasse diagram* of a partial order is a directed graph containing an edge $x \to y$ whenever $x < y$ and there does not exist $z$ with $x < z < y$. That is, we connect a pair of items in the partial order by an edge whenever the pair belongs to the *covering relation* of the partial order. In the partial order, $x < y$ if and only if there exists a directed path from $x$ to $y$ in the Hasse diagram.

To derive a quasi-ordinal space from a partially ordered set of concepts, we interpret the edges of the Hasse diagram as prerequisite relations between concepts. That is, if $x$ and $y$ are concepts, represented as vertices in a Hasse

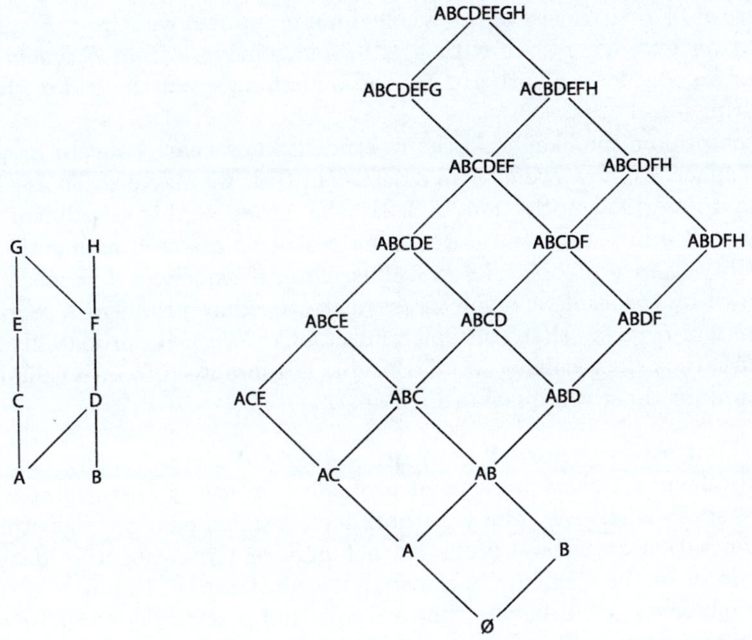


**Figure 13.1.** Left: a partial order, shown as a Hasse diagram in which each edge is directed from the concept at its lower endpoint to the concept at its upper endpoint. Right: the learning space derived from the partial order on the left.

diagram containing the edge $x \to y$, then we take it as given that $y$ may not be learned unless $x$ has also already been learned. For instance, in elementary arithmetic, one cannot perform multi-digit addition without having already learned how to do single digit addition, so a learning space involving these two concepts should be represented by a Hasse diagram containing a path or edge from the vertex representing single-digit addition to the vertex representing multi-digit addition. A set $S$ forms a valid state of knowledge in quasi-ordinal spaces if and only if it is a *lower set*: that is, for every edge $x \to y$ of the Hasse diagram, either $x \in S$ or $y \notin S$. Figure 13.1 shows a Hasse diagram on eight concepts, and the 19 states in the quasi-ordinal space derived from it.

The *fringe* of a state $S$ in a learning space is defined (Cf. Section 1.2.6) to be the set of concepts that, when added to or removed from $S$, lead to another state in the learning space. The *outer fringe* consists of the concepts $x \notin S$ such that $S \cup \{x\}$ is also a state in the learning space; the *inner fringe* consists of the concepts $x \in S$ such that $S \setminus \{x\}$ is also a state; the fringe is the union of the outer and inner fringes. In a quasi-ordinal space, the inner fringe of $S$ consists of the maximal concepts in $S$, and the outer fringe consists of the minimal concepts not in $S$; these properties allow the fringe of $S$ to be calculated easily in a single pass through the edges of the Hasse diagram. We initialize a set $F$ to consist of the whole domain; we remove $x$ from $F$ whenever we discover an edge $x \to y$ with $y \in S$, and remove $y$ from $F$ whenever we discover an edge $x \to y$ with $x \notin S$. The remaining set at the end of this scan is the fringe.

To determine the likelihood that a student knows each concept in a quasi-ordinal space, from answers to an assessment test, we may use the assessment algorithm described in Section 13.3. By the design of this assessment procedure, if $x < y$ in the partial order, then $x$ will be assessed as having higher probability than $y$ of being known. Therefore, if we assess a concept as being known by the student if the assessment procedure produces a probability greater than 50% for that concept, and as unknown if its probability is less than 50%, then the resulting set of concepts is guaranteed to be a valid knowledge state for the given quasi-ordinal space.

**13.4.2 Fast state generation in quasi-ordinal spaces.** The main computational difficulty in this assessment problem is in listing all the states of the learning space efficiently. Many authors have studied efficient algorithms for state generation in quasi-ordinal space (equivalently, listing the ideals of a partial order or the elements of a distributive lattice); cf. Habib et al. (2001) and its references. We briefly outline a simple and practical method for solving this problem using *reverse search* (Avis and Fukuda, 1996), a general technique for generating the elements of many types of combinatorial structure. This will serve as a warm-up for our reverse search algorithm for generating the states of more general learning spaces.

Reverse search depends on having a predecessor relationship among the states of the given quasi-ordinal space, which we define as follows. Choose arbitrarily a linear extension of the concepts in a partial order (or equivalently a topological ordering of its Hasse diagram); that is, a sequence of the concepts such that, if $x < y$ in the order, then $x$ must appear prior to $y$ in the sequence. For every state $S$ of the quasi-ordinal space derived from the partial order, let $x$ be the concept in $S$ that is latest in the linear extension; then $S \setminus \{x\}$ is necessarily also a state, which we call the *predecessor* of $S$. (The existence of this element proves the accessibility property of the learning space.) Repeated removal of elements will eventually reach the empty set, so the graph formed by connecting each state to its predecessor is a tree, rooted at the empty set. Reverse search, applied to this predecessor relationship, amounts to performing a depth first traversal of this tree.

To perform this traversal efficiently, we maintain for each state $S$ of the traversal a set children($S$) of concepts that may be added to $S$ to form a child of $S$. When the recursion steps from $S$ to $S \cup \{x\}$, we calculate children($S \cup \{x\}$) from children($S$) by removing from children($S$) any $y$ occurring prior to $x$ in the topological order, and adding to children($S$) any concept $y$ reachable from $x$ by a Hasse diagram edge $x \to y$ such that all other prererequisites of $y$ already belong to $S$. Once we have calculated children($S \cup \{x\}$), we may output $S \cup \{x\}$ as one of our states and continue recursively to each state $S \cup \{x, y\}$ for each $y$ in children($S \cup \{x\}$), and so on.

Very little is needed in the way of data structures beyond the Hasse diagram itself to implement this recursive traversal efficiently. Primarily, we need a way of quickly determining whether all prerequisites of some concept $y$ belong to a set $S$ visited by the traversal. This may be done efficiently by maintaining, for each $y$, a count of the prerequisites of $y$ that do not yet belong to the current state of the traversal. Whenever the traversal steps from a state $S$ to another state $S \cup \{x\}$, and $x$ is a prerequisite of $y$, we decrement this count, and whenever the traversal returns from $S \cup \{x\}$ to $S$ we increment this count. In this way, we may test whether all prerequisites of $y$ belong to the current state $S$ in constant time, simply by checking whether the count for $y$ is zero. The time spent updating these counts, whenever we step from a state $S$ to a state $S \cup \{x\}$ or vice versa, is proportional to the number of Hasse diagram edges that go out of $x$.

There is a different way of checking whether all prerequisites of $y$ belong to the current state $S$, based on manipulation of bitvectors (sequences of 0s and 1s packed into computer words). We may represent the state $S$ as a bitvector with a 1 in position $i$ if the $i$th concept (in some numbering of the concepts) belongs to $S$ and with a 0 in that position otherwise. Whenever the traversal steps from a state $S$ to another state $S \cup \{x\}$ or vice versa, we need only change a single bit of this bitvector, the one corresponding to concept $x$. To test whether all prerequisites of $y$ belong to the state $S$, we perform a bitwise Boolean or function between two bitvectors, one of which represents $S$ and the other of which has a 0 in each prerequisite position and a 1 in each other bit, and test whether the resulting bitvector is the all-ones bitvector (the binary number $-1$). This alternative representation may be updated in constant time when the algorithm steps from one state to another and allows us to check the prerequisites of $y$ in a constant number of bitvector operations. Theoretically each prerequisite test using the bitvector method requires time proportional to the number of concepts in the learning space, slower than the counting method, but in practice it is fast because typical modern computer architectures allow for the testing of prerequisite relations for 32 or more concepts simultaneously in a single machine instruction.

If there are $n$ concepts in the quasi-ordinal space, each step from a state $S$ to another state $S \cup \{x\}$ takes $O(d_{\text{in}}(x))$ time to check which members of children($S$) need to be removed, and $O(d_{out}(x))$ time to decrement the prerequisite counts and find concepts whose count has reached zero, where $d_{\text{in}}$ and $d_{\text{out}}$ represent the indegree and outdegree of $x$ in the Hasse diagram. In typical examples one may expect the indegree and outdegree to both be small, on average, but even in the worst case (Figure 13.2) the time is no more than $O(n)$ per state in a learning space with $n$ concepts. It is this level of state generation efficiency that we hope to approach or meet with our more general learning space representation.

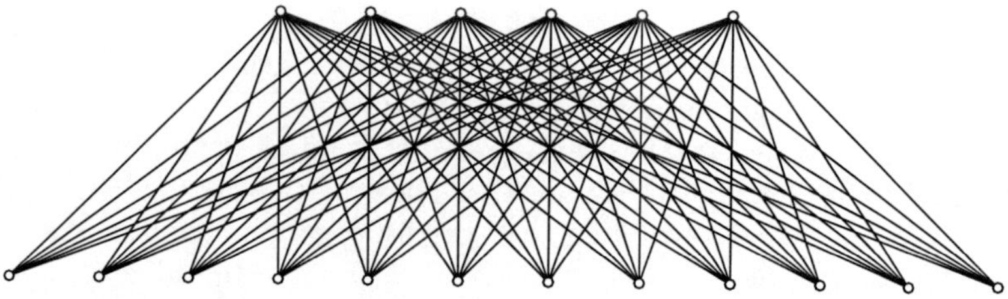

**Figure 13.2.** The Hasse diagram of a worst case example for the partial order state generation algorithm. Each of $2n/3$ items (bottom) is connected by an edge to each of $n/3$ items (top). The corresponding learning space has $2^{2n/3} + 2^{n/3} - 1$ states, $2^{2n/3}$ of which cause the state generation algorithm to perform $n/3$ prerequisite checks per state, so the total time is $\Omega(n)$ per state.

## 13.5 Learning Spaces from Learning Sequences

Given any learning space, there may be many orderings through which a student, starting from no knowledge, could learn all the concepts in the space. We call such an ordering a *learning sequence*; these have also been called *learning paths* by Falmagne and Doignon (2011), while in combinatorics they are also known as *basic words*. We define them more formally below.

**13.5.1 Definition.** A *learning sequence* is a one-to-one map $\sigma$ from the integers $\{0, 1, 2, \ldots n-1\}$ to the $n$ concepts forming the domain of a learning space, with the property that each *prefix* $P_i(\sigma) = \sigma(\{0, 1, 2, \ldots i-1\})$ is a valid knowledge state in the learning space.

**13.5.2 Example.** The learning space depicted in Figure 13.3 has four learning sequences: (1) $A, B, C$, (2) $A, C, B$, (3) $C, A, B$, and (4) $C, B, A$. Learning sequences may also be interpreted as shortest paths from the empty set to the set of all concepts, in the graph formed from the states of a learning space by connecting pairs of states that differ by a single concept. The sequence of concepts by which each two consecutive states differ in such a path forms a learning sequence, and the sequence of prefixes of a learning sequence forms a path of this type. Thus, in Figure 13.3, the learning sequence $A, C, B$ corresponds to the path $\emptyset, \{A\}, \{A, C\}, \{A, B, C\}$ and vice versa. For a quasi-ordinal learning space, such as the one in Figure 13.1, learning sequences are the same thing as linear extensions of the underlying partial order. However, Definition 13.5.1 can be applied to any learning space, and Figure 13.3 provides an example which does not come from a partial order.

If we are given a set $\Sigma$ of learning sequences, we can immediately infer that all prefixes of $\Sigma$ are states in the associated learning space. But we can also infer the existence of other states by using the union-closure property of learning spaces: any set formed from a union of prefixes of $\Sigma$ must also be a

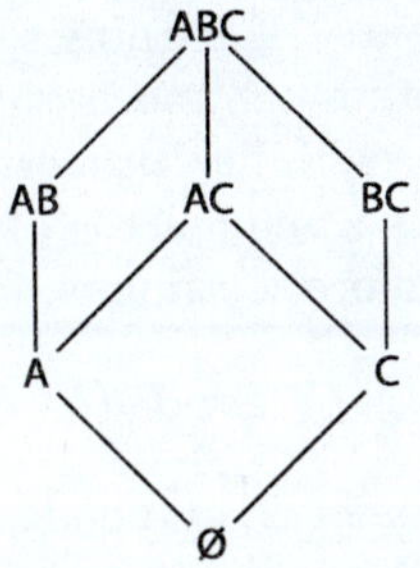


**Figure 13.3.** A learning space that is not quasi-ordinal, as it does not contain the intersection $\{A,B\} \cap \{B,C\} = \{B\}$ of two of its states.

state in the associated learning space. For instance, suppose we have the two sequences $A, B, C$ and $C, B, A$ from the learning space in Figure 13.3. The prefixes of these sequences are the six sets $\emptyset$, $\{A\}$, $\{A,B\}$, $\{A,B,C\}$, $\{C\}$, and $\{B,C\}$. However, by forming unions of prefixes we may also form the seventh set $\{A\} \cup \{C\} = \{A,C\}$. All seven states of the learning space can be recovered in this way from unions of the prefixes of these two sequences.

**13.5.3 Definition.** Let $\Sigma$ be any set of sequences over a finite set of concepts. We define $\mathcal{L}_\Sigma$ to be the family of sets that are unions of prefixes of $\Sigma$.

**13.5.4 Theorem.** *For any set $\Sigma$ of sequences, $\mathcal{L}_\Sigma$ is a learning space.*

PROOF. If two sets $S$ and $T$ both belong to $\mathcal{L}_\Sigma$, then by the definition of $\mathcal{L}_\Sigma$ both $S$ and $T$ are unions of prefixes of sequences in $\Sigma$. Then $S \cup T$ is also a union of prefixes of sequences in $\Sigma$, so it belongs to $\mathcal{L}_\Sigma$. Thus, $\mathcal{L}_\Sigma$ is closed under unions.

If a nonempty set $S$ belongs to $\mathcal{L}_\Sigma$, then it is a union of prefixes of strings in $\Sigma$; among all sets of prefixes whose union is $S$, we may choose one whose total length is minimum. Let $P_i(\sigma_j)$ be one of the prefixes in this set. Consider the set of prefixes formed from the ones defining $S$ by replacing $P_i(\sigma_j)$ by $P_{i-1}(\sigma_j)$, and let its union be $T$. Then $T \neq S$, for otherwise we would have a set of prefixes whose union is $S$ with smaller total length than the set with minimum possible length, a contradiction. Thus, $T$ differs from $S$, but it can only differ by the removal of a single element $\sigma_j(i-1)$. We have shown that for every nonempty set $S$ there exists a valid state $T$ formed from $S$ by removing a single element, so $\mathcal{L}_\Sigma$ is accessible.

As family of sets that is both closed under unions and accessible, $\mathcal{L}_\Sigma$ is by definition a learning space. □

**13.5.5 Theorem.** *Let $S$ be a state of a learning space $\mathcal{L}$. Then there is a learning sequence of $\mathcal{L}$ containing $S$ as one of its prefixes.*

PROOF. The accessibility property of learning spaces, applied repeatedly starting with the set of all concepts in $\mathcal{L}$, shows that $\mathcal{L}$ has at least one learning sequence $\sigma$. Similarly, the accessibility property applied repeatedly starting from $S$ shows that there exists a sequence $\rho$ of the concepts in $S$ each prefix of which is a state of $\mathcal{L}$. Therefore, by union closure, the sequence of sets

$$P_0(\rho), P_1(\rho), \ldots, S, S \cup P_0(\sigma), S \cup P_1(\sigma), \ldots$$

starts from the empty set, includes $S$, and ends at the set of all concepts, with each set in the sequence being a valid state in $\mathcal{L}$. Each two consecutive sets in this sequence either differ by a single element or are the same. Removing the duplicates from the sequence leads to a sequence of sets that corresponds to a learning sequence having $S$ as one of its prefixes. □

**13.5.6 Definition.** An *atom* of a learning space is a nonempty state $S$ for which there exists only one concept $x \in S$ such that $S \setminus \{x\}$ is a valid state. In the antimatroid literature, atoms are called paths, but we avoid this terminology as it may cause confusion with paths in the state graph.

**13.5.7 Lemma.** *A nonempty state $S$ of a learning space $\mathcal{L}$ is an atom if and only if it cannot be formed as the union of two smaller states.*

PROOF. If $S = T \cup U$ where $T$ and $U$ are both smaller than $S$, then by an argument similar to that in Theorem 13.5.5 there is a learning sequence containing both $T$ and $S$ as prefixes; the last concept in this sequence that belongs to $S$ is removable from $S$ and does not belong to $T$. Symmetrically, there is a concept that is removable from $S$ and does not belong to $U$. These two removable concepts cannot be equal, so $S$ is not an atom.

In the other direction, if $S$ is not an atom, then it has two removable concepts $x$ and $y$, and is the union of the two states $S \setminus \{x\}$ and $S \setminus \{y\}$. □

**13.5.8 Theorem.** *Let $\mathcal{L}$ be any learning space, and let $\Sigma$ be a set of learning sequences of $\mathcal{L}$. Then $\mathcal{L}_\Sigma \subseteq \mathcal{L}$, with equality if and only if every atom of $\mathcal{L}$ is one of the prefixes of $\Sigma$.*

PROOF. By the definition of learning sequences, each prefix of a sequence in $\Sigma$ is a state of $\mathcal{L}$, so their unions are also states of $\mathcal{L}$ and $\mathcal{L}_\Sigma \subseteq \mathcal{L}$.

If an atom is not one of the prefixes of $\Sigma$, then by Lemma 13.5.7 it is also not a union of prefixes of $\Sigma$, and therefore does not belong to $\mathcal{L}_\Sigma$, so $\mathcal{L}_\Sigma$ differs from $\mathcal{L}$. However, if every atom is a prefix of $\Sigma$, then it follows by induction on the size of the states that every state of $\mathcal{L}$ belongs to $\mathcal{L}_\Sigma$, for either it is an atom or it is the union of two smaller states that were shown earlier in the induction to belong to $\mathcal{L}_\Sigma$. □

This theorem may be viewed as a restatement of the standard equivalence between the definition of antimatroids as set systems and antimatroids as formal languages; cf. Theorem III.1.4 of Korte et al. (1991). In Chapter 14 we will show how to find a set of as few learning sequences as possible that includes every atom of a given learning space, and therefore that represents the learning space as concisely as possible.

## 13.6 Representing States by Numerical Vectors

It is possible to name states in $\mathcal{L}_\Sigma$ by vectors in $\mathbb{Z}^{|\Sigma|}$, in such a way that each state has a unique name and the concepts in each state can be reconstructed easily from the state's name. This naming scheme will be useful in our state enumeration algorithm.

**13.6.1 Definition.** Given a state $S$ in a learning space $\mathcal{L}$, and a set $\Sigma = \{\sigma_0, \sigma_1, \ldots, \sigma_{k-1}\}$ of $k$ learning sequences within the space, define $\text{mex}_i(S)$ to be the minimum index in $\sigma_i$ of a concept excluded from $S$; that is, $\text{mex}_i(S) = \min\{j \mid \sigma_i(j) \notin S\}$. Define $\text{mex}(S)$ to be the vector

$$\text{mex}(S) = (\text{mex}_0(S), \text{mex}_1(S), \ldots, \text{mex}_{k-1}(S)).$$

**13.6.2 Example.** The indices $\text{mex}(S)$ for an example of a learning space $\mathcal{L}_\Sigma$ are depicted in Figure 13.4. The learning space shown in the figure cannot be derived from a partial order, as it has states $\{B, D, F\}$ and $\{B, C, E, F\}$ but not their intersection $\{B, F\}$.

Note that $\text{mex}_i(S)$ depends, not just on $i$ and $S$, but also on $\mathcal{L}$ and $\Sigma$; we will assume for the rest of this chapter that $\mathcal{L}$ and $\Sigma$ are fixed and we omit them from the notation for the sake of simplicity. If $S$ is the whole domain, we define for completeness $\text{mex}_i(S) = n$. Equivalently, therefore, $\text{mex}_i(S)$ is the size of the largest prefix of $\sigma_i$ that is a subset of $S$. Thus, the mex function maps states in the learning space to vectors in $\mathbb{Z}^k$.

**13.6.3 Definition.** Define $\text{up}(v)$, mapping vectors in $\mathbb{Z}^k$ to states in the learning space, by

$$\text{up}(v) = \bigcup_{0 \le i < k} P_{v_i}(\sigma_i).$$

That is, we interpret the coordinates of $v$ as lengths of prefixes of each of the sequences in $\Sigma$, and form the union of these prefixes. For any set $S$, $\text{up}(\text{mex}(S)) \subset S$, with $\text{up}(\text{mex}(S)) = S$ if and only if $S$ is a state in the learning space $\mathcal{L}_\Sigma$.

If $\text{mex}(S)$ is known, $\text{mex}(S \setminus \{x\})$ may easily be calculated, by taking the coordinatewise minimum of $\text{mex}(S)$ and the positions of $x$. However, calculating $\text{mex}(S \cup \{x\})$ from $\text{mex}(S)$ appears to be more complicated. As part of

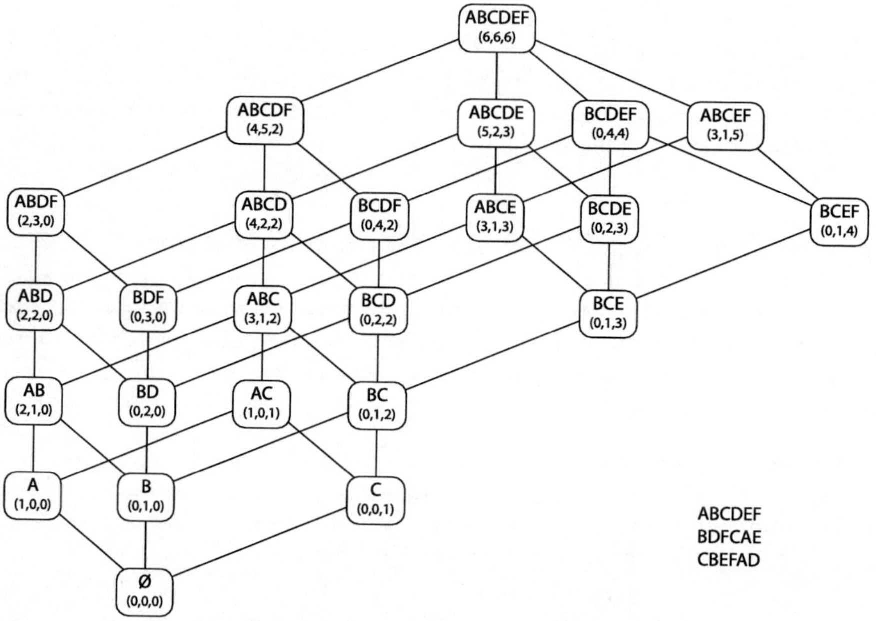


**Figure 13.4.** Learning space $\mathcal{L}_\Sigma$ generated from three sequences $ABCDEF$, $BDFCAE$, and $CBEFAD$. Each state is shown with its index mex($S$).

the state generation procedure in Section 13.7, we include a bitvector-based data structure that allows us to maintain a set subject to both insertions and deletions and calculate its mex indices, more quickly than recalculating these indices from scratch each time they are needed.

**13.6.4 The two fringes of a knowledge state.** As with learning spaces derived from partial orders, we will also need to calculate fringes of states. The *outer fringe* of $S$, the items that may be added to state $S$ to form a new state, is particularly easy to describe: Each state $S \cup \{x\}$ may be formed as the union of the prefixes defining $S$ itself, together with one more prefix $P_{\text{mex}_i(S)}(\sigma_i)$, so the outer fringe of $S$ is

$$\{\sigma_i(\text{mex}_i(S)) \mid 0 \le i < k\}.$$

The *inner fringe* of $S$, those items in $S$ that may be removed to form another state, is a little more complicated, but still easily expressible using our indexing scheme: it is

$$\{x \in S \mid \text{up}(\text{mex}(S \setminus \{x\})) = S \setminus \{x\}\}.$$

The formula $\text{up}(\text{mex}(S \setminus \{x\})) = S \setminus \{x\}$ is true if and only if $S \setminus \{x\}$ is a state of the learning space, so we may construct the lower fringe by testing whether this formula is true for each member $x$ of $S$.

## 13.7 Generating the States of a Learning Space

We wish to generalize knowledge assessment algorithms from quasi-ordinal spaces to learning spaces derived from learning sequences. The keys to this generalization are methods for efficiently listing all states of a learning space, and for projecting learning spaces onto smaller subsets of concepts in order to speed up the algorithm even further by reducing the size of the state space (cf. Chapter 14). As we describe, a representation based on learning sequences allows us to perform these operations efficiently.

**13.7.1 A tree of states.** As with partial orders, we can list all states of a sequence-based learning space by an algorithm based on exploring a tree in which the nodes are the states of the learning space and the parent of each node is found by a predecessor operation that chooses in some canonical way an element to remove from the corresponding state. To generate the predecessor of a state, we repeatedly reduce the last nonzero coordinate of its mex vector until the up function maps the reduced vector to a set different from the state itself. In order to reverse this, and find the successors of a state $S$ (that is, the states having $S$ as its predecessor), we need merely perform the following steps:

1. Set $p$ to the smallest value of $i$ such that $S$ is the union of prefixes from the first $i$ sequences.
2. For each $i \geq p$, such that $\sigma_i(\text{mex}_i(S)) \neq \sigma_j(\text{mex}_j(S))$ for all $j < i$ (that is, such that the first excluded concept in $\sigma_i$ differs from the first excluded concept in all earlier sequences):
   a) Let $v$ be formed from $\text{mex}(S)$ by adding one to its $i$th coordinate.
   b) Let $T = \text{up}(v) = S \cup \{\sigma_i(\text{mex}_i(S))\}$ and output $T$.

This procedure generates exactly the states that have $S$ as their predecessor.

**13.7.2 Recursive tree search.** To generate all states in the learning space, we perform a depth first traversal of the states of the space, starting from the empty set, by applying this procedure to find the successors of each state and then recursively searching each successor that it finds. If $T$ is a successor of $S$, the value of $p$ needed within the search for the successors of $T$ equals the value of $i$ used when we generated $T$ from its parent $S$; therefore, we may pass these $p$ values through the recursive traversal and avoid calculating them at each step. However, the mex values of $S$ and $T$ may differ not just in the $i$th coordinate but also in other coordinates greater than $i$, and must be recalculated. If the learning space is generated by $k$ learning sequences, the time per state is $O(k)$, except for the time to calculate the mex value of each new state.

**13.7.3 Efficient mex recalculation.** When we defined the mex operation we showed that removing an element from a state allows the new mex to be calculated easily, but, unfortunately, in the case of the state generation procedure, each new state is generated not by removing an element but by adding one. To speed up this part of the state generation, we maintain as part of the recursive traversal a collection of bitvectors, with one bitvector per learning sequence defining the learning space and one bit in each bitvector for each concept of the learning space. For notational convenience, we describe these bitvectors here as binary numbers. In each bitvector $B_i$ we store the number $\sum\{2^j \mid \sigma_i(j) \in S\}$ representing the set of positions of learning sequence $\sigma_i$ that are present in the current state $S$. Each of these bitvectors may be updated in constant time for each step forwards and backwards in the state traversal, and each mex calculation may be performed by finding the first zero bit in the binary representation of each integer, which may be performed efficiently using a combination of arithmetic and bitwise boolean operations. Specifically, the bitwise exclusive or of $B_i$ and $B_i + 1$ is an integer of the form $2^j - 1$, where $j - 1$ is the position of the first zero in $B_i$ and should be used as the value of $\text{mex}_i(S)$. Using this method, calculating the mex value of each new state can be performed in a total of $O(k)$ bitvector operations.

As a theoretically efficient but less practical alternative to this bitvector method, it would instead be possible to maintain, for each learning sequence, a priority queue of the excluded elements in that sequence, prioritized by their position in the sequence. Each step from state to state in the state generation procedure could be performed using $O(k)$ priority queue operations; using van Emde Boas trees for each priority queue (van Emde Boas et al., 1976) would give a time of $O(k \log\log n)$ per state, where $n$ is the number of concepts in the learning space.

**13.7.4 Application to assessment.** We can use this state generation procedure to assess the likelihood that a student has mastered each concept in a learning space, using the Bayesian procedure described in Section 13.3. Specifically, we calculate a likelihood for each state of the space as a product of the prior probability of the state with a set of terms, one term per question asked of the student. As we generate all states, we maintain in top-down fashion the likelihoods of each state and in bottom-up fashion the sums of likelihoods in each subtree of the tree of states. The likelihood of a concept is the sum of the likelihoods of the states containing it, which is the sum of the likelihoods of subtrees whose parent edge is labeled by that concept, and the probability of knowing a concept is this sum normalized by dividing by the sum of likelihoods of all states. Like the assessment procedure for quasi-ordinal spaces, this assessment procedure takes constant additional time per state, beyond the time required to generate all states.

## 13.8 Future Work

Although we have made significant progress in learning space implementation, we believe there is plenty of scope for additional investigation, particularly on the following topics:

**Counting states.** Is there an efficient method for counting or estimating the number of states in a learning space, without taking the time to generate all states? This would have implications for our ability to set sample sizes appropriately in knowledge assessment algorithms, as well as for calculating more accurate priori probability distributions on projected states when using projections of learning spaces to speed up the assessment algorithm. From past experience with similar combinatorial counting problems (e.g., Jerrum et al., 2001) we expect that the complexity of a randomized approximation scheme for the counting problem is likely equivalent to the complexity of sampling states from the learning space uniformly at random, which seems to be of interest independently.

**Inference of error rates.** Rather than assuming that careless mistakes and lucky guesses each occur with fixed rates, one could envision a more sophisticated Bayesian assessment procedure that treats the chance of careless errors as a variable with an a priori probability distribution, and attempts to infer a maximum likelihood value of this variable based on the student's answers. Such a procedure would likely be based on an EM-algorithm approach in which one alternates applications of the likelihood calculation described here with an algorithm for estimating the careless error probability given the calculated likelihoods, and would allow the system to better fit its model to each student's performance. However the details of such an approach still need to be worked out.

**Faster state space generation.** Generating the states of a quasi-ordinal space that has $n$ concepts takes time $O(n)$ per generated state, without assumption, and may often be faster, while we can only show a similarly fast time bound of $O(k)$ per generated state for a learning space generated by $k$ learning sequences with the additional assumption (reasonable for small learning spaces but not for large ones) that the bitvector operations for maintaining and updating the mex values of the generated states take constant time per operation. Can more efficient worst case guarantees on the performance of state generation in learning spaces can be proven?

**Question selection strategy.** As described, the assessment procedure selects the next question to ask the student as the one with likelihood closest to 50% of being known, but this may not be an optimal selection strategy. The effect of an improved strategy could be to reduce the number of questions needed to assess each student's knowledge, over and above the reduction afforded by more accurately defining the learning space on which the assessment is based.

## 13.9 Conclusions

We have shown that a computer representation of learning spaces by learning sequences can approach the efficiency of the existing quasi-ordinal space representation for knowledge assessment, while allowing a broader class of learning spaces that may more easily be adapted by adding and removing states. We believe that the algorithms described here are sufficiently detailed and efficient to be suitable for implementation.

# 14

# Projection, Decomposition, and Adaption of Learning Spaces

David Eppstein[1]

In Chapter 13 we described *learning sequences*, a powerful tool for the computer representation of learning spaces, and we showed how to use learning sequences as the basis for computer algorithms that efficiently perform the state generation necessary for knowledge assessment in a learning space. In this chapter we show how to use learning sequences as the basis of several different methods for deriving one learning space from another:

- We show how to *project* a learning space with a large number of concepts onto a smaller learning space that uses a sampled subset of the concepts, by identifying two states of the larger learning space whenever their differences are disjoint from the sample. This concept of projection is an essential part of efficient assessment for learning spaces that are too large to generate all states (cf. Section 8.4).
- We show how to *decompose* a learning space into a *join* of simpler learning spaces, and in particular how to find the smallest possible set of learning sequences that defines a given learning space. This number is directly related to the efficiency of the assessment procedure described in Chapter 13.
- We show how to *adapt* a learning space by adding or removing individual states that belong to its *inner and outer fringe*, allowing it to more accurately model the possible states of knowledge of students.

## 14.1 Projection

The assessment procedure of for learning spaces and quasi-ordinal spaces of Falmagne and Doignon (1988a), described in Section 13.3, involves listing every state of a given learning space and performing some simple calculations for each state. Although this works well for learning spaces with 50 to 100 concepts, it becomes too slow for larger learning spaces due to the combinatorial explosion in the number of states that those spaces have. Therefore, the

[1]Dept. of Computer Science, University of California, Irvine. Supported in part by NSF grants 0830403 and 1217322 and by the Office of Naval Research under grant N00014-08-1-1015.

J.-C. Falmagne et al. (eds.), *Knowledge Spaces: Applications in Education*,
DOI 10.1007/978-3-642-35329-1_14, 

ALEKS system (as of its 2006 implementation) resorts to a sampling scheme that allows its assessment procedure to run at interactive speeds for much larger learning spaces. This sampling scheme uses a distance measure between pairs of concepts to find a small sample of the concepts, projects the given learning spaces onto smaller learning spaces using only the sampled concepts, and bounds the assessed likelihood that a student knows a concept in the original learning space by the likelihoods of nearby concepts in the sample.

We first outline these methods as they have already been applied to quasi-ordinal spaces, and then describe how to extend them to more general learning spaces. To define the distance between two elements $x$ and $y$ in a partial order, let $\Delta_{x,y}$ be the set of items whose comparison to $x$ differs from its comparison to $y$:

$$\begin{aligned}\Delta_{x,y} = \{x,y\} &\cup \{z \mid z < x \wedge z \not< y\} \cup \{z \mid z < y \wedge z \not< x\} \\ &\cup \{z \mid x < z \wedge y \not< z\} \cup \{z \mid y < z \wedge x \not< z\}.\end{aligned}$$

Define the distance between $x$ and $y$ to be $d(x,y) = |\Delta_{x,y}| - 1$. This distance satisfies the mathematical axioms defining a metric space: positivity ($d(x,y) \geq 0$ with equality only when $x = y$), symmetry ($d(x,y) = d(y,x)$), and transitivity ($d(x,z) \leq d(x,y) + d(y,z)$). Positivity follows from the inclusion of $\{x,y\}$ in $\Delta_{x,y}$, and the symmetry of the distance function follows from the symmetry of the definition of $\Delta_{x,y}$. To see that transitivity necessarily holds for this distance, observe that, for any $x$, $y$, and $z$, $\Delta_{x,z} \subset \Delta_{x,y} \cup \Delta_{y,z}$, and that the union is not disjoint as $y$ belongs to both sides.

For a suitably chosen distance threshhold $\delta$, we may find a sample $S$ of the concepts of the given quasi-ordinal space space such that every concept is within distance $\delta$ of a member of $S$. Although there is no mathematical proof of such a fact, the intent of this sampling technique is that assessment on a nearby sample concept is likely to be informative about the assessment of each unsampled concept.

Once a sample of concepts has been chosen, we may form a quasi-ordinal space on the sample by restricting the partial order defining the whole space to the sample. Finally, the assessment of likelihoods on the sampled concepts is used to bound the likelihoods of the remaining unsampled concepts, to determine which ones the student is likely to know or not to know. If $x < y$, $y$ belongs to the sample, and the student knows $y$ with probability $p$, then the student is taken to know the easier concept $x$ with probability at least $p$. Similarly, if $x < y$, $x$ belongs to the sample, and the student knows $x$ with probability $p$, then the student is taken to know the harder concept $y$ with probability at most $p$. Thus, in order for this sampling technique to be informative about the unsampled concepts, it is necessary for each unsampled concept to be bounded above and below by nearby sampled concepts; it is not sufficient to have concepts in the sample that are nearby but incomparable.

This sampling process, sample learning space construction, and likelihood bound, are used together repeatedly to refine the portion of the learning space that is relevant for the student. Initially, all states are considered relevant, and a sample with a high distance threshhold is chosen. After several steps of refinement, a larger number of concepts have likelihoods that can be bounded away to one side or another of 50%, and a sample is chosen with a smaller distance threshhold among only those remaining informative concepts. Eventually, this refinement process converges with all concepts having likelihoods bounded away from 50%, which we may use to construct a most likely knowledge state for the student.

We now discuss how to generalize these ideas from quasi-ordinal spaces to learning spaces defined by learning sequences. We may define the distance between two concepts to be the sum (or some other combination) of the distances between them in each of the learning sequences defining the given learning space. Because this definition depends on the sequences and not just on the learning spaces, it is not as intrinsic as the definition of distance for partial orders, but it may still suffice as the basis of a sampling method for which every unsampled concept is near to at least one sampled concept that will be informative for it.

If $\mathcal{F}$ is any family of sets, and $P$ is any subset of $\cup\mathcal{F}$, we may define the *projection*

$$\mathcal{F}_P = \{S \cap P \mid S \in \mathcal{F}\}$$

by intersecting each state in $\mathcal{F}$ with $P$.

**14.1.1 Theorem.** *If $\mathcal{F}$ is a learning space, and $P$ is any subset of $\cup\mathcal{F}$, then $\mathcal{F}_P$ is also a learning space.*

PROOF. The projection inherits the union closure property of learning spaces from $\mathcal{F}$, as $(S_1 \cap P) \cup (S_2 \cap P) = (S_1 \cup S_2) \cap P$.

To prove the accessibility property of learning spaces for a set $S \cap P$, $S \in \mathcal{F}$, repeatedly use accessibility in $\mathcal{F}$ to remove concepts from $S$ until a concept in $P$ is removed, and let the predecessor of $S \cap P$ be formed by removing the same concept from $S \cap P$. □

This notion of a projection agrees with the restriction of a partial order, used for quasi-ordinal space. It may also be constructed easily from a set of learning sequences defining an arbitrary learning space:

**14.1.2 Theorem.** *Let $\mathcal{L}$ be a learning space, defined from a collection of learning sequences $\sigma_i$, and let $P$ be a subset of the concepts in $\mathcal{L}$. For each learning sequence $\sigma_i$, let $\sigma'_i$ be the subsequence of $\sigma_i$ consisting only of the concepts in $P$, in the same order as they appear within $\sigma_i$. Then the set of learning sequences $\sigma'_i$ define the learning space $\mathcal{L}_P$.*

PROOF. The set of concepts appearing in each prefix of $\sigma'_i$ is the intersection with $P$ of the set of concepts appearing in a corresponding prefix of $\sigma$, which is a valid state of $\mathcal{L}$. Thus, each prefix of $\sigma'_i$ defines the intersection with $P$

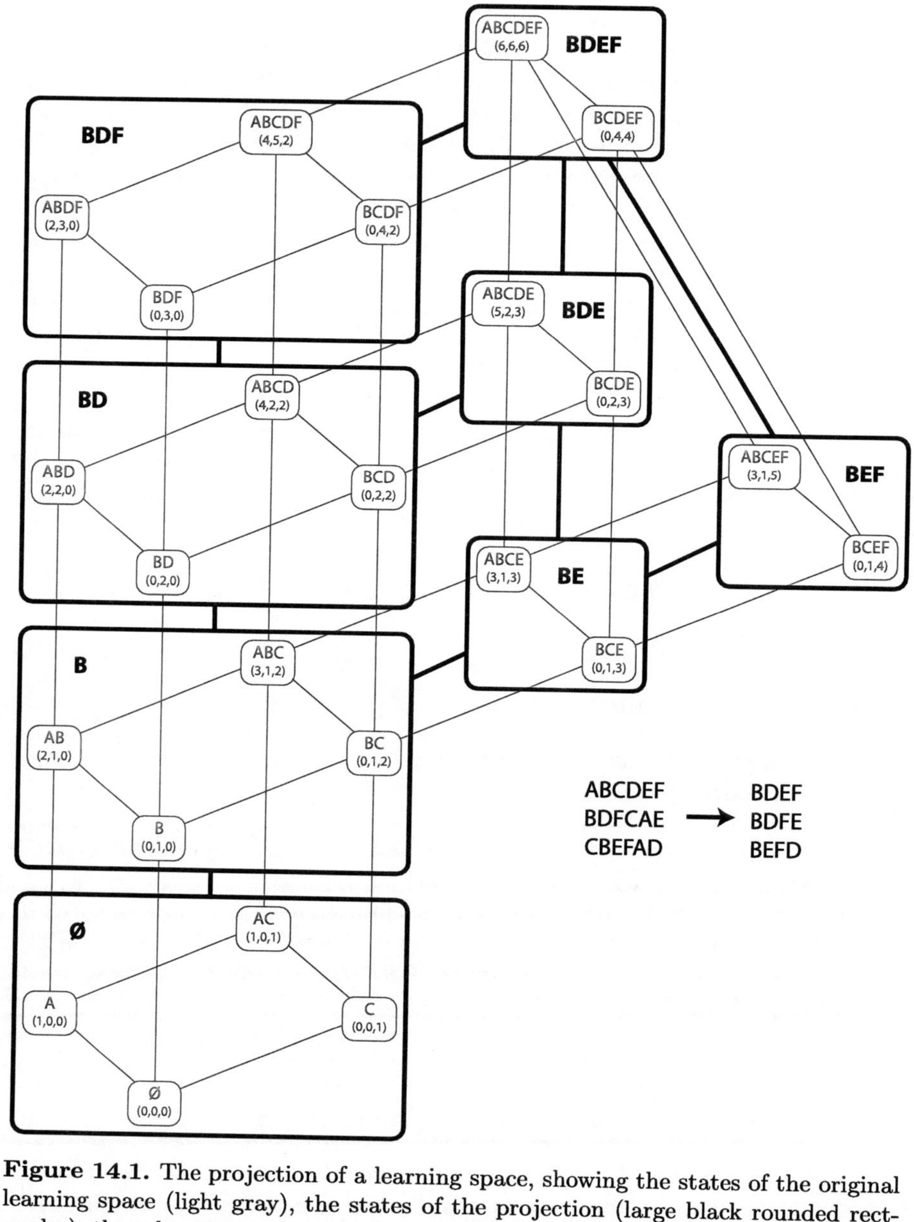


**Figure 14.1.** The projection of a learning space, showing the states of the original learning space (light gray), the states of the projection (large black rounded rectangles), three learning sequences that define the original learning space (left of the arrow), and the corresponding three learning sequences in the projection (right of the arrow).

of a state of $\mathcal{L}$, which is a state of $\mathcal{L}_P$. Because $\mathcal{L}_P$ is a learning space, it is closed under unions, so every union of prefixes of sequences $\sigma'_i$ is also a state of $\mathcal{L}_P$. These unions of prefixes are exactly the states in the learning space defined by the sequences $\sigma'_i$, and by this argument every such state is also a state of $\mathcal{L}_P$.

In the other direction, let $S$ be any state of $\mathcal{L}_P$; then $S = T \cap P$ where $T$ is a state of $\mathcal{L}$. Because $\mathcal{L}$ is defined from the learning sequences $\sigma_i$, $T$ may be formed as a union of prefixes of these learning sequences. Intersecting each of these prefixes with $P$ gives a representation of $S$ as a union of prefixes of the learning sequences $\sigma'_i$. Thus, every state of $\mathcal{L}_P$ is also a state of the learning space defined by the sequences $\sigma'_i$. □

As well as the states of the projection, Figure 14.1 also shows a set of learning sequences defining the original learning space $\mathcal{L}$ and their restriction to a set of learning sequences defining the projected space.

The choice of how many concepts to use in modeling knowledge as a learning space is somewhat arbitrary: we may model some areas of knowledge coarsely, as a relatively small set of concepts, and others more finely using many concepts. We do not expect our assessment of the likelihood that a student knows a given concept to vary significantly based on the choice of how coarsely the space is modeled. Therefore, though there is little mathematical justification for this simplification, we suggest performing the assessment algorithm in a projected space $\mathcal{L}_P$ as if it were the whole space. The results of the likelihood calculation in this projection may differ from those of the calculation in the whole space $\mathcal{L}$, but there seems little reason in principle to view one of the results as more reliable than the other, and if $P$ is small then the projected assessment may be much more efficient than the assessment in the original space.

The remaining component of the sample-based assessment algorithm is the extension from an assessment of a sample back to an assessment within the original unprojected learning space. After performing an assessment within the sampled concepts, we may partition the sample into two subsets, a set $K$ of concepts that the student is assessed to know, and a set $U$ of concepts that the student is assessed not to know. The subset of learning space states $\mathcal{L}(K, U)$ consistent with those beliefs are exactly the states that project to state $K$ in the projected learning space $\mathcal{L}_{K \cup U}$. Therefore, we can view $\mathcal{L}(K, U)$ as a *fiber* of the projection, the inverse image of $K$.

Within $\mathcal{L}(K, U)$ there is a unique maximal state $\neg U$ that may be constructed as the union of a set of prefixes of the learning sequences defining $\mathcal{L}$, where each prefix is taken up to but not including the first member of $U$. Note that the maximal state $\neg U$ of $\mathcal{L}(K, U)$ does not depend on $K$, and that a set $K$ is a valid state of the projection $\mathcal{L}_{K \cup U}$ if and only if $K \subset \neg U$. We may use this construction to determine the assessment of a concept $x \notin K \cup U$ as follows:

- If $x \notin \neg U$, then there do not exist any states of $\mathcal{L}$ consistent with the possibility that the student knows $K \cup \{x\}$ and does not know $U$. In this case, we can conclude that the student does not know $x$.
- If $K \not\subset \neg(U \cup \{x\})$, then there do not exist any states of $\mathcal{L}$ consistent with the possibility that the student knows $K$ and does not know $U \cup \{x\}$. In this case, we can conclude that the student knows $x$.
- In the remaining case, there are states of $\mathcal{L}$ consistent with the possibility that the student knows $K \cup \{x\}$ and does not know $U$, and other states consistent with the possibility that the student knows $K$ and does not know $U \cup \{x\}$. In this case, we can make no assessment of whether the student knows $x$, and must leave it for a later round of the sampling algorithm.

## 14.2 Fibers of Projections

In the previous section we defined a *fiber* $\mathcal{L}(K, U)$ to be the inverse image of state $K$ in the projection $\mathcal{L}_{K \cup U}$. Within a knowledge assessment algorithm based on sampling and projection, the fiber determines the set of states in a learning space that is compatible with an assessment performed on a sample of its concepts. Therefore it is of some importance to understand what combinatorial structures fibers might have and how to work with them algorithmically.

**14.2.1 Theorem.** *If $\mathcal{L}$ is a quasi-ordinal space, then so is the fiber $\mathcal{L}(K, U)$.*

PROOF. We observe that, more generally, if $\mathcal{F}$ is any family of sets, and $\mathcal{F}(K, U)$ denotes the subfamily of sets in $\mathcal{F}$ that are supersets of $K$ and disjoint from $U$, then closure under unions or intersections of $\mathcal{F}$ leads to the same property of $\mathcal{F}(K, U)$. For, every set in $\mathcal{F}(K, U)$ is a superset of $K$ and taking the union or intersection of two supersets of $K$ cannot form a set that is not a superset of $K$. Symmetrically, every set in $\mathcal{F}(K, U)$ is disjoint from $U$, and unions and intersections of sets that are themselves disjoint from $U$ must result in sets that are still disjoint from $U$. Thus, every union or intersection of sets in $\mathcal{F}(K, U)$ is a superset of $K$ that is disjoint from $U$, and therefore it belongs to $\mathcal{F}(K, U)$ itself.

Additionally, if $\mathcal{F}$ is any well-graded family of sets, then $\mathcal{F}(K, U)$ inherits the same property: if $X$ and $Y$ are any two sets in $\mathcal{F}$ that both belong to $\mathcal{F}(K, U)$, then the well-graded path in $\mathcal{F}$ from $X$ to $Y$ must remain within $\mathcal{F}(K, U)$, for a path from $X$ to $Y$ that involved insertions or deletions of members of $K \cup U$ would have to have equally many insertions as deletions of those elements, violating the requirement of of a well-graded path that each element is inserted or deleted only once.

The quasi-ordinal spaces can be characterized as well-graded set families that are closed under unions and intersections. Thus, by the above reasoning, if $\mathcal{L}$ is a quasi-ordinal space, then so is the fiber $\mathcal{L}(K, U)$. □

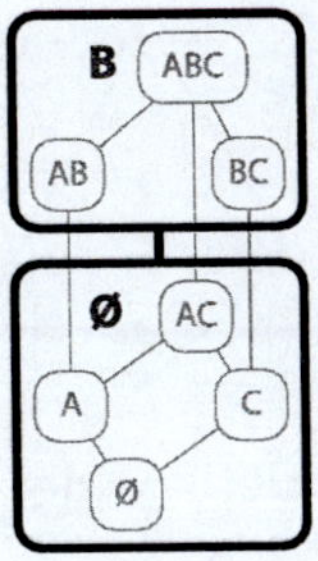


**Figure 14.2.** A learning space $\mathcal{L}$ and projection $\mathcal{L}_{\{B\}}$ for which the fiber $\mathcal{L}(B, \emptyset)$ (the set of three states within the top box of the figure) is not itself a learning space.

In contrast, the fiber of a learning space might not be a learning space. For instance, for the learning space $\mathcal{L}$ of Figure 14.2, the projection $\mathcal{L}_{\{B\}}$ has two states, $\emptyset$ and $\{B\}$, but the inverse image $\mathcal{L}(B, \emptyset)$ of $\{B\}$ is the family of three states $\{\{A, B\}, \{A, B, C\}, \{B, C\}\}$, which does not form a learning space as it fails the accessibility property. However, by the same reasoning as in the proof of Theorem 14.2.1, the fiber of a learning space projection is well-graded and closed under unions, and therefore forms a *closed medium* (Falmagne and Ovchinnikov, 2002).

If $\mathcal{L}$ is a learning space, we define an *upper subfamily* $\mathcal{L}^+$ of $\mathcal{L}$ to be a subset of the states of $\mathcal{L}$, such that if $S \subset T$ are two states of $\mathcal{L}$ with $S \in \mathcal{L}^+$ then $T \in \mathcal{L}^+$.

**14.2.2 Theorem.** *Every fiber $\mathcal{L}(K, U)$ forms an upper subfamily of a learning space $\mathcal{L}'$, and every upper subfamily $\mathcal{L}^+$ of a learning space $\mathcal{L}$ can be represented as a fiber $\mathcal{L}'(K, U)$ for some learning space $\mathcal{L}'$ and sets $K$ and $U$.*

PROOF. First, let $\mathcal{L}(K, U)$ be a fiber, and let $S = \cup\mathcal{L}(K, U)$. Then $S$ is the unique maximal state of $\mathcal{L}(K, U)$, and $\mathcal{L}(K, U) = \mathcal{L}_S(K, \emptyset)$. We assert that $\mathcal{L}(K, U)$ is an upper subfamily of $\mathcal{L}_S$. For, if $A \in \mathcal{L}(K, U)$, $B \in \mathcal{L}_S$, and $A \subset B$, then $B$ like $A$ must contain all members of $K$, so $B$ must belong to $\mathcal{L}(K, U)$.

In the other direction, suppose $\mathcal{L}$ is a learning space, and $\mathcal{L}^+$ is an upper subfamily of $\mathcal{L}$. Then, $\mathcal{L}^+$ can be represented as a fiber of a learning space, as follows. Form $\mathcal{L}'$ by adding to $\mathcal{L}$ the sets of the form $S \cup \{x\}$ where $S \in \mathcal{L}^+$ and $x \notin \cup\mathcal{L}$. Project $\mathcal{L}'$ onto $\{x\}$; then $\mathcal{L}^+$ is the inverse image of $\{x\}$ under this projection. □

Figure 14.3 shows a well-graded union-closed set family $\mathcal{F}$ that cannot be an upper subfamily $\mathcal{L}^+$ of a learning space $\mathcal{L}$. To see this, note that any learning space $\mathcal{L}$ that might be a completion of $\mathcal{F}$ necessarily contains a singleton as one of its states; because $\mathcal{F}$ is symmetric we can assume without loss of generality that this singleton is $\{a\}$. Then, by closure under unions, $\mathcal{L}$ would also contain $\{a\} \cup \{d, e\} = \{a, d, e\}$. But $\{a, d, e\}$ is not in $\mathcal{F}$ while its subset

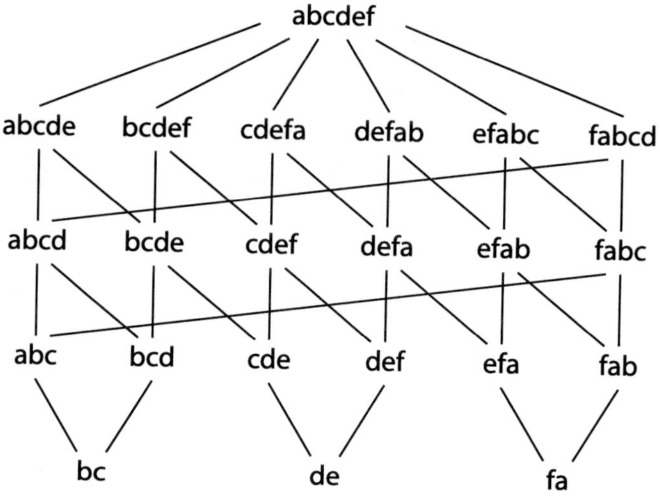


**Figure 14.3.** A well-graded and union-closed set family that cannot be the fiber of a projected learning space.

$\{d, e\}$ is, contradicting the assumption that $\mathcal{F}$ is an upper subfamily of $\mathcal{L}$. Therefore, the family of fibers of projections of learning spaces is a proper subclass of the family of closed media.

**14.2.3 Theorem.** *Let $\mathcal{B}$ be a given family of sets, the closure under union of which generates a family $\mathcal{F}$. Then in time polynomial in the total size of the sets in $\mathcal{B}$ we can determine whether $\mathcal{F}$ is an upper subfamily $\mathcal{L}^+$ of a learning space $\mathcal{L}$, and if so find a collection of learning sequences defining $\mathcal{L}$.*

PROOF. First, observe that we may test membership of a set $S$ in $\mathcal{F}$ in polynomial time, as $S \in \mathcal{F}$ if and only if $S = \cup\{T \in \mathcal{B} \mid T \subset S\}$. We define a set $S$ to be *safe* if, for any $T \in \mathcal{F}$, the union $S \cup T$ also belongs to $\mathcal{F}$. Equivalently, $S$ is safe if and only if, for any $T \in \mathcal{B} \cup \{\emptyset\}$, the union $S \cup T$ belongs to $\mathcal{F}$, so we may test safety in polynomial time. If two sets $S$ and $T$ are safe, so is their union. If $\mathcal{F} \subset \mathcal{G}$, then $\mathcal{F}$ is an upper subfamily of $\mathcal{G}$ if and only if all sets in $\mathcal{G} \setminus \mathcal{F}$ are safe.

We say that a set $S$ is *reachable* if we can find a sequence $\sigma_S$ of all the elements of $S$ such that all prefixes of this sequence are safe. If $S$ and $T$ are both reachable, with $S \subset T$, then we may assume $\sigma_S$ forms an initial subsequence of $\sigma_T$, for if not the concatenation of $\sigma_S$ with the members of $\sigma_T$ not belonging to $\sigma_S$ forms another sequence on the elements of $T$ with the same property that all prefixes are reachable, by the union-closure of safe sets. Therefore, we may test reachability of $S$ in polynomial time, and find a sequence $\sigma_S$ for any reachable $S$, by the following greedy algorithm:

1. Initialize $\sigma$ to empty.
2. While $\sigma$ does not contain all elements of $S$:
   a) If there exists $x$ in $S$ that can be added to $\sigma$ to form a longer safe prefix, do so.
   b) Otherwise, terminate the algorithm and report that $S$ is not reachable.
3. Return $\sigma_S = \sigma$.

If $\mathcal{F}$ is an upper subfamily of a learning space $\mathcal{L}$, and $S \in \mathcal{B}$, then all sets in $\mathcal{L}$ must be safe, so $S$ is reachable via a learning sequence of $\mathcal{L}$ for which $S$ is a prefix. Therefore, if any set in $\mathcal{B}$ is not reachable, then $\mathcal{F}$ is not an upper subfamily of a learning space. On the other hand, if every set $S$ in $\mathcal{B}$ is reachable via a sequence $\sigma_S$, then the unions of prefixes of these sequences form a learning space $\mathcal{L}$ containing $\mathcal{B}$ in which every set is safe; therefore $\mathcal{B}$ is an upper subfamily of this learning space $\mathcal{L}$. The atoms of this learning space consists of certain prefixes of the sequences $\sigma_S$ for $S \in \mathcal{B}$, and we may use our concise representation algorithm to convert these atoms into a collection of learning sequences representing $\mathcal{L}$. □

This result gives us hope that there exists a simple combinatorial description of the set families that form upper subfamilies of learning spaces. However, finding the *smallest* learning space for which $\mathcal{F}$ is an upper subfamily is considerably more difficult.

**14.2.4 Theorem.** *Let $\mathcal{B}$ be a family of sets and let $\mathcal{F}$ be the family of unions of members of $\mathcal{B}$. Then it is NP-complete to determine, given $\mathcal{B}$ and an integer $K$, whether $\mathcal{B}$ is the upper subfamily of a learning space $\mathcal{L}$ with $|\mathcal{L} \setminus \mathcal{F}| \leq K$.*

PROOF. Note that, by our previous construction, $K$ need only be polynomial in the total size of $\mathcal{B}$. Therefore, if there exists a suitably small $\mathcal{L}$, we may exhibit it by listing the additional sets added to $\mathcal{F}$ to form $\mathcal{L}$, and test whether it correctly solves the problem by testing the safety of the additional sets, the union-closure of the added sets, and the reachability of each set in $\mathcal{B}$ by a sequence each prefix of which is a union of the added sets and of other sets already in $\mathcal{B}$. Therefore, the problem is in NP.

To show NP-hardness, we reduce from the Vertex Cover problem, in which we are given a connected undirected graph $G$ and an integer $K$, and must find a set of $K$ or fewer vertices containing at least one endpoint of each edge of the graph. From an instance of Vertex Cover, we define $\mathcal{B}$ as containing a set of the two endpoints of each edge; the family $\mathcal{F}$ generated from $\mathcal{B}$ consists of all sets of two or more vertices from connected subgraphs of $G$. Any learning space containing $\mathcal{F}$ must contain a subfamily of singleton sets for the vertices in a vertex cover of $G$, or some set in $\mathcal{B}$ would not be accessible, and conversely if $C$ is any vertex cover of $G$ we may form a learning space $\mathcal{L}$ with $|\mathcal{L} \setminus \mathcal{F}| = |C|$ by including a singleton set for each member of $C$. □

## 14.3 Finding Concise Sets of Learning Sequences

We have seen that a learning space may be defined from a set $\Sigma$ of learning sequences. The running time of our algorithms depends on the number of sequences in $\Sigma$, so in order for this representation of a learning space to yield efficient algorithms, we need this number to be small.

We outline in this section a method for finding the smallest set of learning sequences that define a given learning space, given as input a description of the space by a (non-optimal) set of learning sequences. The method is based on a standard characterization of the minimum number of learning sequences necessary to define the given space (in antimatroid terminology, the *convex dimension* of the learning space) as the width of an associated partially ordered set; cf. Theorem III.6.9 of Korte et al. (1991). This characterization allows an optimal set of learning sequences to be constructed using a graph matching algorithm. Relatedly, we have provided in Eppstein (2008) a geometric characterization of the learning spaces that may be defined from two learning sequences, from which we described an algorithm for finding a pair of defining sequences in time linear in the number of states of the given learning space. The algorithm described here is more general, in that it works regardless of the number of learning sequences needed to define the space, and it also has the advantage that it runs in an amount of time polynomial in the total size of the set of input sequences, even when the number of states in the learning space that they define may be exponentially larger.

**14.3.1 Definition.** A *chain* in a partial order is a set of items that are all comparable to each other, or equivalently a sequence $x_0, x_1, \ldots$ such that $x_i < x_j$ if and only if $i < j$. A *chain cover* is a set of chains that together include all items in the order. An *antichain* is a set of items no two of which are comparable to each other. The *width* of a partial order is the maximum cardinality of any of its antichains, or equivalently (by Dilworth's theorem) the minimum number of chains in a chain cover.

An optimal chain decomposition of a given partial order may be found in polynomial time via bipartite graph matching, as illustrated in Figure 14.4. The far left of the figure shows a partial order, and the center left shows a bipartite graph with two vertices for each item in the partial order, one on each side. The graph has an edge from a vertex $x$ on the left to a vertex $y$ on the right whenever $x < y$. A maximum matching in this graph (a set of edges of maximum cardinality, no two of which share an endpoint) is shown in center right. From any matching in this graph, we may derive a chain cover in which the consecutive pairs of elements in the chains are given by the matched edges; the cover derived in this way is shown on the far right of the figure. In a partial order with $n$ items, matchings with $k$ matched edges correspond to chain covers with $n - k$ chains, so the minimum chain cover can be found from the maximum matching. Hopcroft and Karp (1973) showed how to find

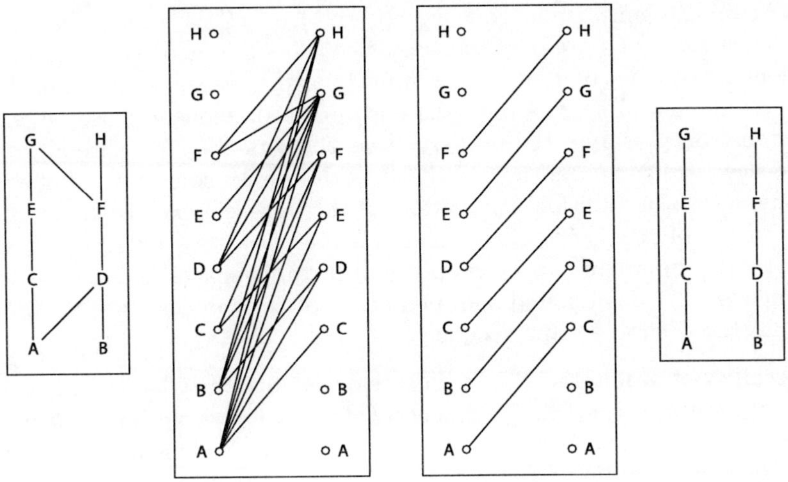


**Figure 14.4.** An optimal chain decomposition of a partial order may be found via bipartite graph matching.

a maximum matching in a bipartite graph with $n$ vertices and $m$ edges in time $O(mn^{1/2})$; for partial orders with $n$ items this translates to an $O(n^{5/2})$ time bound for finding an optimal chain decomposition and therefore finding the width. A maximum antichain may also be found in the same amount of time; we omit the details as they are not necessary for our learning space application.

If we partially order the states of a learning space by set inclusion, then a chain of states consists of a nested family of sets.

**14.3.2 Lemma.** *Let $\mathcal{C}$ be a chain of states in a learning space $\mathcal{L}$. Then there exists a learning sequence $\sigma$ of $\mathcal{L}$ such that each state of $\mathcal{C}$ is a prefix of $\sigma$. If $\mathcal{L}$ is defined from a set $\Sigma$ of learning sequences, then $\sigma$ may be constructed from $\mathcal{C}$ in time polynomial in the total size of $\Sigma$.*

PROOF. We may assume without loss of generality that $\mathcal{C}$ contains $\emptyset$ and $\cup\mathcal{L}$. Sort the sets in $\mathcal{C}$ by size:

$$\emptyset = S_0 \subset S_1 \subset S_2 \subset \cdots \subset S_k = \cup\mathcal{L}.$$

We will build up $\sigma$ in stages, maintaining the invariant that, after the $i$th stage, we will have chosen values for $\sigma(0), \sigma(1), \ldots \sigma(|S_i| - 1)$ such that all sets $S_j$, $j \leq i$, are prefixes of $\sigma$ and such that every prefix of $\sigma$ is a state of $\mathcal{L}$. Initially, the empty sequence satisfies this requirement for $i = 0$. The invariant ensures that, when we have completed this process, $\sigma$ will be a learning sequence with the desired properties.

In stage $i$, we will have already generated a sequence for the elements of $S_{i-1}$. By repeatedly applying the accessibility property of learning spaces to

$S_i$, we can also generate a sequence $\tau(0), \tau(1), \ldots \tau(|S_i| - 1)$ of the elements of $S_i$, such that each prefix of this sequence is a state in $\mathcal{L}$. We then form our new longer sequence of $\sigma_j$ values by concatenating the subsequence of the $\tau_j$'s not already chosen as $\sigma_j$'s onto the end of the previously chosen $\sigma_i$'s. This concatenation preserves the property that the sets $S_j$, $j < i$, are all prefixes of the new sequence, and now $S_i$ itself is the prefix consisting of the whole concatenation. In addition, each prefix of the new sequence is the union of a prefix of the old sequence of $\sigma_j$'s and of a prefix of the sequence of $\tau_j$'s, and therefore by the union closure property of learning spaces is a state of $\mathcal{L}$.

Once we have completed the final stage of this process, for $S_k = \cup\mathcal{L}$, $\sigma$ provides the desired learning sequence. □

Recall (Definition 13.5.6 and Theorem 13.5.8 from Chapter 13) that an *atom* is a state of a learning sequence $\mathcal{L}$ from which only one element may be removed, and that a set $\Sigma$ of learning sequences is sufficient to define $\mathcal{L}$ if and only if every atom of $\mathcal{L}$ is a prefix of a sequence in $\Sigma$.

**14.3.3 Definition.** The *path poset* of a learning space is the set of its atoms, partially ordered by inclusion.

**14.3.4 Lemma.** *Let $\Sigma$ be a (non-optimal) set of learning sequences defining learning space $\mathcal{L}$. Then in time polynomial in the total size of $\Sigma$ we may construct the path poset of $\mathcal{L}$.*

PROOF. By Theorem 13.5.8, each atom must be a prefix of $\Sigma$. We may determine which prefixes are atoms by computing the inner fringe for each prefix, as detailed in Chapter 13. Once the atoms have been found, we may form a partially ordered set from them by testing each pair of states for set inclusion. □

**14.3.5 Theorem.** *Let $\Sigma$ be a (non-optimal) set of learning sequences defining learning space $\mathcal{L}$. Then in time polynomial in the total size of $\Sigma$ we may construct a minimum-cardinality set $\Sigma'$ of learning sequences that also defines $\mathcal{L}$.*

PROOF. We find the path poset of $\mathcal{L}$ by Lemma 14.3.4, decompose it into a minimum number of chains by using the bipartite matching algorithm, and construct a learning sequence for each chain by Lemma 14.3.2. Every representation of $\mathcal{L}$ by learning sequences must give rise to a cover of the path poset by chains, with one chain per learning sequence, and the representation $\Sigma'$ constructed as above uses the minimum possible number of chains for such a decomposition, so it has the minimum possible number of learning sequences of any representation of $\mathcal{L}$. □

## 14.4 Decomposition of Learning Spaces

The problem of finding the minimum cardinality set of learning sequences defining a learning space can be interpreted as a form of decomposition of the space, and the algorithmic success of our learning sequence representation motivates us to seek more powerful forms of decompositions of learning spaces.

**14.4.1 Definition.** If $\mathcal{L}_1$ and $\mathcal{L}_2$ are two learning spaces on the same set of concepts, we may define a *join* operation combining the two into a single larger learning space:

$$\mathcal{L}_1 \sqcup \mathcal{L}_2 = \{S_1 \cup S_2 \mid S_i \in \mathcal{L}_i\}.$$

The join operation is commutative, associative, and *idempotent* (that is, $\mathcal{L} \sqcup \mathcal{L} = \mathcal{L}$ for any learning space $\mathcal{L}$); therefore it forms a *semilattice* although not, in general, a medium. The problem of computing the convex dimension of a learning space $\mathcal{L}$, and of representing $\mathcal{L}$ via a minimum number of learning sequences, can be viewed in this way as expressing $\mathcal{L}$ as the join of a small number of simpler learning spaces: if we define a *totally ordered learning space* to be a learning space formed from a single learning sequence, then $\mathcal{L}$ has convex dimension at most $K$, and can be represented via $K$ learning sequences, if and only if it is the join of at most $K$ totally ordered learning spaces. The totally ordered learning spaces form the smallest subclass of learning spaces such that all learning spaces can be represented as joins of learning spaces in the subclass, for a learning space is irreducible for the join operation if and only if it is totally ordered.

More generally, if $\mathcal{L}$ can be represented as a join of a small number of learning spaces $\mathcal{L}_i$, each of which in turn has an efficient algorithmic representation, then it seems likely that we may also perform algorithms such as state generation efficiently for $\mathcal{L}$ itself, with an efficiency depending in part on the number $K$ of learning spaces used to represent $\mathcal{L}$. But to use representions based on joins of special classes of learning spaces, we need to be able to find such representations efficiently. We have already shown that, if $\mathcal{F}$ is the family of totally ordered learning spaces, then a representation of any learning space as the join of a minimum number of members of $\mathcal{F}$ can be found efficiently. For which other families $\mathcal{F}$ is such efficient decomposition possible? Unfortunately we have a negative answer to this question.

**14.4.2 Theorem.** *It is NP-complete, given a set $\mathcal{B}$ of atoms that define a learning space $\mathcal{L}$, and given an integer $K$, to determine whether $\mathcal{L}$ is the join of $K$ or fewer quasi-ordinal spaces.*

PROOF. If $\mathcal{L}$ is the join of $K$ or fewer quasi-ordinal spaces, we may exhibit a partial order for each such space, test whether each atom of each such space belongs to $\mathcal{L}$, and test whether each atom of $\mathcal{L}$ belongs to at least one of these spaces; therefore, testing whether $\mathcal{L}$ is the join of $K$ or fewer quasi-ordinal spaces can be done in NP.

To show NP-hardness, we reduce from the known NP-complete problem of graph edge coloring. In this problem we are given an undirected graph $G$ and an integer $K$ and must assign $K$ distinct colors to the edges of $G$ in such a way that no two edges that share an endpoint have the same color. From $G$ we form a learning space $\mathcal{L}$ with one concept per vertex $v$, and two concepts $\ell_e$ and $r_e$ per edge $e$. We include as atoms of $\mathcal{L}$ the singleton sets $\{\ell_e\}$ for each edge, the sets $\{ell_e, v\}$ for each edge $e$ having an endpoint $v$, and the sets $\{\ell_e, v, w, r_e\}$ for each edge $e$ having two endpoints $v$ and $w$.

Then, if $\mathcal{L}' \subset \mathcal{L}$ is a quasi-ordinal learning space containing a state $\{\ell_e, v, w, r_e\}$, the partial order defining $\mathcal{L}'$ must have $\ell_e$ (and no other concept) preceding $v$ and $w$. Therefore, it can contain no other state $\{\ell_f, u, v, r_f\}$ for an edge $f$ sharing an endpoint with $e$. So, if $\mathcal{L}$ is the join of $K$ quasi-ordinal spaces, we can find a coloring of the edges of $G$ with $K$ colors, by coloring each edge according to which member of the join contains state $\{\ell_e, v, w, r_e\}$. Conversely, for any independent set of edges, we can find a quasi-ordinal subspace of $\mathcal{L}$ containing the states $\{\ell_e, v, w, r_e\}$ for each edge in the independent set, so for each $K$-edge-coloring of $G$ we can find a representation of $\mathcal{L}$ as the join of $K$ quasi-ordinal spaces.

Since decomposition into $K$ quasi-ordinal spaces is in NP and a known NP-hard problem can be reduced to it, it is NP-complete. □

The same proof shows that the problem remains hard when $K$ is fixed to any constant greater than two. A similar reduction (omitting the concepts $\ell_e$) shows that it is NP-complete to represent a learning space as a join of three or more *hierarchies*, where a hierarchy is defined as a quasi-ordinal learning space for an order the Hasse diagram of which has at most one outgoing edge per concept. A hierarchy can be characterized as a learning space each two atoms of which are either disjoint or nested, so one can test whether a learning space $\mathcal{L}$ is a join of two hierarchies by forming a graph in which we connect two atoms of $\mathcal{L}$ by an edge whenever they do not satisfy this condition and testing whether this graph is bipartite; we omit the details. We have not yet determined whether it is possible to test efficiently whether a given learning space is the join of two quasi-ordinal spaces.

## 14.5 Adapting a Learning Space

Thiéry (2001) defines the *fringe* of a learning space to be the collection of sets that can be added to the space as new states, or removed from the states of the space, in order to form new learning spaces with more or fewer states. He calculates the fringes of a space as part of an algorithm for adapting a knowledge space to make it more accurately reflect the observed knowledge of students. One of the great advantages of learning sequence based construction of learning spaces, over partial orders, is their ability to allow such adaption: learning sequences can represent any learning space, and in particular can represent the spaces formed by adding states to and removing states from an existing space. In a computerized testing system such as ALEKS, using a learning space that accurately models student knowledge would improve its interactions with students, compared with using a less-accurate model. The student interactions from such a system provide a readily available supply of data that could be used to improve the accuracy of the model. Motivated by this idea, we describe here how to calculate the fringes of a learning space defined from a set of learning sequences.

As with the fringe of an individual state in a learning space, we may distinguish two subsets of the fringe of the learning space itself. The *inner fringe* of $\mathcal{L}$ consists of those states of $\mathcal{L}$ that may be removed, leaving a learning space for the same set of concepts with fewer states, while the *outer fringe* of $\mathcal{L}$ consists of those sets that are not states of $\mathcal{L}$ but may be added as states, resulting in a learning space on the same set of concepts but with more states.

We first consider the inner fringe.

**14.5.1 Theorem.** *State $S$ belongs to the inner fringe of $\mathcal{L}$ if, and only if, it satisfies all of the following requirements:*

1. *$S$ is an atom of $\mathcal{L}$.*
2. *$S \neq \cup\mathcal{L}$.*
3. *No set $S \cup \{x\}$, $x \notin S$, is an atom of $\mathcal{L}$.*

PROOF. If $S$ were not an atom, it could be formed as the union of other sets in $\mathcal{L}$, and removing it would violate union closure. If $S$ is an atom, but equals $\cup\mathcal{L}$, then removing it would violate the requirement that the resulting space have the same set of concepts. And if some set $S\cup\{x\}$ were an atom, removing $S$ would violate accessibility for that set. Therefore, a set that violates one or more of the requirements in the theorem cannot belong to the inner fringe.

In the other direction, suppose that all the requirements of the theorem are all satisfied. Because $S$ is an atom, it cannot be the union of other sets, so its removal cannot affect the closure under unions of $\mathcal{L}$. Because each state $S \cup \{x\}$ is not an atom, it has at least two removable elements, at least one of which is not $x$, so the removal of $S$ preserves the accessibility property of $\mathcal{L}$. And because $S$ is not the top state $\cup\mathcal{L}$ of $\mathcal{L}$, its removal cannot change the set of concepts of the resulting space. Thus, removing $S$ forms a learning space on the same set of concepts, meeting the definition of the inner fringe. □

If $S$ belongs to the inner fringe of $\mathcal{L}$, then the atoms of the learning space formed by removing $S$ consist of the atoms of $\mathcal{L}$ other than $S$, together with any state $S\cup\{x\}$ that becomes an atom by the removal. Therefore, the learning space formed by $S$ may be represented using a small number of atoms, from which in polynomial time we may construct a learning sequence representation of the space following the same outline as in Theorem 14.3.5.

Theorem 14.5.1 leads to an efficient algorithm for finding all of the states in the inner fringe of a learning space $\mathcal{L}$ defined from a set of learning spaces: the atoms of $\mathcal{L}$ must be prefixes of these sequences, and it is straightforward to test for each prefix whether it meets the conditions of the theorem.

The sets in the outer fringe of a learning space $\mathcal{L}$, sets that may be added as new states to a learning space $\mathcal{L}$ to form an augmented learning space, also have a simple characterization, but their algorithmic generation is not as efficient.

**14.5.2 Theorem.** *Set $T$ belongs to the outer fringe of $\mathcal{L}$ if, and only if, it has the form $S \cup \{x\}$ where $S$ is a state of $\mathcal{L}$ and $x$ belongs to the intersection of the outer fringes of all states of the form $S \cup \{y\}$.*

PROOF. If $T$ does not have the form $S \cup \{x\}$ where $S$ is a state of $\mathcal{L}$, then the set system formed by adding $T$ will not be accessible. And if $T = S \cup \{x\}$ but there exists a state $S \cup \{y\}$ for which $x$ does not belong to the outer fringe of the state, then $T \cup (S \cup \{y\})$ does not belong to $\mathcal{L}$, so the set system formed by adding $T$ will not be closed under unions.

In the other direction, if $T = S \cup \{x\}$ for a state $S$ of $\mathcal{L}$, then the set system formed by adding $T$ will be accessible. In forming a union of $T$ with a state $U$ of $\mathcal{L}$, if $U \subset S$ then the union is $T$ itself. Otherwise we may use the properties of learning spaces to find $y \in U \setminus S$ such that $S \cup \{y\}$ is also a state of $\mathcal{L}$; then $T \cup U = (T \cup (S \cup \{y\})) \cup U$. But, if $x$ belongs to the outer fringe of all states $S \cup \{y\}$ then $T \cup (S \cup \{y\}$ is a state of $\mathcal{L}$, and we have refactored $T \cup U$ into the union of two states of $\mathcal{L}$, which must therefore also belong to $\mathcal{L}$. Thus, if $T$ obeys the conditions of the theorem, adding $T$ to $\mathcal{L}$ preserves closure under unions. □

Thus, to generate all sets that may be added to $\mathcal{L}$, we need merely list all states of $\mathcal{L}$, list for each state $S$ the states $S \cup \{y\}$ by computing the outer fringe of $S$, and intersect the outer fringes of the states $S \cup \{y\}$. Each such set is listed only once by this procedure, as the set $S$ for which it is listed must be unique or it would already belong to $\mathcal{L}$ by union closure. However, the time for this procedure is controlled by the number of states in $\mathcal{L}$ rather than being a polynomial function of the total size of the learning sequences describing $\mathcal{L}$.

If we add a set $S \cup \{x\}$ as a new state to a learning space $\mathcal{L}$, the atoms of the new learning space consists of the newly added set, together with all atoms of $\mathcal{L}$ that are not of the form $S \cup \{x, y\}$. (The atoms of $\mathcal{L}$ that are of the form $S \cup \{x, y\}$ are no longer atoms in $L \cup \{S \cup \{x\}\}$, as they are the union of states $S \cup \{x\}$ and $S \cup \{y\}$; however, the addition of $S \cup \{x\}$ does not change whether states not of the form $S \cup \{x, y\}$ are atoms.) As before, by applying the methods of Theorem 14.3.5 it is straightforward to construct a concise set of learning sequences describing the modified learning space.

## 14.6 Future Work

**Visualization of learning spaces.** In designing a computerized curriculum, it may be helpful to provide the designers with visualisations of a learning space; however, realistic learning spaces have too many states to draw them in the standard way as a state-transition diagram. Quasi-ordinal spaces may be visualized instead by drawing their Hasse diagrams as graphs, but this technique does not work so well for more general learning spaces. In earlier work (Eppstein, 2005, 2008) we found algorithms for drawing the state-transition diagrams of small learning spaces and other types of medium, however these methods work best when the space being drawn has a well-behaved embedding into a low-dimensional Euclidean space. Can we generalize these drawing approaches to learning spaces with higher convex dimension? Is there a possibility of a hybrid approach that draws portions of a learning space as a Hasse diagram on concepts and resorts to the more complex state space only when necessary?

**Faster outer fringe construction.** Our algorithm for constructing the family of sets that can be added to a learning space to form new larger learning spaces involves generating all states of the learning space. However, there may be many fewer fringe sets than there are states. In some sense, the atoms of a learning space consist of the minimal sets in the space, while the outer fringe consists of the maximal sets not in the space. Thus, it is plausible that one could adapt hypergraph transversal algorithms (Fredman and Khachiyan, 1996), which can be used to convert minimal sets in a family to maximal sets not in a family for certain other types of set families, to the purpose of finding the outer fringe of a learning space in time pseudopolynomial in the number of sets in the outer fringe. Such a result would also have implications for the computational complexity of inferring a learning space from questions asked of an expert (Dowling, 1993a). However, we have not worked out the details of such an efficient algorithm for listing outer fringe states.

**Reconciliation of expert opinions.** Along with its applications to concise representation and efficient state generation in learning spaces, the join of multiple learning spaces may be useful for a problem arising when constructing learning spaces from the answers of experts (Dowling, 1993a): two different experts may give quite different answers when asked what they believe about the prerequisite structure of a given set of concepts, leading to quite different learning spaces on those concepts. The construction procedure of Dowling (1993a) involves asking experts a series of questions about whether valid knowledge states can exist with certain combinations of concepts, but the answers to these questions have only been found reliable when the combinations involve at most two concepts at a time; the learning spaces generated by limiting the questioning to such combinations are necessarily quasi-ordinal. To reliably generate more complex learning spaces, it seems necessary to combine the results from questioning multiple experts. The join provides a mathematical mechanism for reconciling those answers and finding a common learning space containing as states any set of concepts believed to form a state by any of the experts, but the learning spaces constructed in this way are likely to be much larger than necessary. More research is needed on methods for combining information from multiple experts to generate learning spaces of size comparable to the space that would be constructed by questioning a single expert, while simultaneously taking advantage of the multiplicity of experts to generate spaces that more accurately model the students' knowledge.

**Structure of the family of learning spaces**. As Thiéry (2001) showed, when one learning space forms a subfamily or superfamily of the other, there exists a shortest path from one to the other in which each step adds or removes a state and each set family in this shortest path is also a learning space. That is, the family of learning spaces has a chain property similar to that of individual learning spaces and media. This fact motivates the calculation of the fringes of a learning space, as the sets in the fringe represent potential neighbors in such paths. We also know that the family of learning spaces on a given domain forms a semilattice under the join operation, but this operation is not simply the union of set families. And we know that the family of learning spaces is not in general well-graded, so it does not form a medium under operations that add and remove sets. What other structure does the family of learning spaces have, and how can that structure help us quickly adapt a learning space to changing information about the possible knowledge states of students?

## 14.7 Conclusions

We have shown how to sample from a learning space, form a projected space, and make inferences from the projection back to the original space, and we have studied the characterization and recognition of fibers of a projection. We have shown how to represent learning spaces using concise sets of learning sequences and, generalizing this, have studied the complexity of finding decompositions of learning spaces into small numbers of simpler spaces. We have also characterized the states that may be added to or removed from a learning space, preserving its properties, and provided algorithms for listing these states and for finding learning sequence representations of the resulting modified learning spaces. We believe these methods will prove to be useful in the automated assessment of knowledge for learning spaces that are too large to list all states explicitly, and in the adaption of learning spaces to multiple experts and feedback from student data.

# Bibliography


J. Aczél. *Lectures on Functional Equations and their Applications.* Academic Press, New York and San Diego, 1966.

D. Albert. Surmise relations between tests. Manuscript of a talk held at the 28th Annual Meeting of the Society for Mathematical Psychology, University of California, Irvine, 1995.

D. Albert and T. Held. Establishing knowledge spaces by systematical problem construction. In D. Albert, editor, *Knowledge Structures*, pages 78–112. Springer-Verlag, New York, 1994.

D. Albert and T. Held. Component based knowledge spaces in problem solving and inductive reasoning. In D. Albert and J. Lukas, editors, *Knowledge Spaces: Theories, Empirical Research, Applications*, pages 15–40. Lawrence Erlbaum Associates, Mahwah, NJ, 1999.

D. Albert and C. Hockemeyer. Adaptive and dynamic hypertext tutoring systems based on knowledge space theory. In B. du Boulay and R. Mizoguchi, editors, *Artificial Intelligence in Education: Knowledge and Media in Learning Systems*, volume 39 of *Frontiers in Artificial Intelligence and Applications*, pages 553–555, Amsterdam, 1997. IOS Press.

D. Albert and C. Hockemeyer. Developing curricula for tutoring systems based on prerequisite relationships. In G. Cumming, T. Okamoto, and L. Gomez, editors, *Advanced Research in Computers and Communications in Education: New Human Abilities for the Networked Society*, volume 2 of *Frontiers in Artificial Intelligence and Applications*, pages 325–328, Amsterdam, 1999. IOS Press. Proceedings of the 7th International Conference on Computers in Education (ICCE), Chiba, Japan.

D. Albert and C. Hockemeyer. Applying demand analysis of a set of test problems for developing an adaptive course. In *Proceedings of the International Conference on Computers in Education ICCE 2002*, pages 69–70, Los Alamitos, 2002. IEEE Computer Society Press. doi: 10.1109/CIE.2002.1185866.

D. Albert and J. Lukas, editors. *Knowledge Spaces: Theories, Empirical Research, Applications.* Lawrence Erlbaum Associates, Mahwah, NJ, 1999.



J.-C. Falmagne et al. (eds.), *Knowledge Spaces: Applications in Education*,
DOI 10.1007/978-3-642-35329-1,

D. Albert and J. Musch. Knowledge space modeling of Ravens advanced progressive matrices (APM) test. In V. Kolesaric and D. Ivanec, editors, *Abstracts of the 4th Alps Adria Psychology Symposium, October 3-5, 1996*, page 39, Zagreb, Croatia, 1996. Department of Psychology, Faculty of Philosophy, University of Zagreb.

D. Albert, M. Schrepp, and T. Held. Construction of knowledge spaces for problem solving in chess. In G.H. Fischer and D. Laming, editors, *Contributions to Mathematical Psychology, Psychometrics, and Methodology*, Recent Research in Psychology, pages 123–135. Springer-Verlag, New York, 1994. doi: 10.1007/978-1-4612-4308-3_9.

D. Albert, C. Hockemeyer, Z. Kulcsar, and G. Shorten. Competence assessment for spinal anaesthesia. In A. Holzinger, editor, *HCI and Usability for Medicine and Health Care. Proceedings of the Third Symposium of the Workgroup Human-Computer Interaction and Usability Engineering of the Austrian Computer Society, USAB 2007 Graz, Austria, November, 22, 2007*, volume 4799 of *Lecture Notes in Computer Science*, pages 165–170, Berlin, 2007. Springer-Verlag. doi: 10.1007/978-3-540-76805-0_14.

D. Albert, M.D. Kickmeier-Rust, and F. Matsuda. A formal framework for modelling the developmental course of competence and performance in the distance, speed, and time domain. *Developmental Review*, 28:401–420, 2008. doi: 10.1016/j.dr.2008.05.001.

R.G. Almond, L.V. DiBello, B. Moulder, and J.D. Zapata-Rivera. Modeling diagnostic assessments with Bayesian networks. *Journal of Educational Measurement*, 44:341–359, 2007. doi: 10.1111/j.1745-3984.2007.00043.x.

E.B. Andersen. Latent structure analysis: A survey. *Scandinavian Journal of Statistics*, 9:1–12, 1982.

J. Angrist and V. Lavy. New evidence on classroom computers and pupil learning. *The Economic Journal*, 112:735–765, 2002. doi: 10.1111/1468-0297.00068.

R.D. Arasasingham, M. Taagepera, F. Potter, and S. Lonjers. Using knowledge space theory to assess student understanding of stoichiometry. *Journal of Chemical Education*, 81:1517–1523, 2004. doi: 10.1021/ed081p1517.

R.D. Arasasingham, M. Taagepera, F. Potter, I. Martorell, and S. Lonjers. Assessing the effect of web-based learning tools on student understanding of stoichiometry using knowledge space theory. *Journal of Chemical Education*, 82:1251–1262, 2005. doi: 10.1021/ed082p1251.

T. Augustin, C. Hockemeyer, M.D. Kickmeier-Rust, and D. Albert. Individualized skill assessment in digital learning games: Basic definitions and mathematical formalism. *IEEE Transactions on Learning Technologies*, 4 (2):138–142, 2011. doi: 10.1109/TLT.2010.21.

L. Averell and A. Heathcote. The form of the forgetting curve and the fate of memories. *Journal of Mathematical Psychology*, 55:23–35, 2010. doi: 10.1016/j.jmp.2010.08.009.

D. Avis and K. Fukuda. Reverse search for enumeration. *Discrete Applied Mathematics*, 65:21–46, 1996. doi: 10.1016/0166-218X(95)00026-N.

E.L. Baker, M. Gearhart, and J.L. Herman. Evaluating the Apple classrooms of tomorrow. In E.L. Baker and H.F. O'Neil, Jr., editors, *Technology assessment in education and training*, pages 735–765. Lawrence Erlbaum Associates, Hillsdale, NJ, 1994.

V. Bali and R. Alvarez. The race gap in student achievement scores: Longitudinal evidence from a racially diverse school district. *Policy Studies Journal*, 32(3):393–415, 2004. doi: 10.1111/j.1541-0072.2004.00072.x.

J. Baron and F. Norman. SATs, achievement tests, and high-school class rank as predictors of college performance. *Educational and Psychological Measurement*, 52:1047–1055, 1992. doi: 10.1177/0013164492052004029.

W.M. Bart and P.W. Airasian. Determination of the ordering among seven Piagetian tasks by an ordering-theoretic method. *Journal of Educational Psychology*, 66(2):277–284, 1974. doi: 10.1037/h0036180.

W.M. Bart and D.J. Krus. An ordering-theoretic method to determine hierarchies among items. *Educational and Psychological Measurement*, 33: 291–300, 1973. doi: 10.1177/001316447303300208.

K. Baumunk and C.E. Dowling. Validity of spaces for assessing knowledge about fractions. *Journal of Mathematical Psychology*, 41:99–105, 1997. doi: 10.1006/jmps.1997.1152.

K. Baumunk, C.E. Dowling, and C. Hockemeyer. Indices for detecting invalid prequisite relationships. Talk at the 28th European Mathematical Psychology Group (EMPG) Meeting, Nijmegen, The Netherlands, September 2–5 1997.

I.I. Bejar, R. Chaffin, and S. Embretson. *Cognitive and Psychometric Analysis of Analogical Problem Solving*. Recent Research in Psychology. Springer-Verlag, New York, 1991. doi: 10.1007/978-1-4613-9690-1.

G. Birkhoff. Rings of sets. *Duke Mathematical Journal*, 3:443–454, 1937.

B.S. Bloom. The 2-sigma problem: The search for methods of group instruction as effective as one-to-one tutoring. *Educational Researcher*, 13(6):4–16, 1984. doi: 10.2307/1175554.

A. Boomsma, M.A.J. Van Duijn, and T.A.B. Snijders, editors. *Essays on Item Response Theory*. Lecture Notes in Statistics. Springer-Verlag, New York, 2001. doi: 10.1007/978-1-4613-0169-1.

J. Borges and M. Levene. Data mining of user navigation patterns. In B. Masand and M. Spiliopoulou, editors, *Web Usage Analysis and User Profiling*, volume 1836 of *Lecture Notes in Artificial Intelligence*, pages 92–111. Springer-Verlag, Berlin, 2000. doi: 10.1007/3-540-44934-5_6.

D. Borsboom. The attack of the psychometricians. *Psychometrika*, 71:425–440, 2006. doi: 10.1007/s11336-006-1447-6.

S. Brandt, D. Albert, and C. Hockemeyer. Surmise relations between tests—mathematical considerations. *Discrete Applied Mathematics*, 127(2):221–239, 2003. doi: 10.1016/S0166-218X(02)00207-X.

J.D. Bransford, A.L. Brown, and R.R. Cocking, editors. *How People Learn: Brain, Mind, Experience and School*. National Academy Press, Washington, D.C., 1999.

J. Brooks-Gunn, P. Klebanov, and G. Duncan. Ethnic differences in children's intelligence test scores: Role of economic deprivation, home environment, and maternal characteristics. *Child Development*, 65:346–360, 1996. doi: 10.2307/1131822.

S. Brown, L. Neath, and N. Chater. A temporal ratio model of memory. *Psychological Review*, 114:539–576, 2007. doi: 10.1037/0033-295X.114.3.539.

D. Bullock, J. Callahan, Y. Ban, A. Ahlgren, and C. Schrader. The implementation of an online mathematics placement exam and its effects on student success in precalculus and calculus. In *ASEE Annual Conference and Exposition, Conference Proceedings, 2009*, 2009.

J. Callahan, S.Y. Chyung, J. Guild, W. Clement, J. Guarino, D. Bullock, and C. Schrader. Enhancing precalculus curricula with e-learning: Implementation and assessment. In *Proceedings ASEE Annual Conference and Exposition, June 2008*, 2008.

J. Carpenter and R.E. Hanna. Predicting student preparedness in calculus. In *Proceedings of the 2006 American Society for Engineering Education Annual Conference*, 2006.

P.A. Carpenter, M.A. Just, and P. Shell. What one intelligence test measures: A theoretical account of the processing in the Raven Progressive Matrices Test. *Psychological Review*, 97(3):404–431, 1990. doi: 10.1037/0033-295X.97.3.404.

N. Caspard and B. Monjardet. The lattices of closure systems, closure operators, and implicational systems on a finite set: A survey. *Discrete Applied Mathematics*, 127:241–269, 2003. doi: 10.1016/j.dam.2004.08.001.

N. Caspard and B. Monjardet. Erratum to: The lattices of closure systems, closure operators, and implicational systems on a finite set: A survey. *Discrete Applied Mathematics*, 145:333, 2005. doi: 10.1016/S0166-218X(04)00238-0.

D.R. Cavagnaro. Projection of a medium. *Journal of Mathematical Psychology*, 52:55–63, 2008. doi: 10.1016/j.jmp.2007.08.002.

S. Chaiklin. The zone of proximal development in Vygotsky's analysis of learning and instruction. In J.S. Chipman, L. Hurwicz, M.K. Richter, and H.F. Sonnenschein, editors, *Vygotsky's Educational Theory and Practice in Cultural Context*. Cambridge University Press, Cambridge, MA, 2003.

N. Chomsky. *Aspects of the Theory of Syntax*. MIT Press, Cambridge, MA, 1965.

D.H. Clements. 'Concrete' manipulatives, concrete ideas. *Contemporary Issues in Early Childhood*, 1(1):45–60, 1999. doi: 10.2304/ciec.2000.1.1.7.

P.A. Cohen, J.A. Kulik, and C.C. Kulik. Educational outcomes of tutoring: A meta-analysis of findings. *American Educational Research Journal*, 19: 237–248, 1982. doi: 10.2307/1162567.

E. Cosyn and H.B. Uzun. Note on two sufficient axioms for a well-graded knowledge space. *Journal of Mathematical Psychology*, 53(1):40–42, 2009. doi: 10.1016/j.jmp.2008.09.005.

T.M. Cover and J.A. Thomas. *Elements of Information Theory*. Wiley-interscience, 2006.

L. Crocker and J. Algina. *Introduction to Classical & Modern Test Theory.* Wadsworth–Thomson Learning, 1986.

M.R. Crowley. Cincinnati's experiment in Negro education: A comparative study of the segregated and mixed school. *The Journal of Negro Education*, 1(1):25–33, 1932. doi: 10.2307/2292012.

B.A. Davey and H.A. Priestley. *Introduction to Lattices and Order.* Cambridge Mathematical Textbooks. Cambridge University Press, Cambridge, UK, 1990.

C.M. Dayton and G.B. Macready. A probabilistic model for validation of behavioral hierarchies. *Psychometrika*, 41:189–204, 1976. doi: 10.1007/BF02291838.

P. de Bra. Adaptive hypermedia. In H.H. Adelsberger, Kinshuk, M. Pawlowski, and D. Sampson, editors, *Handbook on Information Technologies for Education and Training*, pages 29–46. Springer-Verlag, Berlin, 2008.

J. de la Torre and J. Douglas. Higher-order latent trait models for cognitive diagnosis. *Psychometrika*, 69:333–353, 2004. doi: 10.1007/BF02295640.

E. Degreef, J.-P. Doignon, A. Ducamp, and J.-Cl. Falmagne. Languages for the assessment of knowledge. *Journal of Mathematical Psychology*, 30:243–256, 1986. doi: 10.1016/0022-2496(86)90032-5.

M. DeLucia. Middlesex County College: An ALEKS Case Study. 2008.

A. Dempster, N. Laird, and D. Rubin. Maximum likelihood for incomplete data via the EM algorithm. *Journal of the Royal Statistical Society, Series B*, 39(1):1–38, 1977.

M.C. Desmarais and M. Gagnon. Bayesian student models based on item to item knowledge structures. In W. Nejdl and K. Tochtermann, editors, *Innovative Approaches for learning and Knowledge Sharing. First European Conference on Technology Enhanced learning, EC-TEL 2006*, number 4227 in Lecture Notes in Computer Science, pages 111–124, Heidelberg, 2006. Springer-Verlag. doi: 10.1007/11876663_11.

L.V. DiBello and W. Stout. Guest editors' introduction and overview: IRT-based cognitive diagnostic models and related methods. *Journal of Educational Measurement*, 44:285–291, 2007. doi: 10.1111/j.1745-3984.2007.00039.x.

J.-P. Doignon. Probabilistic assessment of knowledge. In D. Albert, editor, *Knowledge Structures*, pages 1–56. Springer-Verlag, New York, 1994a.

J.-P. Doignon. Knowledge spaces and skill assignments. In G.H. Fischer and D. Laming, editors, *Contributions to Mathematical Psychology, Psychometrics, and Methodology*, pages 111–121. Springer-Verlag, New York, 1994b. doi: 10.1007/978-1-4612-4308-3_8.

J.-P. Doignon and J.-Cl. Falmagne. Spaces for the assessment of knowledge. *International Journal of Man-Machine Studies*, 23:175–196, 1985. doi: 10.1016/S0020-7373(85)80031-6.

J.-P. Doignon and J.-Cl. Falmagne. Knowledge assessment: A set theoretical framework. In B. Ganter, R. Wille, and K.E. Wolfe, editors, *Beiträge zur Begriffsanalyse: Vorträge der Arbeitstagung Begriffsanalyse, Darmstadt 1986*, pages 129–140, Mannheim, 1987. BI Wissenschaftsverlag.

J.-P. Doignon and J.-Cl. Falmagne. Parametrization of knowledge structures. *Discrete Applied Mathematics*, 21:87–100, 1988. doi: 10.1016/0166-218X(88)90046-7.

J.-P. Doignon and J.-Cl. Falmagne. *Knowledge Spaces.* Springer-Verlag, Berlin, Heidelberg, and New York, 1999. doi: 10.1007/978-3-642-58625-5.

C.E. Dowling. Applying the basis of a knowledge space for controlling the questioning of an expert. *Journal of Mathematical Psychology*, 37:21–48, 1993a. doi: 10.1006/jmps.1993.1002.

C.E. Dowling. On the irredundant generation of knowledge spaces. *Journal of Mathematical Psychology*, 37:49–62, 1993b. doi: 10.1006/jmps.1993.1003.

C.E. Dowling and C. Hockemeyer. Computing the intersection of knowledge spaces using only their basis. In C.E. Dowling, F.S. Roberts, and P. Theuns, editors, *Recent Progress in Mathematical Psychology*, pages 133–141. Lawrence Erlbaum Associates, Mahwah, NJ, 1998.

C.E. Dowling and C. Hockemeyer. Automata for the assessment of knowledge. *IEEE Transactions on Knowledge and Data Engineering*, 13(1):451–461, 2001. doi: 10.1109/69.929902.

C.E. Dowling, C. Hockemeyer, and A.H. Ludwig. Adaptive assessment and training using the neighbourhood of knowledge states. In C. Frasson, G. Gauthier, and A. Lesgold, editors, *Intelligent Tutoring Systems*, volume 1086 of *Lecture Notes in Computer Science*, pages 578–586, Berlin, 1996a. Springer-Verlag. doi: 10.1007/3-540-61327-7_157.

C.E. Dowling, U. Koch, and K.A. Quante. A new interface for querying experts on prerequisite relationships. In J. Grundy and M. Apperley, editors, *Proceedings of the Sixth Australian Conference on Computer–Human Interaction (OzCHI 96)*, pages 320–321, Los Alamitos, California, November 1996b. IEEE Computer Society Press. doi: 10.1109/OZCHI.1996.560150.

F. Drasgow. Polychoric and polyserial correlations. In L. Kotz and N.L. Johnson, editors, *Encyclopedia of Statistical Sciences*, volume 7, pages 69–74. Wiley, New York, 1988. doi: 10.1002/0471667196.ess2014.pub2.

A. Ducamp and J.-Cl. Falmagne. Composite measurement. *Journal of Mathematical Psychology*, 6:359–390, 1969. doi: 10.1016/0022-2496(69)90012-1.

I. Düntsch and G. Gediga. Skills and knowledge structures. *British Journal of Mathematical and Statistical Psychology*, 48:9–27, 1995. doi: 10.1111/j.2044-8317.1995.tb01047.x.

H. Ebbinghaus. *Memory: A Contribution to Experimental Psychology*. Dover, New York, 1885. Eng. trans. 1913.

S.E. Embretson. Multicomponent latent trait models for test design. In S.E. Embretson, editor, *Test Design: Developments in Psychology and Psychometrics*. Academic Press, Orlando, FL, 1995.

D. Eppstein. Algorithms for drawing media. In *Graph Drawing: 12th International Symposium, GD 2004, New York, NY, USA, September 29–October 2, 2004*, volume 3383 of *Lecture Notes in Computer Science*, pages 173–183,

Berlin, Heidelberg, and New York, 2005. Springer-Verlag. doi: 10.1007/978-3-540-31843-9_19.

D. Eppstein. Upright-quad drawing of *st*-planar learning spaces. *J. Graph Algorithms and Applications*, 12(1):51–72, 2008. doi: 10.7155/jgaa.00159.

D. Eppstein, J.-Cl. Falmagne, and S. Ovchinnikov. *Media Theory*. Springer-Verlag, Berlin, Heidelberg, and New York, 2007.

D. Eppstein, J.-C. Falmagne, and H. Uzun. On verifying and engineering the wellgradedness of a union-closed family. *Journal of Mathematical Psychology*, 53:34–39, 2009. doi: 10.1016/j.jmp.2008.09.002.

B.S. Everitt and D.J. Hand. *Finite Mixture Distributions*. Chapman and Hall, London, 1981.

J.-Cl. Falmagne. Probabilistic knowledge spaces: A review. In F. Roberts, editor, *Applications of Combinatorics and Graph Theory to the Biological and Social Sciences*, volume 17 of *The IMA Volumes in Mathematics and Its Applications*. Springer-Verlag, New York, 1989a. doi: 10.1007/978-1-4684-6381-1_4.

J.-Cl. Falmagne. Stochastic learning paths in a knowledge structure. *Journal of Mathematical Psychology*, 37:489–512, 1993. doi: 10.1006/jmps.1993.1031.

J.-Cl. Falmagne. Finite Markov learning models for knowledge structures. In G.H. Fischer and D. Laming, editors, *Contributions to Mathematical Psychology, Psychometrics, and Methodology*, Recent Research in Psychology, pages 75–89. Springer-Verlag, New York, 1994. doi: 10.1007/978-1-4612-4308-3_6.

J.-Cl. Falmagne. Projections of a learning space. Electronic Preprint 0803.0575, arXiv.org, 2008.

J.-Cl. Falmagne and J.-P. Doignon. A class of stochastic procedures for the assessment of knowledge. *British Journal of Mathematical and Statistical Psychology*, 41:1–23, 1988a. doi: 10.1111/j.2044-8317.1988.tb00884.x.

J.-Cl. Falmagne and J.-P. Doignon. A Markovian procedure for assessing the state of a system. *Journal of Mathematical Psychology*, 32:232–258, 1988b. doi: 10.1016/0022-2496(88)90011-9.

J.-Cl. Falmagne and J.-P. Doignon. Meshing knowledge structures. In C.E. Dowling, F.S. Roberts, and P. Theuns, editors, *Recent Progress in Mathematical Psychology*, Scientific Psychology Series, pages 143–153. Lawrence Erlbaum Associates, Mahwah, NJ, 1998.

J.-Cl. Falmagne and J.-P. Doignon. *Learning Spaces*. Interdisciplinary Applied Mathematics. Springer-Verlag, Heidelberg, 2011. doi: 10.1007/978-3-642-01039-2.

J.-Cl. Falmagne and S. Ovchinnikov. Media theory. *Discrete Applied Mathematics*, 121:83–101, 2002. doi: 10.1016/S0166-218X(01)00235-9.

J.-Cl. Falmagne, M. Koppen, M. Villano, J.-P. Doignon, and L. Johanessen. Introduction to knowledge spaces: How to build, test and search them. *Psychological Review*, 97:204–224, 1990. doi: 10.1037/0033-295X.97.2.201.

J.-Cl. Falmagne, E. Cosyn, J.-P. Doignon, and N. Thiéry. The assessment of knowledge, in theory and in practice. In B. Ganter and L. Kwuida, editors, *Formal Concept Analysis, 4th International Conference, ICFCA 2006, Dresden, Germany, February 13–17, 2006*, Lecture Notes in Artificial Intelligence, pages 61–79. Springer-Verlag, Berlin, Heidelberg, and New York, 2006. doi: 10.1007/11671404_4.

J.-Cl. Falmagne, E. Cosyn, C. Doble, N. Thiéry, and H.B. Uzun. Assessing mathematical knowledge in a learning space: Validity and/or reliability. Submitted to *Psychological Review*, 2008.

H. Feger. *Analysis of Co-occurrence Data.* Verlag Shaker, Aachen, 1994.

H. Feger. Configuration frequency analysis and feature pattern analysis: Some comparative observations. *Psychologische Beiträge*, 42:448–468, 2000.

R.F. Ferguson. What doesn't meet the eye: Understanding and addressing racial disparities in high-achieving suburban schools. Technical report, North Central Regional Educational Laboratory, Oak Brook, IL, 2002. URL `http://www.eric.ed.gov/ERICWebPortal/detail?accno=ED474390`.

E.O. Finkenbinder. The curve of forgetting. *The American Journal of Psychology*, 24:8–32, 1913. doi: 10.2307/1413271.

G.H. Fischer and I.W. Molenaar, editors. *Rasch Models: Foundations, Recent Developments, and Applications.* Springer-Verlag, New York, 1995.

C. Flament. *L'Analyse Booléenne de Questionnaire.* Mouton, Paris, 1976.

D. Fletcher. Effectiveness and cost of interactive videodisc instruction in defense training and education. IDA Report R2372, Institute for Defense Analysis, Arlington, VA, 1990. URL `http://www.eric.ed.gov/ERICWebPortal/detail?accno=ED326194`.

M.L. Fredman and L. Khachiyan. On the complexity of dualization of monotone disjunctive normal forms. *Journal of Algorithms*, 21(3):618–628, 1996. doi: 10.1006/jagm.1996.0062.

O. Freitag. Graphische Darstellung von Expertenurteilen [Graphical Presentation of Experts' Judgments]. Diplomarbeit, Technische Universität Braunschweig, Germany, 1999.

R.G. Fryer, Jr. and S.D. Levitt. Understanding the Black-White test score gap in the first two years of school. *Review of Economics and Statistics*, 86 (2):447–464, 2004. doi: 10.1162/003465304323031049.

B. Ganter and R. Wille. *Formale Begriffsanalyse: Mathematische Grundlagen.* Springer-Verlag, Berlin-Heidelber, 1996. English translation by C. Franske: *Formal Concept Analysis: Mathematical Foundations*, Springer-Verlag, 1998.

G. Gediga and I. Düntsch. Skill set analysis in knowledge structures. *British Journal of Mathematical and Statistical Psychology*, 55:361–384, 2002. doi: 10.1348/000711002760554516.

S.G. Ghurye and D.L. Wallace. A convolutive class of monotone likelihood ratio families. *The Annals of Mathematical Statistics*, 30:1158–1164, 1959. doi: 10.1214/aoms/1177706101.

J.P. Gollub, M.W. Bertenthal, J.B. Labov, and P.C. Curtis. *Learning and Understanding: Improving Advanced study of Mathematics and Science in U. S. High Schools*. National Academy Press, Washington DC, 2002.

L.A. Goodman. *Analysing Qualitative/Categorial Variables: Loglinear Models and Latent Structure Analysis*. Cambridge University Press, Cambridge, 1978.

L.A. Goodman and W.H. Kruskal. Measures of association for cross classifications. *Journal of the American Statistical Association*, 49:732–764, 1954. doi: 10.2307/2281536.

L.A. Goodman and W.H. Kruskal. Measures of association for cross classifications, III: Approximate sampling theory. *Journal of the American Statistical Association*, 58:310–364, 1963. doi: 10.2307/2283271.

D.A. Grayson. Two-group classification in latent trait theory: Scores with monotone likelihood ratio. *Psychometrika*, 53:383–392, 1988. doi: 10.1007/BF02294219.

T. Greer, W.P. Dunlap, and G.O. Beaty. Intensity discrimination with gated and continuous sinusoids. *Educational and Psychological Measurement*, 63 (6):931–950, 2003.

L. Guttman. A basis for scaling qualitative data. *American Sociological Review*, 9:139–150, 1944. doi: 10.2307/2086306.

M. Habib, R. Medina, L. Nourine, and G. Steiner. Efficient algorithms on distributive lattices. *Discrete Applied Mathematics*, 110:169–187, 2001. doi: 10.1016/S0166-218X(00)00258-4.

J.A. Hagenaars. *Categorical Longitudinal Data: Loglinear Analysis of Panel, Trend and Cohort Data*. Sage Publications, Newbury Park, 1990.

J.A. Hagenaars and A.L. McCutcheon, editors. *Applied Latent Class Analysis*. Cambridge University Press, Cambridge, 2002. doi: 10.1017/CBO9780511499531.

G.W. Hagerty and S. Smith. Using the Web-based interactive software ALEKS to enhance college algebra. *Mathematics and Computer Education*, 39:183–194, 2005.

P. Hájek and T. Havránek. On generation of inductive hypotheses. *International journal of Man-Machine Studies*, 9:415–438, 1977. doi: 10.1016/S0020-7373(77)80011-4.

P. Hájek, I. Havel, and M. Chytil. The GUHA method of automatic hypotheses determination. *Computing*, 1:293–308, 1966. doi: 10.1007/BF02345483.

F. Halasz and M. Schwartz. The Dexter hypertext reference model. In J. Moline, D. Benigni, and J. Baronas, editors, *Proceedings of the Hypertext Standardization Workshop*, volume 500–178 of *NIST Special Publications*, pages 95–133, Gaithersburg, MD 20899, January 1990. National Institute of Standards and Technology.

F. Halasz and M. Schwartz. The Dexter hypertext reference model. *Communications of the ACM*, 37(2):30–39, 1994. doi: 10.1145/175235.175237.

J. Hampikian. AC 2007-1998: Benefits of a tutorial mathematics program for engeneering students enrolled in precalculus: A template for assessment.

In *Proceedings of the 2007 American Society for Engineering Education Annual Conference & Exposition, Honolulu, HI*, pages 24–27, 2007.

J. Hampikian, J. Gardner, A. Moll, P. Pyke, and C. Schrader. Integrated pre-freshman engineering and precalculus mathematics. In *Proceedings of the 2006 Annual Conference of the American Society for Engineering Education*, 2006.

E.A. Hanushek and S.G. Rivkin. Teacher quality. In E.A. Hanushek and F. Welch, editors, *Handbook of the Economics of Education*, pages 1051–1078. North Holland, Amsterdam, 2006. doi: 10.1016/S1574-0692(06)02018-6.

D.N. Harris and C.D. Herrington. Accountability, standards, and the growing achievement gap: Lessons from the past half century. *American Journal of Education*, 112:209–238, 2006. doi: 10.1086/498995.

L.V. Hedges and A. Nowell. Changes in the Black-White gap in achievement test scores. *Sociology of education*, 2(2):111–135, 1999. doi: 10.2307/2673179.

T. Heinen. *Latent Class and Discrete Latent Trait Models: Similarities and Differences.* Sage Publications, Thousand Oaks, 1996.

T. Held. *Establishment and empirical validation of problem structures based on domain specific skills and textual properties.* PhD thesis, University of Heidelberg, Heidelberg, Germany, 1993.

T. Held. An integrated approach for constructing, coding, and structuring a body of word problems. In D. Albert and J. Lukas, editors, *Knowledge Spaces: Theories, Empirical Research, Applications*, pages 67–102. Lawrence Erlbaum Associates, Mahwah, NJ, 1999.

T. Held and K. Korossy. Data analysis as a heuristic for establishing theoretically founded item structures. *Zeitschrift für Psychologie*, 206:169–188, 1998.

J. Heller and C. Repitsch. Distributed skill functions and the meshing of knowledge structures. *Journal of Mathematical Psychology*, 52:147–157, 2008. doi: 10.1016/j.jmp.2008.01.003.

J. Heller, C. Steiner, C. Hockemeyer, and D. Albert. Competence-based knowledge structures for personalised learning. *International Journal on E–Learning*, 5(1):75–88, 2006.

J. Heller, M. Levene, K. Keenoy, D. Albert, and C. Hockemeyer. Cognitive aspects of trails: A stochastic model linking navigation behaviour to the learner's cognitive state. In J. Schoonenboom, M. Levene, J. Heller, K. Keenoy, and M. Turcsanyi-Szabo, editors, *Trails in Education: Technologies that Support Navigational Learning*, pages 119–146. Sense Publishers, Rotterdam, 2007.

B.T. Hemker, L.A. van der Ark, and K. Sijtsma. On measurement properties of continuation ratio models. *Psychometrika*, 66:487–506, 2001. doi: 10.1007/BF02296191.

C. Hockemeyer. RATH—A relational adaptive tutoring hypertext WWW-environment. Rapport 1997/3, Institut für Psychologie, Karl-Franzens-Universität Graz, Austria, 1997a.

C. Hockemeyer. Using the basis of a knowledge space for determining the fringe of a knowledge state. *Journal of Mathematical Psychology*, 41:275–279, 1997b. doi: 10.1006/jmps.1997.1170.

C. Hockemeyer. KST tools user manual (2nd edition). Technical Report 2001/3, Institut für Psychologie, Karl-Franzens-Universität Graz, Austria, 2001. URL http://wundt.uni-graz.at/kst.html.

C. Hockemeyer. A comparison of non-deterministic procedures for the adaptive assessment of knowledge. *Psychologische Beiträge*, 44:495–503, 2002.

C. Hockemeyer. Competence based adaptive e-learning in dynamic domains. In F.W. Hesse and Y. Tamura, editors, *The Joint Workshop of Cognition and Learning through Media–Communication for Advanced E-Learning (JWCL)*, pages 79–82, Berlin, September 2003.

C. Hockemeyer, T. Held, and D. Albert. RATH—A Relational Adaptive Tutoring Hypertext WWW-Environment Based on Knowledge Space Theory. In C. Alvegard, editor, *CALISCHE'98: Proceedings of the Fourth International Conference on Computer Aided Learning in Science and Engineering*, pages 417–423, Göteborg, Sweden, June 1998. Chalmers University of Technology. URL http://wundt.kfunigraz.ac.at/rath/publications/calisce/.

C. Hockemeyer, O. Conlan, V. Wade, and D. Albert. Applying competence prerequisite structures for eLearning and skill management. *Journal of Universal Computer Science*, 9:1428–1436, 2003.

C. Hockemeyer, A. Nussbaumer, E. Lövquist, A. Aboulafia, D. Breen, G. Shorten, and D. Albert. Applying a web and simulation-based system for adaptive competence assessment of spinal anaesthesia. In M. Spaniol, Q. Li, R. Klamma, and R. Lau, editors, *Advances in Web-Based learning—ICWL 2009*, volume 5686 of *Lecture Notes in Computer Science*, pages 182–191, Berlin, 8 2009. Springer-Verlag. doi: 10.1007/978-3-642-03426-8_23.

T.G. Holzman, J.W. Pellegrino, and R. Glaser. Cognitive variables in series completion. *Journal of Educational Psychology*, 75(4):603–618, 1983. doi: 10.1037/0022-0663.75.4.603.

J.E. Hopcroft and R.M. Karp. An $O(n^{5/2})$ algorithm for maximum matchings in bipartite graphs. *SIAM J. on Computing*, 2(4):225–231, 1973. doi: 10.1137/0202019.

H. Huynh. A new proof for monotone likelihood ratio for the sum of independent Bernoulli random variables. *Psychometrika*, 59:77–79, 1994. doi: 10.1007/BF02294266.

M. Jerrum, A. Sinclair, and E. Vigoda. A polynomial-time approximation algorithm for the permanent of a matrix with non-negative entries. In *Proc. 33rd ACM Symp. on Theory of Computing*, pages 712–271, 2001. doi: 10.1145/380752.380877.

W.H. Jeynes. A meta-analysis: The effects of parental involvement on minority children's academic achievement. *Education and Urban Society*, 35:202–218, 2003. doi: 10.1177/0013124502239392.

S. Judge. The impact of computer technology on academic achievement of young African American children. *Journal of Research in Childhood Education*, 20:91–101, 2005. doi: 10.1080/02568540509594554.

S. Judge, K. Puckett, and S.M. Bell. Closing the digital divide: Update from the early childhood longitudinal study. *Journal of Education Research*, 100 (1):52–60, 2006. doi: 10.3200/JOER.100.1.52-60.

B.W. Junker and K. Sijtsma. Cognitive assessment models with few assumptions, and connections with nonparametric item response theory. *Applied Psychological Measurement*, 25(3):258–272, 2001. doi: 10.1177/01466210122032064.

M. Kambouri, M. Koppen, M. Villano, and J.-Cl. Falmagne. Knowledge assessment: Tapping human expertise by the QUERY routine. *International Journal of Human-Computer Studies*, 40:119–151, 1994. doi: 10.1006/ijhc.1994.1006.

R.A. Karabinus. The $r$-point biserial limitation. *Educational and Psychological Measurement*, 35:277–282, 1975. doi: 10.1177/001316447503500205.

J.G. Kemeny and J.L. Snell. *Finite Markov Chains*. Van Nostrand, Princeton, N.J., 1960.

M.D. Kickmeier-Rust, N. Peirce, O. Conlan, D. Schwarz, D. Verpoorten, and D. Albert. Immersive digital games: The Interfaces for next-generation e-learning? In *Universal Access in Human-Computer Interaction. Applications and Services*, volume 4556 of *Lecture Notes in Computer Science*, pages 647–656, New York, Heidelberg, 2007. Springer-Verlag. doi: 10.1007/978-3-540-73283-9_71.

M.D. Kickmeier-Rust, C. Hockemeyer, D. Albert, and T. Augustin. Micro adaptive, non-invasive assessment in educational games. In M. Eisenberg, Kinshuk, M. Chang, and R. McGreal, editors, *Proceedings of the second IEEE International Conference on Digital Game and Intelligent Toy Enhanced Learning*, pages 135–137, Banff, Canada, 2008a. doi: 10.1109/DIGITEL.2008.10.

M.D. Kickmeier-Rust, B. Marte, S.B. Linek, T. Lalonde, and D. Albert. Learning with computer games: Micro level feedback and interventions. In M.E. Auer, editor, *Proceedings of the International Conference on Interactive Computer Aided Learning (ICL)*, Kassel, 2008b. Kassel University Press. CD–ROM publication.

K.H. Kim and F.W. Roush. Group relationships and homomorphisms of Boolean matrix semigroups. *Journal of Mathematical Psychology*, 28:448–452, 1984. doi: 10.1016/0022-2496(84)90011-7.

K.J. Klauer. Induktives Denken: Definition, Theorie und Training. *Zeitschrift für Experimentelle Psychologie*, 44(2):213–219, 1997.

M. Koppen. Extracting human expertise for constructing knowledge spaces: An algorithm. *Journal of Mathematical Psychology*, 37:1–20, 1993. doi: 10.1006/jmps.1993.1001.

M. Koppen. The construction of knowledge spaces by querying experts. In G.H. Fischer and D. Laming, editors, *Contributions to Mathematical Psychology, Psychometrics, and Methodology*, pages 137–147. Springer-Verlag, New York, 1994. doi: 10.1007/978-1-4612-4308-3_10.

M. Koppen and J.-P. Doignon. How to build a knowledge space by querying an expert. *Journal of Mathematical Psychology*, 34:311–331, 1990. doi: 10.1016/0022-2496(90)90035-8.

C. Körner and D. Albert. Speed of comprehension of visualized ordered sets. In K.W. Kallus, N. Posthumus, and P. Jiménez, editors, *Current Psychological Research in Austria. Proceedings of the 4th Scientific Conference of the Austrian Psychological Society (ÖGP)*, pages 179–182, Graz, 2001. Akademische Druck- u. Verlagsanstalt.

K. Korossy. *Modellierung von Wissen als Kompetenz und Performance. Eine Erweiterung der Wissensstruktur-Theorie von Doignon und Falmagne [Modelling knowledge as competence and performance. An extension of the theory of knowledge structures by Doignon and Falmagne]*. PhD thesis, University of Heidelberg, Heidelberg, Germany, 1993.

K. Korossy. Extending the theory of knowledge spaces: A competence–performance approach. *Zeitschrift für Psychologie*, 205:53–82, 1997.

K. Korossy. Modeling knowledge as competence and performance. In D. Albert and J. Lukas, editors, *Knowledge Spaces: Theories, Empirical Research, Applications*, pages 103–132. Lawrence Erlbaum Associates, Mahwah, NJ, 1999.

B. Korte, L. Lovász, and R. Schrader. *Greedoids*. Number 4 in Algorithms and Combinatorics. Springer-Verlag, 1991.

R.B. Kozma and J. Russell. Multimedia and understanding: Expert and novice responses to different representations of chemical phenomena. *Journal of Research in Science Teaching*, 34:949–968, 1997. doi: 10.1002/(SICI)1098-2736(199711)34:9¡949::AID-TEA7¿3.0.CO;2-U.

C.H. Kraemer. Correlation coefficients in medical research: From product moment correlation to the odds ratio. *Statistical Methods in Medical Research*, 15:525–545, 2006. doi: 10.1177/0962280206070650.

J.A. Kulik. Meta-analytic studies of findings on computer-based instruction. In E.L. Baker and H.F. O'Neil, Jr., editors, *Technology assessment in education and training*. Lawrence Erlbaum Associates, Hillsdale, NJ, 1994.

S. Kullback. *Information Theory and Statistics*. John Wiley and Sons, New York, 1959.

S. Kullback and R.A. Leibler. On information and sufficiency. *Annals of Mathematical Statistics*, 22:79–86, 1951. doi: 10.1214/aoms/1177729694.

P.F. Lazarsfeld and N.W. Henry. *Latent Structure Analysis*. Houghton Mifflin, Boston, 1968.

K. Lewin. Problems of research in social psychology. In D. Cartwright, editor, *Field Theory in Social Science: Selected Theoretical Papers*, pages 155–169. Harper & Brothers, New York, 1951.

T. Ley and D. Albert. Kompetenzmanagement als formalisierbare Abbildung von Wissen und Personalwesen. *Wirtschaftspsychologie*, 5(3):86–93, 2003.

T. Ley, D. Albert, and S. Lindstaedt. Competence management using the competence performance approach: Modeling, assessment, validation, and use. In M.A. Sicilia, editor, *Competencies in Organizational E-learning: Concepts and Tools*, pages 83–119. Idea Group Inc., Hershey, PA, USA, 2007. doi: 10.4018/978-1-59904-343-2.ch004.

S.T. Lubienski. A closer look at black-white mathematics gaps: Intersections of race and SES in NAEP achievement and instructional practices data. *Journal of Negro Education*, 71(4):269–287, 2002. doi: 10.2307/3211180.

S.T. Lubienski. On "gap gazing" in mathematics education: The need for gaps analyses. *Journal for Research in Mathematics Education*, 39(4):350–356, 2008.

J. Lukas. Modellierung von Fehlkonzepten in einer algebraischen Wissensstruktur [Modeling misconceptions in an algebraic knowledge structure]. *Kognitionswissenschaft*, 6(4):196–204, 1997. doi: 10.1007/s001970050042.

J. Lukas and D. Albert. Knowledge assessment based on skill assignment and psychological task analysis. In G. Strube and K. F. Wender, editors, *The Cognitive Psychology of Knowledge*, pages 139–159. Elsevier Science Publishers B.V., Amsterdam, 1993.

I.L. MacDonald and W. Zucchini. *Hidden Markov Models and Other Types of Models for Discrete-valued Time Series.* Chapman and Hall, London, 1997.

E. Maris. Estimating multiple classification latent class models. *Psychometrika*, 64(2):187–212, 1999. doi: 10.1007/BF02294535.

J.L. Martin and J.A. Wiley. Algebraic representations of beliefs and attitudes II: Microbelief models for dichotomous belief data. *Sociological Methodology*, 30(1):123–164, 2000. doi: 10.1111/0081-1750.00077.

A.L. McCutcheon. *Latent Class Analysis.* Sage Publications, Newbury Park, 1987.

G.J. McLachlan and K.E. Basford. *Mixture Models: Inference and Application to Clustering.* Marcel Dekker, New York, 1988.

G.J. McLachlan and D. Peel. *Finite Mixture Models.* Wiley, New York, 2000.

E. Melis and J. Siekmann. ActiveMath: An intelligent tutoring system for mathematics. In *Proceedings of Seventh International Conference on Artificial Intelligence and Soft Computing (ICAISC)*, volume 3070 of *Lecture Notes in Computer Science*, pages 91–101. Springer-Verlag, 2004. doi: 10.1007/978-3-540-24844-6_12.

R. Mislevy. Test theory reconceived. *Journal of Educational Measurement*, 33:379–416, 1996. doi: 10.1111/j.1745-3984.1996.tb00498.x.

A. Mitra, S. Joshi, K.J. Kemper, C. Woods, and J. Gobble. Demographic differences and attitudes toward computers among healthcare professionals earning continuing education credits on-line. *Journal of Educational Computing Research*, 35(1):31–43, 2006. doi: 10.2190/8123-3KU2-648J-7593.

R.J. Mokken. *A Theory and Procedure of Scale Analysis.* De Gruyter, The Hague, Mouton/Berlin, 1971.

R.J. Mokken. Nonparametric models for dichotomous responses. In W.J. van der Linden and R.K. Hambleton, editors, *Handbook of Modern Item Response Theory*, pages 351–367. Springer-Verlag, New York, 1997. doi: 10.1007/978-1-4757-2691-6_20.

J. Moore. Do we really value learning? *Journal of Chemical Education*, 76:5, 1999. doi: 10.1021/ed076p5.

C.E. Müller. A procedure for facilitating an expert's judgments on a set of rules. In E.E. Roskam, editor, *Mathematical Psychology in Progress*, Recent Research in Psychology, pages 157–170. Springer-Verlag, Berlin, Heidelberg, and New York, 1989. doi: 10.1007/978-3-642-83943-6_10.

M.B. Nakleh, K.A. Lowrey, and R.A. Mitchell. Narrowing the gap between concepts and algorithms in freshman chemistry. *Journal of Chemical Education*, 8:758–762, 1996. doi: 10.1021/ed073p758.

J. Nunnally and I. Bernstein. *Psychometric Theory.* McGraw-Hill, New York, 1994.

A.J. Orr. Black-white differences in achievement: The importance of wealth. *Sociology of Education*, 76(4):281–304, 2003. doi: 10.2307/1519867.

E. Packard. It's fun, but does it make you smarter? *Monitor on Psychology*, 38(10), 2007. URL `http://www.apa.org/monitor/nov07/itsfun.aspx`.

M. Phillips, J. Brooks-Gunn, G. Duncan, P. Klebanov, and J. Crane. Family background, parenting practices, and the Black-White test score gap. In C. Jencks and M. Phillips, editors, *The Black-White test score gap*, pages 103–145. Brookings Institution Press, Washington, DC, 1998.

M. Pilgerstorfer, D. Albert, and D.G. Camhy. Considerations on personalized training in philosophy for children. In D.G. Camhy and R. Born, editors, *Encouraging Philosophical Thinking: Proceedings of the International Conference on Philosophy for Children in Graz, Austria*, volume 17 of *Conceptus–Studien*, pages 85–89, Sankt Augustin, Germany, 2006. Academia–Verlag.

I. Ponocny and K. Waldherr. A nonparametric goodness-of-fit procedure for unidimensional polytomous Rasch models. Talk at the 33rd European Mathematical Psychology Group Meeting, Bremen, Germany, August 21–24 2002.

S. Pötzi and G. Wesiak. SRbT tools user manual. Technical Report 2004/1, Institut für Psychologie, Karl-Franzens-Universität Graz, Austria, 2004. URL `http://wundt.uni-graz.at/kst.html`.

M. Prensky. *Digital Game-Based Learning.* McGraw-Hill, New York, 2001.

C.H. Proctor. A probabilistic formulation and statistical analysis of Guttman scaling. *Psychometrika*, 35:73–78, 1970. doi: 10.1007/BF02290594.

R Development Core Team. *R: A Language and Environment for Statistical Computing*. R Foundation for Statistical Computing, Vienna, Austria, 2012.

H. Roediger. Relativity of remembering: Why the laws of memory vanished. *Annual Review of Psychology*, 59:225–254, 2008. doi: 10.1146/annurev.psych.57.102904.190139.

A. Rusch and R. Wille. Knowledge spaces and formal concept analysis. In H.-H. Bock and W. Polasek, editors, *Data Analysis and Information Systems*, pages 427–436, Berlin, Heidelberg, and New York, 1996. Springer-Verlag. doi: 10.1007/978-3-642-80098-6_36.

A. Sargin and A. Ünlü. Inductive item tree analysis: Corrections, improvements, and comparisons. *Mathematical Social Sciences*, 58:376–392, 2009. doi: 10.1016/j.mathsocsci.2009.06.001.

A. Sargin and A. Ünlü. The R package DAKS: Basic functions and complex algorithms in knowledge space theory. In H. Locarek-Junge and C. Weihs, editors, *Classification as a Tool for Research: Proceedings of the 11th IFCS Biennial Conference and 33rd Annual Conference of the Gesellschaft fr Klassifikation e.V., Dresden, March 13-18, 2009*, Studies in Classification, Data Analysis, and Knowledge Organization, pages 263–270, Berlin, 2010. Springer-Verlag. doi: 10.1007/978-3-642-10745-0_28.

H. Scheiblechner. Nonparametric IRT: Testing the bi-isotonicity of isotonic probabilistic models (ISOP). *Psychometrika*, 68:79–96, 2003. doi: 10.1007/BF02296654.

M. Schrepp. *Über die Beziehung zwischen kognitiven Prozessen und Wissensräumen beim Problemlösen [On the Relationship between Cognitive Processes and Knowledge Spaces for Problem Solving]*. PhD thesis, University of Heidelberg, Heidelberg, Germany, 1993.

M. Schrepp. A generalization of knowledge space theory to problems with more than two answer alternatives. *Journal of Mathematical Psychology*, 41:237–243, 1997. doi: 10.1006/jmps.1997.1169.

M. Schrepp. Extracting knowledge structures from observed data. *British Journal of Mathematical and Statistical Psychology*, 52:213–224, 1999a. doi: 10.1348/000711099159071.

M. Schrepp. On the empirical construction of implications between bi-valued test items. *Mathematical Social Sciences*, 38:361–375, 1999b. doi: 10.1016/S0165-4896(99)00025-6.

M. Schrepp. An empirical test of a process model for letter series completion problems. In D. Albert and J. Lukas, editors, *Knowledge Spaces: Theories, Empirical Research, Applications*, pages 133–154. Lawrence Erlbaum Associates, Mahwah, NJ, 1999c.

M. Schrepp. A method for comparing knowledge structures concerning their adequacy. *Journal of Mathematical Psychology*, 45:480–496, 2001. doi: 10.1006/jmps.2000.1329.

M. Schrepp. Explorative analysis of empirical data by Boolean analysis of questionnaires. *Zeitschrift für Psychologie*, 210(2):99–109, 2002. doi: 10.1026//0044-3409.210.2.99.

M. Schrepp. A method for the analysis of hierarchical dependencies between items of a questionnaire. *Methods of Psychological Research—Online*, 19: 43–79, 2003.

M. Schrepp. About the connection between knowledge structures and latent class models. *Methodology: European Journal of Research Methods for the Behavioral and Social Sciences*, 1(3):92–102, 2005. doi: 10.1027/1614-2241.1.3.92.

M. Schrepp. ITA 2.0: A program for classical and inductive item tree analysis. *Journal of Statistical Software*, 16, 2006a.

M. Schrepp. Properties of the correlational agreement coefficient: A comment to Ünlü and Albert (2004). *Mathematical Social Sciences*, 51:117–123, 2006b. doi: 10.1016/j.mathsocsci.2005.07.005.

M. Schrepp. On the evaluation of fit measures for quasi-orders. *Mathematical Social Sciences*, 53:196–208, 2007. doi: 10.1016/j.mathsocsci.2006.11.002.

M. Schrepp, T. Held, and D. Albert. Component-based construction of surmise relations for chess problems. In D. Albert and J. Lukas, editors, *Knowledge Spaces: Theories, Empirical Research, Applications*, pages 41–66. Lawrence Erlbaum Associates, Mahwah, NJ, 1999.

K. Sijtsma. Methodology review: Nonparametric IRT approaches to the analysis of dichotomous item scores. *Applied Psychological Measurement*, 22: 3–31, 1998. doi: 10.1177/01466216980221001.

K. Sijtsma and I.W. Molenaar. *Introduction to Nonparametric Item Response Theory*. Sage Publications, Thousand Oaks, 2002.

H.A. Simon. Information-processing theory of human problem solving. In W.K. Estes, editor, *Handbook of Learning and Cognitive Processes, Vol. 5: Human Information Processing*, pages 271–295. Lawrence Erlbaum Associates, Hillsdale, NJ, 1978.

J. Sivin-Kachala. Report on the effectiveness of technology in schools, 1990–1997. Technical report, Software Publisher's Association, 1998.

C. Stahl and D. Meyer. Package 'KST'. 2009. URL http://cran.r-project.org/web/packages/kst/kst.pdf.

L. Stefanutti. A logistic approach to knowledge structures. *Journal of Mathematical Psychology*, 50:545–561, 2006. doi: 10.1016/j.jmp.2006.07.003.

L. Stefanutti and D. Albert. Efficient assessment of organizational action based on knoweldge space theory. In K. Tochtermann and H. Maurer, editors, *2nd International Conference on Knowledge Management*, Journal of Universal Computer Science, pages 183–190, 2002.

L. Stefanutti and D. Albert. Skill assessment in task simulation. In K. Tochtermann and H. Maurer, editors, *3rd International conference on Knowledge Management*, Journal of Universal Computer Sciences, pages 174–180, 2003.

L. Stefanutti and M. Koppen. A procedure for the incremental construction of a knowledge space. *Journal of Mathematical Psychology*, 47(3):265–277, 2003. doi: 10.1016/S0022-2496(02)00022-6.

L. Stefanutti and E. Robusto. Recovering a probabilistic knowledge structure by constraining its parameter space. *Psychometrika*, 72:83–96, 2009. doi: 10.1007/s11336-008-9095-7.

L. Stefanutti, D. Albert, and C. Hockemeyer. Derivation of knowledge structures for distributed learning objects. In P. Ritrovato, C. Allison, S.A. Cerri, T. Dimitrakos, M. Gaeta, and S. Salerno, editors, *Towards the Learning Grid: Advances in Human Learning Services*, pages 105–112. IOS Press, Amsterdam, 2005.

C. Steiner and D. Albert. Personalising learning through prerequisite structures derived from concept maps. In H. Leung, F. Li, R. Lau, and Q. Li, editors, *Advances in Web Based learning—ICWL 2007*, volume 4823 of *Lecture Notes in Computer Science*, pages 45–54, 2008. doi: 10.1007/978-3-540-78139-4_5.

R.J. Sternberg and M.K. Gardner. Unities in inductive reasoning. *Journal of Experimental Psychology: General*, 112(1):80–116, 1983. doi: 10.1037/0096-3445.112.1.80.

R. Suck. A dimension-related metric on the lattice of knowledge spaces. *Journal of Mathematical Psychology*, 43:394–409, 1999. doi: 10.1006/jmps.1998.1229.

R. Suck. Parsimonious set representations of orders, a generalization of the interval order concept, and knowledge spaces. *Discrete Applied Mathematics*, 127:373–386, 2003. doi: 10.1016/S0166-218X(02)00255-X.

R. Suck. Set representations of orders and a structural equivalent of saturation. *Journal of Mathematical Psychology*, 48:159–166, 2004. doi: 10.1016/j.jmp.2004.03.001.

R. Suck. Knowledge spaces regarded as set representations of skill structures. In E. Dzhafarov and L. Perry, editors, *Descriptive and Normative Approaches to Human Behavior*, pages 249–270. World Scientific, 2011. doi: 10.1142/9789814368018_0010.

M. Taagepera and S. Noori. Mapping students' thinking patterns in learning organic chemistry by the use of the Knowledge Space Theory. *Journal of Chemical Education*, 77:1224–1229, 2000. doi: 10.1021/ed077p1224.

M. Taagepera, F. Potter, G. E. Miller, and K. Lakshminarayan. Mapping students' thinking patterns by the use of Knowledge Space Theory. *International Journal of Science Education*, 19:283–302, 1997. doi: 10.1080/0950069970190303.

M. Taagepera, R.D. Arasasingham, F. Potter, A. Soroudi, and G. Lam. Following the development of the bonding concept using knowledge space theory. *Journal of Chemical Education*, 79:756–762, 2002. doi: 10.1021/ed079p756.

M. Taagepera, R.D. Arasasingham, S. King, F. Potter, I. Martorell, D. Ford, J. Wu, and A.M. Kearney. Integrating symmetry in stereochemical analysis in introductory organic chemistry. *Chemical Education Research and Practice*, 12:322–330, 2011. doi: 10.1039/c1rp90039k.

R.F. Tate. Correlation between a discrete and a continuous variable. Point-biserial correlation. *The Annals of Mathematical Statistics*, 25(3):603–607, 1954. doi: 10.1214/aoms/1177728730.

C. Tatsuoka. Data analytic methods for latent partially ordered classification models. *Journal of the Royal Statistical Society: Series C (Applied Statistics)*, 51(3):337–350, 2002. doi: 10.1111/1467-9876.00272.

P. Theuns. A dichotomization method for Boolean analysis of quantifiable co-occurence data. In G.H. Fischer and D. Laming, editors, *Contributions to Mathematical Psychology, Psychometrics and Methodology*, Scientific Psychology Series, pages 173–194. Springer-Verlag, New York, 1994. doi: 10.1007/978-1-4612-4308-3_28.

P. Theuns. Building a knowledge Space via Boolean analysis of co-occurrence data. In C.E. Dowling, F.S. Roberts, and P. Theuns, editors, *Recent Progress in Mathematical Psychology*, Scientific Psychology Series, pages 173–194. Lawrence Erlbaum Associates, Mahwah, NJ, 1998.

N. Thiéry. *Dynamically Adapting Knowledge Spaces*. PhD thesis, Univ. of California, Irvine, School of Social Sciences, 2001.

D.M. Titterington, A.F.M. Smith, and U.E. Makov. *Statistical Analysis of Finite Mixture Distributions*. Wiley, New York, 1985.

H. Toivonen. Sampling large databases for association rules. In *22th International Conference on Very Large Databases (VLDB96)*, pages 134–145. Morgan Kaufmann Publishers, 1996.

A. Ünlü. Estimation of careless error and lucky guess probabilities for dichotomous test items: A psychometric application of a biometric latent class model with random effects. *Journal of Mathematical Psychology*, 50: 309–328, 2006. doi: 10.1016/j.jmp.2005.10.002.

A. Ünlü. Nonparametric item response theory axioms and properties under nonlinearity and their exemplification with knowledge space theory. *Journal of Mathematical Psychology*, 51:383–400, 2007. doi: 10.1016/j.jmp.2007.07.002.

A. Ünlü. A note on monotone likelihood ratio of the total score variable in unidimensional item response theory. *British Journal of Mathematical and Statistical Psychology*, 61:179–187, 2008. doi: 10.1348/000711007X173391.

A. Ünlü. *The Correlational Agreement Coefficient CA and an Alternative kappa*. Cuvillier Verlag, Göttingen, Germany, 2009.

A. Ünlü. A note on the connection between knowledge structures and latent class models. *Methodology: European Journal of Research Methods for the Behavioral and Social Sciences*, 7:63–67, 2011. doi: 10.1027/1614-2241/a000023.

A. Ünlü and D. Albert. The correlational agreement coefficient $\mathrm{CA}(\leq, D)$ – a mathematical analysis of a descriptive goodness-of-fit measure. *Mathematical Social Sciences*, 48:281–314, 2004. doi: 10.1016/j.mathsocsci.2004.03.003.

A. Ünlü and W.A. Malik. Modified item tree analysis of inductive reasoning data. *Journal of Interdisciplinary Mathematics*, 11:641–652, 2008a.

A. Ünlü and W.A. Malik. Psychometric data analysis: A size/fit trade-off evaluation procedure for knowledge structures. *Journal of Data Science*, 6: 491–514, 2008b.

A. Ünlü and W.A. Malik. Interactive glyph graphics of multivariate data in psychometrics: The software package Gauguin. *Methodology: European Journal of Research Methods for the Behavioral and Social Sciences*, 7: 134–144, 2011. doi: 10.1027/1614-2241/a000031.

A. Ünlü and A. Sargin. Interactive visualization of assessment data: The software package Mondrian. *Applied Psychological Measurement*, 33:148–156, 2009. doi: 10.1177/0146621608319511.

A. Ünlü and A. Sargin. DAKS: An R package for data analysis methods in knowledge space theory. *Journal of Statistical Software*, 37(2):1–31, 2010. URL www.jstatsoft.org/v37/i02.

A. Ünlü and A. Sargin. Mosaic displays for combinatorial psychometric models. In P. Cerchiello and C. Tarantola, editors, *Proceedings of the Classification and Data Analysis Group*, Pavia, Italy, 2011. Italian Statistical Society.

A. Ünlü, S. Brandt, and D. Albert. Test surmise relations, test knowledge structures, and their characterizations. Submitted for publication, 2004.

A. Ünlü, S. Brandt, and D. Albert. Corrigendum to "Surmise relations between tests—mathematical considerations". *Discrete Applied Mathematics*, 155:2401–2402, 2007. doi: 10.1016/j.dam.2007.08.018.

A. Vaarik, M. Taagepera, and M. Tamm. Following the logic of student thinking patterns about atomic orbital structures. *Journal of Baltic Science Education*, 7:27–36, 2008.

J. van Buggenhaut and E. Degreef. On dichotomization methods in Boolean analysis of questionnaires. In E.E. Roskam and R. Suck, editors, *Progress in Mathematical Psychology I*, pages 447–453. Elsevier Science Publishers B.V., North Holland, 1987.

L.A. van der Ark. Relationships and properties of polytomous item response theory models. *Applied Psychological Measurement*, 25:273–282, 2001. doi: 10.1177/01466210122032073.

L.A. van der Ark. Stochastic ordering of the latent trait by the sum score under various polytomous IRT models. *Psychometrika*, 70:283–304, 2005. doi: 10.1007/s11336-000-0862-3.

W.J. van der Linden and R.K. Hambleton, editors. *Handbook of Modern Item Response Theory*. Springer-Verlag, New York, 1997.

P. van Emde Boas, R. Kaas, and E. Zijlstra. Design and implementation of an efficient priority queue. *Mathematical Systems Theory*, 10(1):99–127, 1976. doi: 10.1007/BF01683268.

J.F.J. van Leeuwe. Item tree analysis. *Nederlands Tijdschrift voor de Psychologie*, 29:475–484, 1974.

J.K. Vermunt. *Log-linear Models for Event Histories.* Sage Publications, Thousand Oaks, 1997.

J.K. Vermunt and J. Magidson. *Latent GOLD 2.0 User's Guide.* Statistical Innovations Inc., Belmont, 2000.

J.K. Vermunt and J. Magidson. Latent class analysis. In M.S. Lewis-Beck, A.E. Bryman, and T.F. Liao, editors, *The Sage Encyclopedia of Social Science Research Methods*, pages 549–553. Sage Publications, Thousand Oaks, 2004.

M. Villano. Probabilistic student models: Bayesian belief networks and knowledge space theory. In *Intelligent Tutoring System. Second International Conference*, Lecture Notes in Computer Science, pages 491–498, New York, 1992. Springer-Verlag. doi: 10.1007/3-540-55606-0_58.

S.B. Vincent. The function of the vibrissae in the behavior of the white rat. *Behavioral Monographs*, 1(5), 1912.

L.S. Vygotsky. *Mind and Society: The Development of Higher Mental Processes.* Harvard University Press, Cambridge, MA, 1978.

H. Wainer, N.J. Dorans, D. Eignor, R. Flaugher, B.F. Green, R.J. Mislevy, L. Steinberg, and D. Thissen. *Computerized Adaptive Testing: A Primer.* Lawrence Erlbaum Associates, New Jersey and London, 2000.

J.H. Wandersee, J.J. Mintzes, and J.D. Novak. Research on alternative conceptions in science. In D.L. Gabel, editor, *Handbook of Research on Science Teaching and Learning*, pages 177–210. Macmillan Publishing Company, New York, 1993.

D.J.A. Welsh. Matroids: Fundamental concepts. In R.L. Graham, M. Grötschel, and L. Lovász, editors, *Handbook of Combinatorics*, volume 1. The M.I.T. Press, Cambridge, MA, 1995.

H. Wenglinsky. Does it compute? The relationship between educational technology and student achievement in mathematics. Technical report, Educational Testing Service Policy Information Center, 1998.

G. Wesiak. *Ordering Inductive Reasoning Tests for Adaptive Knowledge Assessments: An Application of Surmise Relations between Tests.* PhD thesis, Karl-Franzens-Universität Graz, 2003. URL http://psydok.sulb.uni-saarland.de/volltexte/2004/380.

G. Wesiak and D. Albert. Knowledge spaces for inductive reasoning tests. In K.W. Kallus, N. Posthumus, and P. Jiménez, editors, *Current Psychological Research in Austria. Proceedings of the 4th Scientific Conference of the Austrian Psychological Society (ÖGP)*, pages 157–160, Graz, 2001. Akademische Druck- u. Verlagsanstalt.

R. Wille. Restructuring lattice theory: An approach based on hierarchies of concepts. In I. Rival, editor, *Ordered sets*, volume 83 of *NATO Advanced Study Institutes Series*, pages 445–470. Reidel, Dordrecht, 1982. doi: 10.1007/978-94-009-7798-3_15.

V.M. Williamson and M.R. Abraham. The effects of computer animation on the particulate mental models of college students. *Journal of Research in Science Teaching*, 32:521–534, 1995. doi: 10.1002/tea.3660320508.

B.P. Woolf, I. Arroyo, C.R. Beal, and T. Murray. Gender and cognitive differences in help effectiveness during problem solving. *Technology, Instruction, Cognition, and Learning*, 3:89–95, 2006.

W.L. Yarroch. Student understanding of chemical equation balancing. *Journal of Research in Science Teaching*, 22:449–459, 1985. doi: 10.1002/tea.3660220507.

A. Zaluski. Knowledge spaces Mathematica package. In *PrimMath 2001—Mathematica u znanosti, tehnologiji i obrazovanju [Mathematica in Science Technology, and Education] Conference, Zagreb, Sept. 27–28*, pages 287–325, 2001.

D. Zhang, D. Albert, C. Hockemeyer, D. Breen, Z. Kulcsar, G. Shorten, A. Aboulafia, and E. Lövquist. Developing competence assessment procedure for spinal anaesthesia. In S. Puuronen, M. Pechenizkiy, A. Tsymbal, and D.-J. Lee, editors, *Proceedings of the 21st IEEE International Symposium on Computer-Based Medical Systems*, pages 397–402, Los Alamitos, CA, 2008. IEEE Computer Society Press. doi: 10.1109/CBMS.2008.25.

# Index



J.-C. Falmagne et al. (eds.), *Knowledge Spaces: Applications in Education*,
DOI 10.1007/978-3-642-35329-1, © Springer-Verlag Berlin Heidelberg 2013